Autonomous and Connected Vehicles

Network Architectures from Legacy Networks to Automotive Ethernet

Dominique Paret and Hassina Rebaine

Translated by
Benjamin A. Engel

WILEY

This edition first published 2022

Originally published in France as:
Véhicules autonomes et connectés. Technologies, architectures et réseaux : du multiplex à l'Ethernet By Dominique PARET & Hassina REBAINE

Registered Office
John Wiley & Sons, Inc., 111 River Street, Hoboken, NJ 07030, USA
John Wiley & Sons Ltd, The Atrium, Southern Gate, Chichester, West Sussex, PO19 8SQ, UK

Editorial Office
The Atrium, Southern Gate, Chichester, West Sussex, PO19 8SQ, UK

For details of our global editorial offices, customer services, and more information about Wiley products visit us at www.wiley.com.

Wiley also publishes its books in a variety of electronic formats and by print-on-demand. Some content that appears in standard print versions of this book may not be available in other formats.

Library of Congress Cataloging-in-Publication Data
Names: Paret, Dominique, author. | Rebaine, Hassina, author. | Engel, Benjamin A., translator.
Title: Autonomous and connected vehicles : network architectures from legacy networks to automotive ethernet / Dominique Paret and Hassina Rebaine ; translated by Benjamin A. Engel.
Other titles: Véhicules autonomes et connectés. English
Description: Hoboken, NJ : Wiley, 2022. | Translation of: Véhicules autonomes et connectés : techniques, technologies, architectures et réseaux : du multiplex à l'ethernet automobile | Includes bibliographical references and index.
Identifiers: LCCN 2021033078 (print) | LCCN 2021033079 (ebook) | ISBN 9781119816126 (hardback) | ISBN 9781119816317 (pdf) | ISBN 9781119816133 (epub) | ISBN 9781119816140 (ebook)
Subjects: LCSH: Automated vehicles. | Driver assistance systems. | Intelligent transportation systems. | Local area networks (Computer networks)
Classification: LCC TL152.8 .P3713 2022 (print) | LCC TL152.8 (ebook) | DDC 629.2--dc23
LC record available at https://lccn.loc.gov/2021033078
LC ebook record available at https://lccn.loc.gov/2021033079

Cover image: © Zapp2Photo/Shutterstock
Cover design by Wiley

Set in 9.5/12.5pt STIXTwoText by Integra Software Services Pvt. Ltd, Pondicherry, India

10 9 8 7 6 5 4 3 2 1

Contents

Foreword

To begin with, I wish to thank Dominique Paret and Hassina Rebaine for their initiative in writing this excellent book about the technologies needed for the autonomous vehicles of the future. The technical literature on the subject is scant as yet, but we are seeing numerous new developments, almost on a daily basis.

France will have to take up a position in relation to these technologies by 2030, which represents an extraordinary opportunity, because the automotive industry is currently undergoing a period of immense and intense change. Vehicles will be (partially) autonomous, connected, electrified, shared, etc. The ways in which they are used are also evolving, and the new forms of transport are expanding the range of what is possible. In addition, users are hungry for new experiences, as a result of the shifting of their way of life and their environment.

Every day, new technologies and associated skills are emerging, to understand, develop and implement these experiences. the traditional engineering Sciences of the twentieth and early twenty-first centuries will also become human sciences, and the engineers of tomorrow will/will have to be designers, marketers, psychologists, lawyers, philosophers, etc.

The usual terminology is also evolving. Amongst other shifts, "comfort" is becoming "wellbeing;" "ergonomics" is giving way to "human–machine interface;" "onboard electronics" are being replaced by "sensors, big data, and algorithms;" the regulation is expanding to apply to personal data; and automation is becoming a matter of ethics.

The training of engineers is no exception to this rapidly changing context. There are profound changes happening in all areas: new schools, new directions, the combination of schools; school/university; the size of promotions; dual-honors courses; international collaboration; apprenticeships; project-based coursework; spin-offs; digital engineering; and new technologies. However, this is not yet enough. The industry must provide its technicians and engineers with the requisite expertise, and work with them to build the academic world, its content, its courses, internships, conferences, projects, professorships, and so on. Today, it is universally accepted that the vehicles of the future will draw heavily on the concept of artificial intelligence, but in actual fact, such AI must first be based on the expanded intelligence of the engineers who designed it.

With that in mind, this extremely thorough book addresses readers wishing to understand the complexity of autonomous vehicles (for all applications), their use in various applications (whether connected or not), and those who design these systems.

To my mind, the most special thing about this work is that it sets out the positions of the fundamental aspects of automotive engineering – functional, software and hardware aspects of the technological building blocks – describing the architectures, examining the different protocols that are used, making designers aware of the regulations and norms in force, discussing the processing of sensitive data and finally, talking about safety and end-to-end cybersecurity. Other points are highlighted in the discussion of the numerous elements in the overall design chain of secure autonomous and connected vehicles, and the technical/economic implementation of these systems in the real world.

Dominique and Hassina have, for many years, been internationally recognized experts in technologies for automobile architectures and networks, both "multiplexed" (LIN, CAN, CAN FD, FlexRay, etc.) and "point-to-point switched" (100 Mbit/s, 1 Gbit/s and multi-Gbit/s Ethernet) in relation to the protocols, physical layers, software and tool development. The highly technical nature of the topics addressed in this book attests to the caliber of their expertise.

I sincerely hope, and firmly expect, that this book will help construct the skills necessary for the development of autonomous vehicles!

Philippe Aumont
Vice-President, SIA – Société des Ingénieurs de l'Automobile

Acknowledgments

As usual, there are many people to whom thanks are due for their kindness, their willingness to listen, their remarks, and their constructive comments. They all know who they are, and to all, we say a heartfelt "Thank you!"

Now we offer thanks to a few, more specific, friends and colleagues, for lengthy informal discussions that have contributed to this book:

- In the automotive and industrial worlds: Jean-Philippe Dehaene and his team (Vector France), Karim El Attachi (Stellantis – formerly PSA + Fiat), Robert Chen (Faurecia), Denis Bugnot, and Muriel Partouche (NXP);
- at the SIA: Philippe Aumont.

We also wish to thank three members and friends of the panel of experts "GDPR – Associates," co-founded by Dominique Paret and Pierre Crégo, for their continuous support:

- Maître Gaëlle Kermorgant – a barrister at the Paris Bar – and Isabelle Pottier – a barrister with Alain Bensoussan-Lexis – for their valuable assistance and involvement in relation to the legalities surrounding present and future automobiles, and Personal Data;
- Mr. Jean-Paul Huon, the CEO of Z#bre, and the co-author of *Secured Connected Objects* (which was published in both French and English by ISTE).

Finally, we warmly thank Ms. Sandra Grayson at John Wiley, for her tenacity and patience throughout this project; our long-time translator, Ben Engel, for the consistently high quality of his work; and numerous other friends who, each in their own way, have brought us good times and joy.

Dominique Paret – Hassina Rebaine
Meudon, 9 May 2021

About the Authors

The two authors of this book have, for many years, been working together on hardware and software for onboard systems, protocols, structures, and architectures of communication networks in the automotive and aeronautical fields, and all variants thereof.

Dominique Paret

Dominique Paret is an electronics engineer (who qualified at the Breguet-ESIEE school of engineering) and the holder of a Master of Advanced Studies (DEA) in physics from the UPMC, Paris VI. For many years, prior to his recent retirement, he was a technical expert and provided technical support in the fields of Contactless/RF technology (contactless chip cards, RFID, NFC, Geoloc, Zigbee, BT, BLE, UWB, UNB, IEEE 804-xxx, etc.), automobile technologies, and multiplexed networks with a major international group producing electronic components (Philips–NXP). He was also a member of and delegate to AFNOR, the ISO, BNA, and the CEN in relation to these domains. In parallel to this career, he taught at some 15 engineering schools, both in his native France and abroad.

Dominique Paret is the founder and CEO of the consultancy and technical expertise company *dp-Consulting*, which he ran for 12 years, and the co-founder of GDPR Associates. He has also authored over 35 technical books, published in French (by Dunod), English (ISTE – John Wiley), Spanish (Paraninfo), Korean, and Chinese.

In short, nobody's perfect, but technology can be!

Hassina Rebaine

A general engineer in electronics who graduated from the Algiers National Polytechnic College, Hassina Rebaine holds a doctorate in electronics on VLSI CAD systems and a DEA in information processing from the INSTN, UPMC, Paris.

Specializing, initially, in the design of simulation tools at SAT (Société Anonyme des Télécommunications) for VLSI ASICs, she then shifted her focus to the design of FPGAs/ASICs written in VHDL, for Verilog (Europe Technologies).

Today, Hassina is Training Manager at Vector, with respect to solutions for automobile onboard systems. Her areas of interest include tools to validate the use of the communication protocols CAN, LIN, FlexRay, and Ethernet. She also teaches at a range of engineering schools and universities.

Preface

Having had a long career as the technical support officer in the semiconductor industry, at Philips Semiconductors/NXP, I had the good fortune to play a direct and active role in the birth, design, development, standardization committees such as the ISO, BNA, Consortium, etc., of the protocols CAN (with Bosch), LIN (with Motorola), FlexRay (with BMW and Freescale), CAN FD and CAN XL (with Bosch – CiA), and Ethernet BroadR-Reach (with BroadCom). In addition, as I have always had a passion for conveying *knowledge*, at the same time, I trained both professionals and students (final-year engineering students) – hundreds of people, in all – in "Embedded Systems and Networks" and "IVNs – In-Vehicle Networks." In parallel, I have published dozens of books over my career, including several in collaboration with Hassina Rebaine, the Technical Support Officer at Vector.

Why this book?

Over recent years, we have published (with Dunod and John Wiley) numerous highly technical books on "CAN," "FlexRay and its applications," and "Multiplexed networks for embedded systems." The latter book notably described the imminent advent, in the industry, of CAN FD and CAN XL, the tsunamis of future ADASs, and the underpinnings of the earliest applications of Ethernet in vehicles that are beginning to have small glimpses of pseudo-autonomy. Since then, we have trawled the market for a book of a sensible level, clear, simple, accurate, and easily accessible, regarding the foundations, the why and the how, and other factors in communication network architectures for autonomous vehicles (such networks are the very "backbone" of a vehicle). Truth to tell, we found there was a gap in the market for such a book – what we found were either overly simplistic books or highly specialized works and university theses focusing on a particular facet of the discipline. With the exception of the few articles, books, journals, etc., cited in the bibliography, the area appears to be a gaping hole. However, over a period of three years, we attended a great many generic (and expensive) high-level conferences (equally expensive), marketing symposia spanning a range of domains and subjects on autonomous vehicles, transport, etc. When we sought to really get into the nitty gritty (true hardware and software architectures, true datarates, true problems – in summary, the daily concerns of automakers, OEMs, SMEs, startups, etc.), we found a similar lack of coverage. In addition, having spent a (very) long time in the field, we realized there was a lack of technical support on the basis of their definitions and applications in the world of intelligent and/or autonomous vehicles. Finding this academic state extremely unsatisfactory, following numerous discussions with a number of colleagues and friends, we decided to once more screw up our courage to the sticking place, mining this domain, and, in the hope of filling some small part of the void, opted to write this essentially technical book designed around this specific facet of "autonomous (or nearly autonomous) vehicles," whose release to the general public is now highly foreseeable – imminent, even.

How this book is constructed, and how to approach it

We have carefully reconstructed and reshuffled this book many times over, to ensure it is coherent and readable, and that readers can easily orientate themselves. The end result is that these pages are divided into five major parts.

Part One clearly defines the parameters of the discussion and outlines the vast subject with which we are dealing, including integral parts of the technology and aspects that are tangentially connected to it. This part offers:

- A **general introduction** to the world of autonomous vehicles, including the precise definitions of the different *levels of autonomy* and connections/connectivity of an automobile, the terminology used and the likely future trends (Chapter 1);
- A description and a detailed breakdown of the numerous aspects, contexts, constraints, and problems (regulatory, legal, normative, moral, ethical, etc.) that weigh upon the design of the autonomous vehicles expected to enter into circulation between 2022 and 2035. At first glance, these matters may seem ancillary, but must be considered in relation to the technology, whatever the autonomous vehicle project (Chapter 2).

Part Two is **more technical**. It is divided into two main parts, illustrated with numerous examples of applications. It includes:

- A detailed technical review of the extremely numerous sensors that are directly or indirectly related to a vehicle's autonomous properties (infrared, sonar, cameras, radar, lidar, inertial navigation systems, etc.) (Chapter 3);
- A detailed technical review of many possible ADASs (Advanced Driver Assistance Systems) and, in particular, of the data fusion in these systems. We examine the integration of AI (artificial intelligence) in the system to make decisions compatible with the desired level of autonomy. The discussion touches on problems relating to mobility, comfort, and security of data transport (Chapter 3).

We then move on to **Part Three**, which is **technical and technological**, and relates to:

- The different possible architectures (hardware and software) used for implementing the different networks in the various zones of the vehicle – power train, chassis, comfort, infotainment, ADAS, etc., – to serve the needs of autonomous and/or connected vehicles, in terms of operational safety and cybersecurity (Chapter 4);
- The increasing power of these networks, reflecting the datarates needed for the automation functions. There is also a detailed discussion of the CAN FD, CAN XL, and FlexRay protocols.

Part Four, which is **highly technical and technological**, describes the possibilities of Ethernet in the industrial world, the features specific to the automobile market, and the new "backbone" structures in "switched" Ethernet networks at 100 Mbit/s, 1 Gbit/s, and several Gbit/s, peculiar to automobiles (Chapter 5).

Finally, **Part Five** gives a detailed description of the **software and tools** needed, which are becoming increasingly important during the **simulation**, **development**, **testing**, **calibration**, etc., of all the devices in the future "autonomous supercomputers on wheels" (Chapter 6).

Target audience

This book is intended for all those who are curious about this new (or nearly new) and vast domain, encapsulating multiple physical, technological, technical, industrial, and marketing aspects. Of course, it is also written for students, professionals in the discipline, and new arrivals to it.

Technical level

Readers need not have a specific level of technical knowledge in order to follow the discussion. The book is intended to be universally accessible, but, throughout, the aim is to satisfy readers' curiosity and provide technical knowledge up to a high level fairly quickly.

Teaching style

As both of us have, for many years, been teaching and training experts in this field, the language used, and the tone, are deliberately accessible and agreeable, without compromising on precision. To provide a complete view of the field, many examples of industrial applications are presented. In addition, running through this book is an unerring desire to impart knowledge, because, to our minds, there is little point writing for oneself. For the curious and intrepid, we have included numerous summary tables, secrets, and anecdotes throughout the text. Simply put, this book is for you, for the pleasure of understanding, learning, and enjoying. We remain "Autonomously Yours!"

N.B

Of course, this book has certain points in common with our earlier publications, which are either identical or similar. That being the case, there will inevitably be a certain amount of repetition. This, we feel, is a price worth paying in order for this book to present a contiguous picture of this novel discipline. We therefore ask our faithful readers to bear with us.

To recap, Ethernet was designed in 1975, I^2C in 1979, D2B (the forebear of MOST) around 1981, CAN in 1983, etc., so all of these technologies have been maturing for around 40 years. They represent the *vintage* in our profession!

Introduction

Warning

This book is not intended to be an encyclopedia on autonomous and/or connected vehicles. Its sole purpose is to explain the types, choices, operation, properties, and architectures of the networks that can be used in autonomous vehicles, depending on a great many external parameters. Thus, one part of the book (the earlier chapters, essentially) details these parameters, with the aim of quantifying their technical implications in concrete terms – for example, in terms of the choice of network topology, datarate, latency, level of security, performance, compliance with norms, standards and regulations, etc.

In this book, various worlds collide – mainly, the world of automobiles, its connected domains, and numerous entities in the worlds of electronics, mechanics, and communication. All of these disciplines have their own specific sets of vocabulary, their own ways of being and acting, design methods, marketing techniques, and commercial approaches, which are generally very different – and this is perfectly normal.

Often, the thinking in electronics can be viewed as linear and Cartesian, taking one step at a time in logical succession. However, in the automotive world, it becomes much more serpentine and interlinked with other factors, because everything reacts to (and influences) everything else, and often we need to view the product that the end client wants as a contiguous whole, rather than a collection of subsets.

Before we begin...

Before setting out on this long journey, let us make two specific points.

On the subject of autonomous and connected vehicles, the Internet offers hundreds of articles (some better than others), presenting some complex and marvellous theories, all sorts of varied and vast future markets, fabulous forms of encryption, etc. As we are not fond of unproductive redundancy, we have focused solely on subjects about which there are not as many articles – i.e. the down-to-earth, day-to-day of this domain, offering a concrete and technical discussion of the vast range of applications and designs. The aim, in so doing, is to guide readers, to overlook nothing, and to avoid the

Autonomous and Connected Vehicles: Network Architectures from Legacy Networks to Automotive Ethernet, First Edition. Dominique Paret and Hassina Rebaine.

pitfalls that may be encountered in the process of designing and implementing secure, autonomous, and connected vehicles. It is all very well to talk about such things, giving speeches and lectures and demonstrations (as we have seen and heard many times). However, to concretely and physically realize a connected solution for commercial purposes, and to sell it in large quantities at a sensible and reasonable price, is far better. Otherwise, it would be as well to do nothing, and forego the unnecessary fuss. This book describes the procedures that must be observed to avoid the usual pitfalls in a project, and facilitate the transition from the virtual to the real and concrete world. Thus, we propose an approach based on due consideration of the technical, financial, ergonomic, etc. standards, rather than on false promises.

On this subject, in late 2018, Bernard Favre, an expert at Inria and the former Head of Research at Volvo-Renault Trucks and of the LUTB Transport and Systems industrial research program, wrote the following. “Autonomous vehicles are a highly complex technology, in which it is probably harder to bring artificial intelligence to bear than in any other application. In no other sector is the technology faced with such a diverse range of situations. At present, we are in the full throes of innovation “in the lab.” As yet, there is no real proven market .... The number of tests that automakers require in order to validate an autonomous vehicle’s performances is soaring. They include physical experiments in real conditions, and digital simulations. ... Having a certain amount of experience of the disparity between what automakers’ projections and announcements say about when new technologies will be available, and when they actually become available for commercialization (for various reasons: maturity, regulations, market acceptability, cost, real performance, etc.), I fear that autonomous vehicles will be no different to what I have seen time and time again in my career”. He concludes by projection that “autonomous cars will be operating on private circuits by 2025. In relation to autonomous vehicles on public/open roads, it is likely to be 2040. ...”

This is a view which we, the authors, have long shared.

This, then, is the explicit aim of this book, which should therefore remain on your coffee tables as a reference until 2035 at least – and that should be enough!

1

The Buzz about Autonomous and Connected Vehicles

This book begins with a two-part chapter, directly connected to the technical wizardry that must be implemented in vehicles in general and, by extension, in autonomous and/or connected vehicles.

By way of a general introduction, this first part offers a brief overview of the vocabulary used in the field. This will help to avoid the common confusions arising on the ground, and offer clarity about the various terms used under the umbrella of autonomous vehicles.

The second describes the vast world of vehicles, the surrounding topics, the media buzz, coverage in the ordinary and specialized press, and the concrete reality of defining, designing, manufacturing, fine-tuning and industrially producing a product, and, in particular, successfully selling it at market.

In 2021:

- There are already over a billion automobiles in the world (source: *Comité des constructeurs français d'automobiles* [CCFA – French Automaker Committee]);
- In 2016, in Paris, drivers spent more than 65 hours stuck in traffic jams. The situation in Moscow was worse still (91 hours) and in Los Angeles (104 hours) (source: INRIX research, 2016);
- Each year, worldwide, 1.3 million people die in traffic accidents (source: WHO);
- Every year, globally, 2.6 million deaths are caused by air pollution, which is partly linked to automotive traffic;
- In 2030, it is projected that 2.3 million people will die as a result of a road traffic accident (source: WHO);
- By 2050, according to predictions, 70% of the world population will live in urban areas (source: WHO).

In addition, the world population is continually increasing, leading to:

- Increased traffic;
- Congestion in city centers;
- Soaring CO_2 emissions;
- The upsurge in road accidents.

Autonomous and Connected Vehicles: Network Architectures from Legacy Networks to Automotive Ethernet, First Edition. Dominique Paret and Hassina Rebaine.

In the short and medium terms, all these subjects raise the question of *urbi et orbi* (in cities and out of them) transport in the future. In addition, in the early twenty-first century, the automotive industry is experiencing major technological changes, and as mentioned in the Foreword, in time, vehicles will come to be (partially) autonomous, connected, often electric, shared, etc. Their uses will evolve, and new forms of mobility, technologies, and skills will extend the range of possibilities.

As stated previously, this technical book is merely a single stone in the understanding of the vast edifice that is autonomous and connected vehicles. We have therefore restricted our field of study to a specific part of that edifice.

1.1 The reasons behind this book and its limitations

Autonomous and/or connected vehicles represent an enormous and highly complex subject, including a great many concepts that must be understood. Thus, we shall begin by briefly presenting the fields we have decided to cover in this book. Note that, while we have chosen to focus on the technical and technological aspects only, each of the subjects has its own accompanying philosophy and technical consequences.

1.1.1 Architectures

In a vehicle, there are a wide range of architectures, which clash, overlap, coexist, etc. We shall examine various architectures here. For example:

- **Functional architecture**, which governs the overall organization of all the system functions in a vehicle. Here, functional architecture is discussed only briefly;
- **Network architecture**, which governs the way in which protocols and communications between the functions and ECUs (computers) in the vehicle are chosen and structured. This will be the main focus of the book, as we move progressively from multiplexed network systems to automotive Ethernet architectures;
- **Hardware architecture**, whose purpose is to structure and define the choices of ECUs, the types of electronics, sensors, actuators, etc. We shall also discuss these in some depth, as they are directly involved in the different types of data that need to be transmitted;
- **Software architecture**, which controls the structure and management of the different software modules in a vehicle. At the end of the book, we shall examine the software architectures that are dedicated to networking;
- **Organic architecture**, which is in charge of the implementation of the different functions in the vehicle's electrical and electronic components;
- **Topological architecture**, which manages the physical arrangement of the different systems and components within a vehicle. The topological architecture is of crucial importance in estimating and minimizing network lengths, which are closely connected to the achievable datarates;
- **Cabling architecture**, which governs the way in which the networks and cabling harnesses are physically divided and implemented in the vehicle, their performances, diameters, weight, and so on.

1.1.2 Communication networks

As we shall see, for many years, numerous types of communication network have been installed in vehicles. Each network is specifically suited to particular application typologies.

The majority of this book focuses on analyzing their quality and performance, with a view to making suitable, safe applications, carried in vehicles with high levels of autonomy and connectivity. Until recently, such networks were largely based on "multiplexed" modes of operation, and many are in the process of shifting towards modes of operation oriented around the Ethernet philosophy, tailored for use in automobiles. The main goal of this book is to guide readers through that technological transition.

1.2 Terminology

It may be unusual to begin a book with a terminology section. However, in order to discuss autonomous vehicles, it is necessary to clearly define and describe the different levels of vehicle autonomy, to overcome the many potential misunderstandings, without the generalization and obfuscation that are typical of mainstream media coverage of this subject.

1.2.1 Autonomous and/or connected vehicles

To begin with, readers must appreciate the profound distinction between "autonomous" vehicles and "connected" vehicles. These two terms represent two completely different things, and must, under no circumstances, be confused.

- By definition, a (true) *autonomous* vehicle must be capable of traveling unassisted, alone, anywhere and at any time, etc., with no restrictions, without the help or even the presence of a driver.To be absolutely clear, either a vehicle is autonomous or it is not. It cannot be nearly autonomous or semi-autonomous, etc. – that makes no sense.

  Nevertheless, in order to rate a vehicle's performances, we can speak of the *levels of autonomy* (dictionary: its "capacity to be autonomous"), taking care to clearly indicate the specific domains and references in question;
- A connected vehicle is a vehicle that is linked to other systems by means of telecommunication systems, telephones, etc.;
- An "autonomous" vehicle is not necessarily "connected," or vice versa.

On the other hand, frequently, an autonomous vehicle often does need to be connected in order to carry out other functions and other tasks (for example: uploading or downloading information about the road on which the vehicle is traveling, etc.) – this is why confusion so often arises.

Autonomous vehicles

The terms "autonomous" and "autonomous vehicles" are much too broad and too imprecisely defined. Again, to prevent confusion, in this book describing the habits and customs of the automotive profession, focusing on vehicles from those of the past to those of the (perhaps distant) future, we shall use precise levels to define these types of autonomy, specified below.

Connected vehicles

It is all very well to speak of connected vehicles – but connected to what, why, and how? Figure 1.1 illustrates some of the possible links and connections. These will be discussed in greater detail in Chapter 4.

To complete this brief general introduction, Figure 1.2 illustrates a vehicle solution whose functions facilitate a certain degree of autonomy, and which also has a number of connections.

1.2.2 Terms and vocabulary relating to autonomous driving

The changes taking place in the automotive sector have made their way into the lexicon – there is a range of terminology dedicated to autonomous vehicles.

Terms and definitions

This vocabulary includes terms such as "ADASs" (advanced driver assistance systems), "driverless taxi," "participatory geo-navigation," etc. Consider a few other examples:

- *Pay-how-you-drive (PHYD) insurance*: "a vehicle insurance contract whose premiums are based on the driver's conduct at the wheel, and the way in which the vehicle is used." Note that driver behavior and vehicle usage are assessed using data transmitted to the insurance company by onboard sensors;
- *Dashboard camera, dashcam, dash camera, scene recorder*: "an onboard camera that records the scene in front of the vehicle." Note that often, only the last few minutes of a recording are actually kept in the memory. These recordings may, for example, be used to document the cirucmstances of an accident;

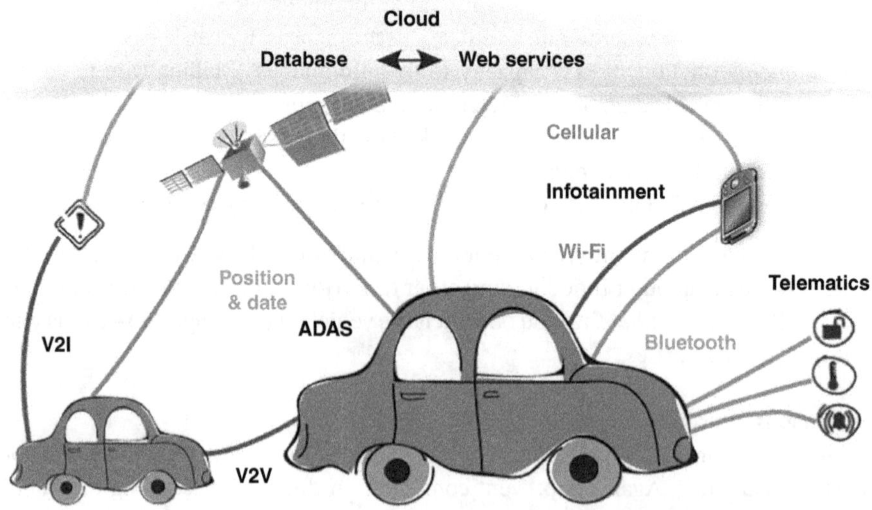

Figure 1.1 Example of links and connections in a connected vehicle.

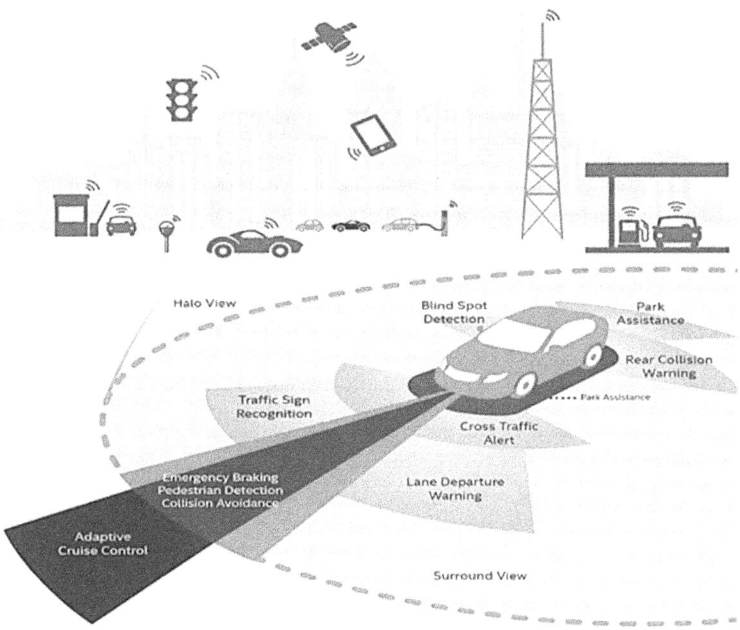

Figure 1.2 Example of functions that help make a vehicle autonomous.

- Autonomous driving or *automated driving*: "a method of automated driving of a vehicle, which does not require input from its users; and, by extension, a system that facilitates this kind of driving;"
- *Traffic jam assist, traffic jam chauffeur, traffic jam pilot*: "a system that allows a vehicle to move independently in traffic jams." The simplest forms of traffic jam autonomous driving systems merely follow the vehicle in front in the same lane (which does represent a certain degree of autonomy); the most complex are also able to change lanes.
- *Driver alert, driver alert system, driver monitoring, driver monitoring system*: "an onboard system that uses sensors and analyzes the driver's behavior to detect any reduction in their alertness, and warn them about it." The most advanced form is an alertness monitoring system:
  - The sensors used may be cameras, which analyze the driver's head and eye movements. There are also systems that analyze the rotation of the steering wheel to assess driver alertness;
  - *Attention assist* is a registered trademark, so the term should not be used in any other context.

For information, Figure 1.3 offers a list of equivalent terms:

Terms
Driverless cab, autonomous taxi, driverless taxi
Mirroring, screen mirroring

Figure 1.3 Other examples.

Let us now look at the levels of autonomy.

Autonomy levels

In the automobile industry, the gradual trend towards autonomous driving follows a scaled technological progression, defined by a classification, which itself is established on the basis of multiple autonomy levels. Level 0 corresponds to a 100% manual vehicle, and the highest level (4 or 5, depending on the standards used) corresponds to a fully autonomous vehicle (restricted to specific use cases), which has no need of a driver.

Use cases

> **Note**
>
> At the time of writing (2021), these different autonomy levels correspond only to applications in specific environments and use cases.

The three defined use cases and their specific features are as follows:

- For private vehicles:
  - In a traffic jam, without changing lane;
  - On the highway, without changing lane;
  - Autonomous parking.
- For industrial vehicles:
  - Speed regulation by infrastructure;
  - Platooning;
  - Garbage trucks;
  - Agricultural sprayers.
- For public transport:
  - Free service on a private site;
  - Shuttle bus services on a sheltered site.

> **Note**
>
> All of these use cases are highly restrictive. Nowhere in these three use cases is "open" circulation of autonomous vehicles mentioned.

We shall now look at the classification of autonomy levels.

NHTSA classification (United States)

In the United States, the Department of Transportation and the National Highway Traffic Safety Administration (NHTSA) have created a five-level vehicle autonomy scale (Figure 1.4).

Let us briefly recap the definitions of these levels.

NHTSA classification	
Autonomy level	**Functionality**
Level 4	Fully autonomous driving
Level 3	Limited autonomous driving
Level 2	Automation of combined functions
Level 1	Automation of certain functions
Level 0	No automation

Figure 1.4 NHTSA vehicle autonomy scale.

Level 0 – no automation
"The driver has total control, at all times, of the vehicle's main functions (engine, accelerator, steering, brakes)."

Level 1 – automation of certain functions
"The automation systems, which apply only to certain vehicle functions, merely assist the driver, who retains overall control."

Level 2 – automation of combined functions
"Control of at least two major functions is combined in the automation system, to replace the driver in certain situations."

Level 3 – limited autonomous driving
"The driver can hand over complete control of the vehicle to the automated system, which will then be in charge of the critical safety functions. However, autonomous driving can only take place in certain environmental conditions and traffic conditions (e.g. only on the highway). The driver must be able to re-assume control reasonably quickly when asked to do so by the system – notably when the requisite conditions for autonomous driving are no longer met (exiting the highway, tailback, etc.)."

Level 4 – full autonomous driving
"The vehicle is designed to perform, unassisted, all the critical safety functions for the whole of the journey. The driver enters a destination or directions, but is not required to be available to resume control. Thus, the driver can leave the steering wheel unattended, and the vehicle is capable of traveling without any occupants."

OICA-SAE classification
The International Organization of Automakers (known as OICA, for its French acronym), is founded on the SAE J3016 classification. The classification system set forth by OICA-SAE is very similar to that of the NHTSA, except that it has six levels instead of five – the equivalent to level 3 in the American classification system is split into two sub-levels.

The functional breakdown of the autonomy levels in the OICA framework is presented for information purposes and is not normative (see Figure 1.5).

	OICA-SAE J3016 Classification					
	0	1	2	3	4	5
	No automation	Driver assistance	Partial automation	Conditional automation	High automation	Full automation
	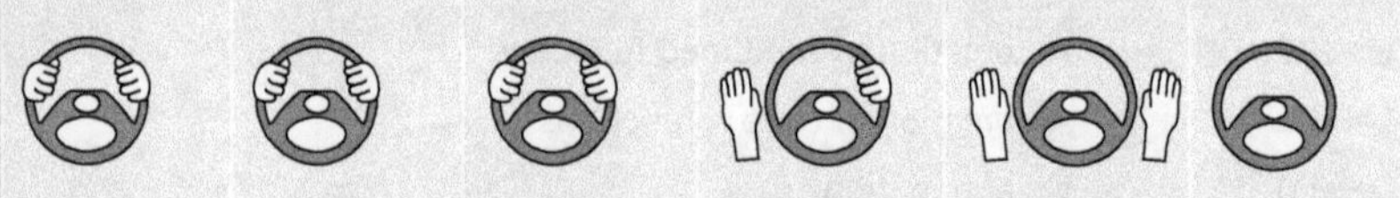					
VDA	Driver only	Assisted	Partially automated	High automated	Fully automated	Driverless
NHTSA	0	1	2	3	3/4	

Figure 1.5 *Levels of vehicle autonomy according to OICA-SAE.*

Level	Functions	Driver	Vehicle
Level 5	Full autonomous driving	No longer needed at all.	Capable of performing all dynamic driving functions, whatever the circumstances.
Level 4	High automation	Should no longer be needed, either for driving or for backup.	Can handle almost all driving tasks in various scenarios.
Level 3	Limited autonomous driving	No longer needed to drive the vehicle, but must be prepared to take over the controls within a specific length of time if asked to by the system.	Steers, accelerates, and brakes automatically, but only in certain situations. The system recognizes its own limitations and alerts the driver to resume control.
Level 2	Automation of combined functions	Must remain alert, even when the vehicle is performing basic driving tasks, and must be able to take over immediately.	Steers, accelerates, and brakes automatically, but only in certain situations.
Level 1	Automation of certain functions	Must drive constantly, but with some basic support in certain situations. Must be able to take over the controls immediately.	Provides basic assistance, such as automatic emergency braking and lane-keeping.
Level 0	No automation	Must always be in charge of driving.	Merely responds to the driver's instructions, but can provide alerts about the vehicle's surroundings.

Figure 1.6 Details of the vehicle autonomy levels according to OICA-SAE.

The OICA-SAE classification offers the distinct advantage of more precisely defining the respective roles of the driver and of the vehicle (see Figure 1.6).

Let us now look at these six levels, starting with the lowest, and offer a little more detail and background explanation, as well as some examples.

Level 0 – no automation

Driver At all times, the driver has total control over the vehicle and its main functions (engine, accelerator, steering, and brakes).

Vehicle The vehicle merely responds to the driver's commands and does not raise any alerts about the environment.

Of course, this is the most conventional level, and the easiest to understand – everything is manual. At all times, the driver has total and exclusive control of the vehicle's primary functions (brakes, steering, acceleration, and driving force), and, logically, there are no automated functions. However, even when the driver does everything themselves, and none of the main functions is automated, the vehicle may still have alert mechanisms. Thus, all old cars can be described as being *level 0* autonomous!

Consider the example of a 4×4 convertible, with hands-free entry, keyless ignition – in short, a 1942 Army Jeep – is a level-0 autonomous car (see Figure 1.7).

Example of level-0 assistance: the audio alert of a reversing sonar sensor, lane departure warning, etc.

This level includes a great many vehicles on the road today.

Let us now examine levels 1 and 2, which hinge on advanced driver assistance systems (ADASs).

Level 1 – Driver assistance

Driver The driver retains overall control, with certain basic supports in place in certain situations.

Vehicle The vehicle can provide some basic assistance, such as emergency braking and lane-keeping support.

Figure 1.7 A level-0 "autonomous" car.

In level-1 autonomy:

- The automation systems, which are in place for certain functions of the vehicle, merely assist the driver;
- The driver can temporarily hand over a driving task to the vehicle, on condition that it only handles one of the two dimensions of steering (longitudinal or transverse);
- The human operator always remains responsible for maneuvers, delegating a portion of the tasks to the systems. The driver must always be able to take over full control of the driving, if the situation requires.

Example of level-1 assistance:

- Adaptive cruise control: the vehicle takes control of the longitudinal dimension (the acceleration), but the driver is responsible for keeping the most appropriate lane position (transverse dimension);
- The anti-lock braking system (ABS) and the electronic stability program (ESP) automatically act on the brakes to help the driver maintain control of the vehicle.

Level 2 – partial automation of combined functions

Driver The driver must remain fully alert while the vehicle performs certain basic driving tasks.

Vehicle The vehicle controls at least two main functions, which are combined in the automated system, to replace the driver in certain situations. It can automatically turn, accelerate, or brake in certain situations.

At level 2, the driver ought only to need to play a supervisory role as the vehicle drives itself. In the event of an accident, the driver is still entirely liable for the failure of the system, because they have not been sufficiently attentive to their environment and neglected their supervisory duty. In all cases, it is important to remember that "the safety of the system as a whole is independent of the autonomy level." At this stage, any action by the driver, however small, will supersede the car's own behavior, in all circumstances. The human remains responsible for maneuvers at all times, and must be ready to take over fully if the situation requires. Note that certain automated functions have been in place in consumer vehicles for a number of years – such as automatic braking, electronic stability control and cruise control – but the human driver is still in charge. In addition to the electronic stability control system, level-1 and level-2 systems, such as AEB and LKA (see below), have begun to be rolled out. In summary:

- Level 1 means that only one automated system can be active at any given time in the vehicle;
- Level 2 means that multiple automated functions can work in tandem.

Examples of level 2:

- Adaptive cruise control combined with lane centering would qualify a vehicle as level 2;
- Automatic emergency braking (AEB) – the vehicle can apply the emergency brake automatically, of its own accord if a collision is imminent – plus *lane keeping assist*

(LKA), which alerts the driver if the vehicle begins to stray – and/or change direction if the vehicle deviates from its lane;
- *Park assist* is a good example of a level-2 function, which is able to park the car without input from the driver on the steering wheel or pedals. The car itself handles all steering and motion parameters, under the driver's supervision. The driver is able to reassume control and correct trajectory at any time.

Today, safety managers all over the world are beginning to include level-1 and level-2 technologies as an essential precondition for a vehicle to aspire to levels 4 or 5. For example, in Europe, all vehicle models launched after 2018 must have automatic emergency braking, *road edge detection* and *lane keeping assist* in order to eventually attain a level-4 or level-5 rating. Given that around 85% of vehicles are intended to eventually reach level 4 – that is, basic cars that are equipped with more advanced ADASs, where the driver must remain fully alert at all times, level-1 and level-2 technology is likely to gain a greatly increased presence over the next three to five years. In good conditions, such a vehicle can take care of a great many normal aspects of driving.

At present, this represents the legal limit, as it is not possible to transfer liability to a machine, even if that machine's decision-making is said to be better than that of a human being.

At levels 3 to 5, we enter into the realm of autonomous driving – to a certain degree.

Level 3 – autonomous driving under certain circumstances

Driver The driver can cede complete control of the vehicle to the automated system, which will then be in charge of the critical safety functions. However, upon request from the autonomous driving system, when it finds itself unable to continue, the driver must always be ready and able to resume control within an acceptable, specified window of time (especially when the conditions for autonomous driving are no longer met – e.g. exiting the highway, approaching a tailback, etc.).

In 2019, Tesla and Google Car were at this stage of automation (see Figure 1.8).

Vehicle The vehicle can take complete control of the steering, acceleration, and brakes in certain conditions. However, autonomous driving is only allowable in certain, specified environmental and traffic conditions (for example, only on an open highway).

In this case, the human driver can delegate a portion of the two-dimensional task of driving, and let their attention slip to carry out other tasks "briefly." However, in certain circumstances, the vehicle will have to ask the driver to take over once more. The driver must be capable of resuming control if conditions require.

Because the driver presumably cannot *instantly* assume control of the vehicle, the system must ensure continued safety for a period of time when the driver has still not re-engaged (for example: in this case, pull over and stop the vehicle if the driver does not respond to repeated requests to take over control). This redundancy must be covered by additional sensors, such as radar, lidar (to detect shapes and objects), and map location, to identify safe drivable roads, and be aware of intersections and other signaling or instructions.

Figure 1.8 *Google Car.*

At this level, the driver can disengage completely until the vehicle requests otherwise. During that time, the vehicle demonstrates a true “capacity for autonomy,” and the technology begins to display a number of substantial advantages in addition to safety, such as increased protection.

The transition from level 2 to level 3 requires a substantial improvement of the levels of functional security and redundancy in the system. At this stage, the technical level is the same as level 2, but liability is temporarily transferred to the system during periods of autonomous driving.

This immediately raises the question: what would the legal position be if the driver were to fail to take over the controls? The answer to this question is rather too tricky, but will be discussed in Chapter 2. In brief, the legal problem that embroils drivers, insurers, and automakers is so complex that only the law can decide on the appropriate behavior if the driver does not respond, when they were supposed to resume control of the vehicle within a given period of time.

This level means that, in certain circumstances, the vehicle can take charge of all driving functions:

- Example 1: the least complex driving environment is the highway (all vehicles are traveling in the same direction, with no pedestrians and no complex intersections). All the major functions are automated, including braking, steering, and acceleration;
- Example 2: slow-speed driving in traffic jams, in which the vehicle’s autonomous system takes charge of positioning and keeping the vehicle in its lane, whilst maintaining a speed that is appropriate for the traffic conditions and the speed of other vehicles. In such cases, a driver could, for example, read a newspaper without paying too much attention to the traffic jam. However, when the jam clears, the car will ask the driver to take over control once more.

Whilst it is true that the levels of vehicle autonomy are based primarily on the sharing of responsibility, it is important to remember that the first three levels of autonomy

Figure 1.9 Examples of level-3 vehicles: the Renault Espace (left) and Peugeot 508 First Edition (right).

imply no responsibility (or liability) on the part of the vehicle. The systems are intended to be no more or less than driver supports (see Figure 1.9).

Level-4 and level-5 vehicles are autonomous in all situations and driving environments – no longer just "in certain circumstances," as is the case at level 3.

Level 4 – high automation

Driver The driver can be a passenger who, on request, can take over the controls and driver when the automated driving system is unable to continue.

Vehicle The vehicle can perform all driving tasks in almost all specified conditions, without requiring attention from the driver.

Level 4, which is almost full autonomy of a vehicle, is characterized by the vehicle's ability to override the human agent in relation to specific functions, and in a number of specific cases. At level 4, the human is no longer supposed to be a driver, because the vehicle is prepared for all situations: thus, the human goes from being a driver to being a mere passenger. In these cases, the system takes care of all functions, autonomously, and is responsible for maneuvers, not even requiring the driver to be present. However, it is always the driver who decides to activate or deactivate autonomous driving mode.

Example: *valet parking*, where the vehicle finds a space and parks on its own, and then comes back to collect the driver.

This level of autonomy is only to be found in the European system (though it has no normative framework). It consists of a system with the capacity for full autonomous driving, all of the time, where the human does not intervene at any time, except to set the destination and allow themselves to be transported, and indicate their preferences. Thus, the vehicle acts as a smart robot.

Level 5 – full autonomous driving

The vehicle is designed to perform, unaided, all critical safety functions, over the entire journey (subject to the usage restrictions mentioned at the start of this chapter). The driver inputs a destination or navigation instructions, but is not required to be available to take over control of the vehicle. Thus, the driver may leave their post, and the vehicle is capable of traveling even without occupants.

Driver There is no longer a need for a driver to take the wheel (if indeed there is a wheel in the vehicle), and anyone can be a passenger.

Level-5 systems never have need of a driver. The vehicle genuinely drives itself, in all conditions, and is capable of reacting in the same way as a human driver would (or better). Therefore, the control components such as the steering wheel and pedals are no longer needed.

Vehicle The vehicle is in charge of driving and can operate in all environments (defined use cases), without human help.

Autonomy level 5 is distinguished from level 4 by the idea of "machine certainty", whereby the system is permitted not to obey a human order if the instruction is deemed abnormal, rash, or dangerous, in which case the machine will need to take an initiative based on the measurements of its sensors. In a number of cases, the vehicle systems can carry out a maneuver that the driver did not request, and can even refuse to perform an operation that would endanger either the vehicle or its passengers (for example: opening the door whilst driving on the highway).

For example, the vehicle drives itself, even without a steering wheel, pedals or other controls used by humans (e.g. indicators). The rollout of this level of autonomy begins with fleets of "carpooling" vehicles, in restricted sectors. Driverless vehicles make the carpooling model more financially viable and more attractive, because they eliminate the most significant cost in carpooling fleets: the driver. In addition, the initial rollout of carpooling fleets will offer two further important benefits:

- When first used, they will include a qualified operator in the driving seat, allowing consumers to gain initial experience with the technology, with the assurance that a qualified operator is monitoring the situation;
- The ability to safely generate real-world performance data. Once a sufficient quantity of data has been generated, then the legislature should approve the wider use of the vehicles.

The transition between autonomy "somewhere" and autonomy "everywhere" should ultimately be achievable at the flick of a switch or the touch of a button, because the underlying maps will be constantly updated, everywhere, rather than only in confined areas that are geolocated and geofenced (see Figure 1.10).

We have now seen the lists of "official" titles for the levels of "autonomy" to which we refer in this book. Unfortunately, the press, the audiovisual media, and marketing services often use language that is somewhat different and imprecise, opening the door to significant confusion.

The problem with these classifications

There is, undeniably, a major problem in these classifications. Often, in the above paragraphs, we see the use of little phrases such as "use cases" or "in certain conditions..." that, unfortunately, are incompletely (if at all) defined explicitly, leaving room for a great deal of fuzzy interpretation, on a technical, legal, and commercial level. When the day comes that there is a "truly totally autonomous vehicle, which can drive anywhere, in all countries, all road conditions, free roads, open roads,

Figure 1.10 A typical example of a level-5 vehicle (the NAVYA shuttle).

roads with heavy traffic... so with absolutely no restrictions," it will be a true level-5 vehicle, but that day is still a very long way off, though certain people would imply that it is near.

Other terminology

Commercially, there are almost as many names for these technologies as there are different models ("genuinely" autonomous vehicles, semi-autonomous vehicles, automated vehicles, robot taxis, etc.). Often, the marketing departments of automakers and other technological giants deliberately maintain a degree of fuzz as to the true capabilities of vehicles they call "autonomous". The official classification into six categories by the OICA-SAE International can be used to roughly define what human drivers and/or autonomous systems can (and cannot) do. On the basis of these categories and their limitations, certain automakers introduce wordplay into their communications. Indeed, there are other names that lend themselves well to marketing/commercial activities, and it is highly likely that these terms will enjoy a certain success in the media. For example:

- Eyes on/eyes off;
- Hands on/hands off;
- Hands off/eyes off;
- Mind on/mind off;
- Autopilot; etc.

"Hands off/eyes off" technology corresponds, approximately, to autonomous driving without human supervision. During periods when control is delegated to the automated system, drivers no longer need to keep their eyes on the road or their hands upon the wheel: the driving is given over entirely to the vehicle. This function is envisaged for the most "boring" or "monotonous" types of driving (for example, in traffic jams), and only on authorized roads. In the press, the expression "hands off/eyes off" recurs frequently, and often with the addition of "mind off." No longer is it a case solely of "muscular disengagement, but mental disengagement too." However, the use cases are not defined.

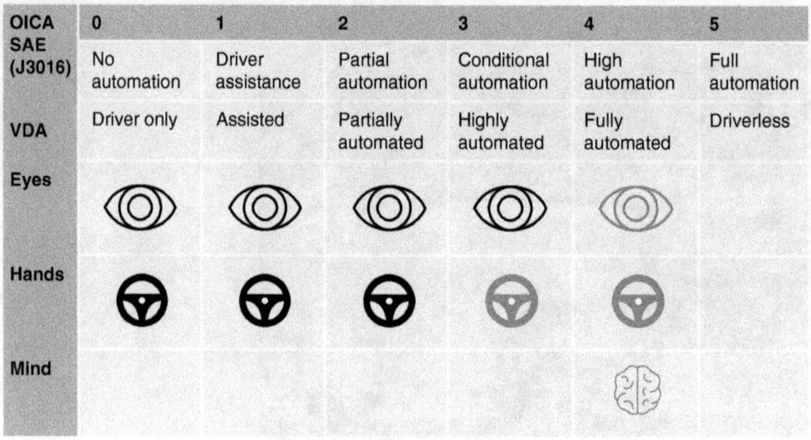

Figure 1.11 Concordance between OICA-SAE levels and eyes/hands/mind on and off.

Thus, when autonomous driving is activated, numerous devices/sensors are responsible for watching the road and monitoring 360° around the vehicle. Certain vehicles include the following:

- Three lidars, or long-range laser scanners (two in the front, one in the rear);
- One long-range forward-facing radar;
- Four medium-range angular radars;
- Three digital cameras with different focal lengths (short/medium/long range) at the top of the windshield;
- Four short-range digital cameras with 180° vision, beneath the rear-view mirrors and in the license plates;
- A strip of 20 short-range ultrasound (sonar) sensors.

The dataset collected by these sensors is analyzed by multiple onboard software "brains," which decide how the vehicle should behave. The driver has no need to look at the road or keep hold of the steering wheel: the vehicle is in "hands off/eyes off" mode. The devices controlling the vehicle's path (steering and brakes), and the associated electronic architecture, are duplicated, in order to obtain the level of safety required for hands off/eyes off autonomous driving. Thus, if there is a fault, the vehicle remains in control of its trajectory, and is capable of achieving safety automatically, in the event that the driver does not take over the controls once more. Figure 1.11 summarizes and illustrates these situations.

Finally, in years to come, in the use of vehicles, we shall see the rise of auto-companies in the areas of "ridesharing" and "shared ownership," where people and organizations share ownership of a vehicle that will take them wherever they want to go. This latter concept is even more transformative than carpooling, because it opens up completely new business models for transport.

In summary

Figure 1.12 offers an overview of this chapter.

Having set out the official backdrop, we are finally ready to launch an in-depth discussion.

1.3 A brief history of the autonomous vehicle

Before diving in to look at the major issues, here is a brief history of the discipline.

The history of autonomous vehicles goes back a long way, before the time of Jules Verne. However, it was in the 1990s that designers began to think seriously about how to achieve such automation, notably devising industrial ADASs that would, gradually, lead to the design of autonomous vehicles.

1.3.1 Autonomous vehicles

The rollout of autonomous vehicles has produced two main models:

- That of a new type of transport, with totally autonomous driving that is easier, more fluid, effortless, auto-adaptive, and so on;
- That of a new type of management of total safety and security (physical, cyber-X, etc.) in our normal, daily environment, in all circumstances.

In addition, it will no longer be necessary to own your own vehicle, or even be able to drive. These two points will have an influence on the vehicles as a whole, and also on public transport and freight.

On the technical aspect, an autonomous vehicle is an extremely complex system. Broadly speaking, it comprises:

- Huge numbers of onboard sensors;
- A great deal of very fast communication technology;
- Onboard, complex "artificial intelligence" systems that are highly sophisticated, notably capable of supporting and deciding on the path the vehicle should take, steering and monitoring the vehicle's different control systems (steering, brakes, accelerator, suspension, etc.) with verifications and redundancy.

In addition, the vehicle must be able to conform to a great many use cases and demonstrate reliability in the face of an infinite number of critical events that it may encounter:

- In ideal conditions, in sunshine, on a clear road with a good surface, this challenge should, in theory, be reasonably easy to meet;
- On the other hand, in rain, snow or on a bad road, road markings are unclear, and the likelihood of encountering an obstacle is high. In this situation, the only way forward is to limit speed or revert to manual driving.

Level		
0	No assistance	The driver has total control over the vehicle's main functions (steering, acceleration, and braking).
1	Driver assistance	The system is able to take charge of either the longitudinal dimension (speed and distance from the vehicle in front) or the lateral dimension of the vehicle's motion (following the white lines), but not both at the same time. It assists the driver, but the human driver is still in charge of driving. **Example**: cruise control, lane departure warning (radar), automatic emergency braking, and collision warning system.
2	Partial automation	In certain situations, the driver can delegate both longitudinal *and* lateral control of the vehicle to the system, but is still required to supervise: they must keep an eye on the system at all times, monitor the vehicle's environment, and take back total control if necessary. **Example**: traffic jam assistance (adaptive cruise control) or parking assist.
3	Conditional automation	The vehicle system, rather than the human driver, is responsible for monitoring the surroundings. However, the driver must remain alert to road conditions, and must be able to immediately resume control if necessary (which implies that the machine must allow them enough time to respond). In this context, driving can be given over entirely to the machine, but only in predefined conditions: on the highway, for example. The system must be capable of recognizing its own limitations – i.e. recognizing when the traffic conditions are no longer compatible with its operation. In this situation, the vehicle asks the driver to take over control by means of a visual and auditory alert, raised several seconds in advance. **Example**: highway driving, automatic parking, platooning.
4	High automation – *Eyes off, hands off, mind off*	No assistance is needed on the part of the driver. They may turn to whatever occupies them, without worrying about the road. However, total autonomy is limited to a specific geographic zone, such as a highway or a parking lot with which the vehicle is compatible. It is also limited to specific weather conditions and visibility (with no fog or snow, or only during the daytime, for example, which is also the case with the above levels of autonomy). When these criteria are fulfilled, the driver is no longer responsible for driving: that responsibility is turned over entirely to the system. However, the human driver is still obliged to take over driving when the vehicle leaves the automated driving zone. Unlike with level 3, if the driver does not respond, the vehicle must be able to react for itself – e.g. by pulling over into a safe position. **Example**: An end to the boredom of long journeys or sitting in traffic jams on the highway.
5	Full automation	A fully autonomous, driverless vehicle. Restricted to known and limited journeys (e.g. shuttle buses, etc.). **Example**: In 2016, Mobileye announced a collaboration with BMW and Intel to produce a fully autonomous, level-5 vehicle by 2021. As things stand in mid-2021, we shall still have to wait and see!

Figure 1.12 *Overview of levels of vehicle autonomy.*

Such demonstrable reliability is all the more important when we consider that, should an accident occur, it is/becomes highly complex to define where the responsibility lies. Though it is widely accepted that autonomous vehicles will be equipped with black boxes (acting as spies, which certain players vociferously decry), it may not be possible to define the reasons for an incorrect trajectory simply by analyzing the data. To demonstrate the performances of these vehicles, it is necessary to conduct numerous tests in environments that are as closely representative of the various use cases as possible. Also, there is no doubt that we shall become accustomed to sharing the road with these vehicles which, it is said, will or should be safer than conventional vehicles.

Beyond the pure technical detail and technology discussed above, the development of these highly complex systems is hampered by a number of technical and non-technical barriers, which we shall discuss at length in Chapter 2, entitled "Aspects relating to autonomous and connected vehicles," and examine:

- The definition of a legal framework and liability in the event of an accident;
- The adaptation of the driving regulations;
- The development of appropriate standards;
- The adaptation of infrastructures to support these new types of vehicles;
- The availability of onboard intelligence technologies, their performance and cost;
- Operational safety, cybersecurity, and guarding against cyberattacks;
- The acceptability of new uses, etc.

1.3.2 Types of propulsion and autonomous vehicles

In principle, there is no direct relation between the terms "propulsion type" and "autonomy." The vehicle may have an internal combustion engine (running on gasoline, diesel, LPG, or similar fuels), be purely electric or hybrid (thermal–electric hybrid, electric fuel cell), etc. It should be of no importance – beyond a few minor adjustments. Vehicle autonomy and connectivity are new functions that are added to the existing vehicle architecture. Inevitably, they consume energy (a few tens or a few hundred watts). Thus, sooner or later, they will impact the vehicle's range – that is, the distance it is able to travel on a specific amount of fuel or energy (see Section 2.14). This is a parameter that needs to be taken into account, because it is clear that we are on the cusp of a total transformation in the auto industry. The French Government has announced its intention, by 2050, to transition from combustion engines to electric vehicles, which echoes the transition to electric trams around 1900. In short, the world has seen this type of transition before. All the technical and economic factors that we see today in the early twenty-first century are similar to those that existed in the early twentieth century: there are numerous small automakers, numerous types of propulsion, distribution networks for new fuels, and types of energy are – quite literally – in the pipeline! and so on.

1.4 Impact of the COVID-19 pandemic

In early 2020, the planet was hit full force by a pandemic, caused by the coronavirus known as COVID-19, and its variants. We could not, in good faith, fail to include a note about the effects the pandemic has had on the autonomous vehicle industry. The financial repercussions of the pandemic will be grave, and sometimes extremely painful for certain automakers, OEMs, and partners (it is likely that the effects will be felt over an extended period of time).

For mainly economic and budgetary reasons, the pandemic had the effect of, for example, greatly accelerating the transitions from combustion engines to hybrid and/or electric engines, to cope with technical demands, leading to immediate (short-term) financial gains.

With respect to the advent of autonomous vehicles, the crisis:

- Lent momentum to projects to develop "robot" vehicles (autonomous mini-trucks to make deliveries, robot taxis, etc.)
- Delayed/postponed certain long-term projects to develop higher-level or top-of-the-range vehicles, which were deemed to be less immediately and less desperately needed. Nonetheless, only part of the work on these projects has been put back until 2022.

These general comments should have no bearing whatsoever on this book, because, as stated in the preface, its content reaches a long way into the future, spanning from 2020 to 2035. Certainly, early on, we should expect a slight delay early on, in the phase of work projected for 2021–2025. However, in our view, there should be little, if any, impact on the subsequent stages in the project. Broadly speaking, the estimated time of arrival of autonomous vehicles that are truly capable of driving on completely open roads, worldwide, ought to be unaffected.

2

Aspects Relating to Autonomous and Connected Vehicles

Before embarking on any kind of project, it is always helpful to know where to tread one's feet, or risk dire disappointment. In this area, not all the scientific "stepping stones" are yet in place. In addition, as previously specified, this book is not intended to be a "cover-all". However, from long experience, we feel it prudent to briefly describe the most common concerns and questions in this area, before delving into the technical "nitty-gritty" of autonomous vehicles.

There are a multitude of constraints – technical, industrial, regulatory/normative, safety-related, etc. – which must be addressed. Unless you fully understand the issues in all of these areas, be aware any project may turn out not to be viable.

That, then, is the framework upon which this chapter is built – a discussion that it is absolutely vital to have![1]

N.B.

This book is a technical reference, and all "related aspects" in this chapter are presented in view of their technical repercussions on the design of autonomous and connected vehicles, the architectures and specifics of their communication networks, etc. The philosophical, financial, political (etc.) aspects are beyond the scope of this book.

Why speak of regulations, laws, legalities, norms, and societal aspects in this book? The answer has to do with the "intelligent" side of autonomy, radio connectivity, and the fact that a great deal of the data being processed will be provided directly by individuals and will be subject to expectations of protection. Consequently, there are an enormous number of regulations, standards, and global constraints that must be satisfied, varying depending on the production and sales sites, numerous types and levels of certification and different levels thereof, and latent protectionism that sometimes (often, even) means vehicles must be produced in the country where they are to be sold.

Autonomous and Connected Vehicles: Network Architectures from Legacy Networks to Automotive Ethernet, First Edition. Dominique Paret and Hassina Rebaine.

The "related aspects" described here are those that are not strictly / overtly technical, but exert a major influence on the technical solutions employed and on the value "1" or "0" of the last bit implemented in a vehicle's onboard computer.

As before, let us look at some of the important vocabulary we shall encounter.

2.1 Vocabulary

Once again, we begin with a recap of the terminology commonly used in the area of autonomous and connected vehicles, which can often lead to confusion:

- **Standard** (not to be confused with a norm): a technical document, proprietary in origin (and thus often including specific patents, licenses and royalties), drawn up by one or more companies in relation to a domain or a product, **with no particular implication of consensus or of national or international agreement.**
  - Example: the CAN protocol, published in 1983 by R. Bosch.
- **Consortium**: a group, partnership or association of actors – particularly of businesses – collaborating on a project or program, with the aim of achieving a particular result.
  - Example: the IEEE is a recognized consortium to achieve a particular end.
- **Norm** (not to be confused with a standard): at national level, a norm is established in the wake of discussions between multiple companies who agree to adopt a certain position. **At international forums, national proposals from multiple States around the world are voted on and adopted by consensus.** A norm allows for RAND (reasonable and non-discriminatory) licensing. It is not mandatory to follow a norm (or a recommendation), but norms may be made legally binding requirements by an act of law or decree.
  - Example: the organizations AFNOR, ETSI, ISO, and CREATE NORMS.
  - Bosch's original CAN (a "standard") became an international norm with the ISO (ISO 11898-X) after a number of modifications were discussed, resolved, and voted upon at international level.
- **Law**: a prescription produced by a State's sovereign authority, universally applicable and defining each player's rights and responsibilities (a behavioral requirement, conventions established by the members of a group, by moral or social mores, etc.).
  - Examples: hospitality laws (the law of the jungle), criminal law, etc.
- **Rules**: a set of measures to which the members of a society, of a group, etc., are collectively subject.
- **Code**: a set of laws and/or rules that govern a specific area of law.
  - Examples: Highway Code, Civil Code, Labor Code, etc.
- **Regulation** (not to be confused with a standard or a norm): a set of legal and regulatory measures that govern a particular field. In short, a "regulation" is a series of documents/official rules issued by an organization attached to a "State" or a "community of States" (e.g. the European Union), which is made mandatory by prescriptions, rules and regulations, laws, decrees, and/or other legal texts governing a social activity.

- **Recommendation** (not to be confused with a norm or a regulation): a recommendation is not mandatory but is highly advisable, especially with a view to interoperability with other technologies (in Brussels, the term used is "soft law").
- **Ethics**: the science of morality, which represents all of the moral concepts held by an individual or a group. Ethics are rules that govern the right way to conduct oneself at a given time, when faced with a specific situation.
- **Morals**: the science of Good and Evil; the theory of human duty, with the objective of doing good. Morals represent the set of behavioral rules considered good practice in relation to the mores, the rules accepted and abided by in a society.
- **Legal**: everything having to do with the law.

This brief interlude now completed, let us drill down into the core of the subject, beginning with the regulations and recommendations in this area.

2.2 Regulatory aspects and recommendations

The following sections detail the regulatory (and therefore compulsory) aspects, and recommendations (non-compulsory) in relation to autonomous vehicles at national, European and world levels. These regulations and recommendations are often wrongly viewed as being barriers to development, but, when correctly applied, they are no such impediment.

2.2.1 Current regulations and those necessary for the introduction of autonomous vehicles

At the time of writing, there are regulatory constraints governing devices in vehicles (autonomous or otherwise) and the worlds surrounding them. Historically, the current rules enforced by the Geneva Convention of 1949 (to which numerous nations are signatories) and article 8, points 1 and 5, of the Vienna Convention on Road Traffic (1968) specify that "Every moving vehicle or combination of vehicles shall have a driver," that the driver must be present in the vehicle, and that there must be a physical connection between the driver and the vehicle's powertrain: "Every driver shall at all times be able to control his vehicle or guide his animals" (see Figure 2.1).

As the regulation currently stands, the phrase "at all times" precludes the circulation of level-3 and level-4 AVs (see the exact definitions and terms for these levels), and certainly not level-5 ones! Thus, these international treaties must be modified before we can hope to legally take hands off the wheel, either partially or completely.

2.2.2 Softening of legislation on autonomous vehicles

Of course, both in Europe and, indeed, throughout the world, discussions are taking place about adapting the regulations. Groups are working on adapting the texts in force, and laying the groundwork for the introduction of high-level autonomous vehicles (more a question of adapting to accommodate an existing situation than

ARTICLE 8

Drivers

1. Every moving vehicle or combination of vehicles shall have a driver.

2. It is recommended that domestic legislation should provide that pack, draught or saddle animals, and, except in such special areas as may be marked at the entry, cattle, singly or in herds, or flocks, shall have a driver.

3. Every driver shall possess the necessary physical and mental ability and be in a fit physical and mental condition to drive.

4. Every driver of a power-driven vehicle shall possess the knowledge and skill necessary for driving the vehicle; however, this requirement shall not be a bar to driving practice by learner-drivers in conformity with domestic legislation.

5. Every driver shall at all times be able to control his vehicle or to guide his animals.

5. bis. Vehicle systems which influence the way vehicles are driven shall be deemed to be in conformity with paragraph 5 of this Article and with paragraph 1 of Article 13, when they are in conformity with the conditions of construction, fitting and utilization according to international legal instruments concerning wheeled vehicles, equipment and parts which can be fitted and/or be used on wheeled vehicles*

Vehicle systems which influence the way vehicles are driven and are not in conformity with the aforementioned conditions of construction, fitting and utilization, shall be deemed to be in conformity with paragraph 5 of this Article and with paragraph 1 of Article 13, when such systems can be overridden or switched off by the driver.

6. A driver of a vehicle shall at all times minimize any activity other than driving. Domestic legislation should lay down rules on the use of phones by drivers of vehicles. In any case, legislation shall prohibit the use by a driver of a motor vehicle or moped of a hand-held phone while the vehicle is in motion

Figure 2.1 Convention on Road Traffic (Vienna, 8 November 1968 – consolidated version).

preparing for it before it arises). Things are moving forward as, in a joint statement, the G7 Ministers of Transport declared their "commitment to seek to identify and remove potential barriers in existing regulations." A range of such barriers have been identified, including legal and ethical issues – notably relating to liability in the event of an accident. At the European level, it is a matter of defining the "major texts and conventions," for example, by reviewing the definition of a "driver". One question, among many others, needs to be addressed: "Does a driver necessarily have to be a human?" At this stage, though, it is too early to produce a final and reliable answer, and a date.

Whilst the Vienna Convention stipulates that a driver (article 1) must have control of his vehicle (articles 8.5 and 13), it specifies nothing as to that driver's physical location. Hence, it is possible for a human being to qualify as a driver even when not inside the vehicle. However, the judiciary in France, for example, has a more restrictive interpretation of the rule, holding that the driver is "the person who is behind the steering wheel or handlebars, who is controlling the vehicle at the time of the accident" (Court of Cassation, second chamber, 14/01/1987, no. 85-14655).

Nevertheless, in Europe, 23 March 2016 was a landmark occasion for all industrialists who have invested in developing autonomous vehicles. The United Nations Economic Commission for Europe (UNECE) introduced an amendment to the Vienna

Convention, stating: "from today (23 March 2016), the automated control systems will be explicitly allowed on public roads, provided they comply with the provisions of the UN Regulations on motor vehicles or so that they can be controlled or deactivated by the driver." The UNECE has also hinted that other advances are being prepared, such as the UNECE vehicle regulation, which would offer a more detailed list of authorized systems: "notably those which, in certain circumstances, can assume control of the vehicle, under the constant supervision of the driver, such as systems to maintain trajectory (to prevent accidental lane deviation), parking assist functions, and the autopilot function for highway driving (the vehicle travels at high speeds under automated control on these axes)." Put slightly differently, this means automatic control of the steering (for example, keeping the car in a given lane). In addition, the speed limit (which was 10 km/h in 2016) imposed on automated driving has been increased to 60 km/h. In due course, it should be increased again to 90 km/h – it is surely just a matter of time.

Thus, as we can see, regulations are beginning to change Europewide, which is a good thing in ensuring Member States' competitive position – in the USA, various individual States have increasing numbers of local and/or private traffic authorizations and dispensations that must be satisfied, with each State aiming to earn or maintain a leading position in the AV market. If Europe wishes to compete with North America, it must facilitate the uptake of such technologies, and continue to ease the regulations on autonomous vehicles. Automakers (CAP), automakers are working with public authorities on two levels: European and national (see the below examples of projects under way in France, Germany, etc.).

2.2.3 Germany and autonomous vehicles

As of 1 January 2021, the regulation adopted by the World Forum for the Harmonization of Vehicle Regulations, organized by the UN, allows the commercial sale of private vehicles (not transport services vehicles) with certain **level-3** autonomous driving capabilities.

More specifically, the regulation:

- Authorizes lane-keeping functions at high speeds (60 km/h maximum);
- Precludes the use of these technologies in urban areas, which are still considered to be too risky in view of the number of potential obstacles;
- Also precludes the use of autonomous driving on a highway at normal speed, because it limits the use of these systems to 60 km/h.

This means that the signatories to this regulation (the majority of industrialized countries, except the USA, China, and Canada) are now entitled to pass standards relating to lane-keeping systems. However, whilst automakers must comply with the strict requirements in this regulation (seatbelt detection, detection of driver absence or inattention, deactivation of infotainment systems during autonomous driving, cybersecurity, etc.), the various States also need to modify their own national highway codes for this purpose.

This regulation takes some of the strain off drivers, by offering automated driving – a function that is hotly anticipated by automakers, who see it as a selling point to drivers who spend a great deal of time stuck in traffic jams each day on their commute to and from work. Functions that are below level 3 are considered more to be driving supports than driving delegation.

Various automakers have already implemented true level-3 autonomous driving functions in vehicles due for release in 2022 – for example, the Mercedes S Class. For the time being, owing to the estimated slow market for business-class models, automakers such as Volkswagen and Mercedes-Benz are ready to launch this type of vehicle, but only for fleets such as taxicabs. However, this is not to say that they will not, in future, expand into the general auto market.

A draft bill to reform the Highway Code and construct a regulatory framework to govern autonomous driving has been in the pipeline since May 2020. In February 2021, the German Federal. Ministry of Transport announced a project to create a legal framework for **level-4** autonomous vehicles. The purpose of this project is to speed up the adoption of a new law, with the idea that, once State authorization has been obtained (by 2022), autonomous vehicles can be driven on German roads in specific areas. Its stated goal is to make Germany "the first country in the world to allow driverless vehicles in regular service across the whole of the national territory". The intention is for drivers to be able to let go of the steering wheel, at speeds of up to 60 km/h, in traffic jams or certain specific areas such as parking lots from 2022. The authorized situations for use of a level-4 autonomous vehicle include mail distribution, autonomous shuttles/buses operated by companies for passengers and employees, and taking them to and from the carpark. This would make Germany – which is the fourth largest auto producer in the world and one of the pioneering players in autonomous driving – the first country to make the transition from research and development to daily reality. At present, Germany is the only country in a position to benefit from this new regulation.

As pointed out above, level-4 autonomous driving is merely one step away from level 5: fully autonomous driving, with vehicles traveling without a driver, in which case the occupants become mere passengers.

The development of autonomous driving at levels 3, 4, and 5 also poses various problems – in particular, relating to the storage of driving data, routes, and liability in the event of an accident. The German association of insurers is pondering the question of where liability would lie in an accident involving an autonomous vehicle: with the owner, the maker, or a programming supplier? Germany, which held the Presidency of the European Union until July 2021, took advantage to advance the European cloud project for mobility – in relation to data protection and, in particular, recording.

For its part, the automobile association VDA – *Verband der Automobilindustrie* – which represents German automakers, believes that (a) if these regulations are passed, it will help the industry advance in the race to develop vehicles with high of autonomy and (b) it will still be several years before we see the introduction of extended level-3 functions or of level 4 (complete delegation of driving, within a predefined environment). The regulatory framework necessary to move on to the next step is currently under construction, but will not be complete for another 2–4 years. Counting the time for standardization, it will be at least 3–5 years before the framework can be implemented, and the VDA does not believe that level-5 vehicles will be operational before 2030. When, one wonders, will we see level-5 personal vehicles?

Thus, there will not be an inundation of autonomous vehicles on the road from 1 January 2022.

2.2.4 France and autonomous vehicles

In order to give a complete picture of the history of this discipline, let us briefly step back in time.

In **February 2009** (a little over 10 years ago at the time of writing), delegated driving vehicles could be "road tested," subject to the Decree of 9 February 2009 relating to the procedures for vehicle registration (JORF no. 0035 of 11 February 2009, page 2402, text no. 29).

In **September 2013**, President François Hollande announced the New Industrial France program.

In **August 2015**, the energy transition law very clearly referred to "delegated driving vehicles" rather than to "autonomous vehicles." In particular, this law paved the way for the Government to pass Order 2016-1057 of 3 August 2016 on the "testing of delegated driving vehicles on public roads".

In **August 2016**, the Ministerial Council passed an order officially allowing the testing, on France's roads, of "delegated driving cars, which are an essential step along the way to calmer mobility, better traffic regulations and traffic safety, and transport that is more efficient and more environmentally friendly, which are the future of the automobile industry." That order introduced an authorization, allowing the use on public roads, "for test purposes, of a vehicle with partial or full driving delegation," upon obtaining authorization from the Transport Ministry. It stipulated that "the driver must always be present, and alert, behind the wheel. For example, he must not be intoxicated." The order was to "serve as the foundation upon which to construct a solid regulatory framework by means of a State Council decree." In addition, the Council noted that the international rules of the Highway Code in force have made it difficult to conduct initial tests, particularly because of the legal and practical issues relating to the uncertainty as to the driver's situation (for example, whether he must keep both hands on the wheel). Thus, the order came as a great relief to enterprises already working in that field, who feared that France would fall behind in relation to this technology because of a sluggish legislative framework that is too slow to evolve.

We can now offer a few comments: remember that since 2016, prototype autonomous vehicles have been allowed to be driven and tested subject to special authorization, and have been tested discreetly (with enormous advertising logos on the bodywork!) on France's roads.

In addition, in view of the multiple players in this field (automakers, OEMs, SMEs, startups, etc.), the French Government harbors great ambitions for France to be a "testing ground," going so far as to envisage the country being a "center of excellence in onboard intelligence, and a leader in security of complex systems." Furthermore, autonomous systems offer benefits in terms of road safety. Such systems are quicker to respond and adapt than human drivers; that reactivity, in combination with the lack of factors that adversely affect human driving (fatigue, intoxication, distraction and inexperience, for example) should help bring down the number of injuries and deaths caused on our roads. One can always dream!

In **May 2018**, the French Government published a lengthy and highly instructive document, entitled *Développement des véhicules autonomes*, which is still current, setting forth the "strategic guidelines for public action" in relation to the development of autonomous vehicles. It defined autonomous vehicle development as a priority in its long-term industrial recovery policy. It also represents a critical policy issue in the areas of transport, digital security, and road safety.

This report provides the strategic framework around which the French Government structures its actions to foster the development of automated or autonomous vehicles. It presents a series of priority actions to facilitate the emergence and development of these technologies, to help French businesses gain a strong position in this singularly buoyant market, and also to address the challenge of the safety, security, and acceptability of these innovations.

In **August 2018**, with a view to having level-3 autonomous vehicles on the road in France in 2020, the State launched a call for (in the sense of a call for tenders) experimental projects, allocating them a budget of around €300 million. In addition, automakers have agreed that, by 2022, AEB should be included as standard on all new vehicles.

In **October 2018**, the schedule attached to the strategy allowed for the gradual roll-out of autonomous vehicles between 2020 and 2022:

- Firstly, level-3 automated vehicles will be authorized (such vehicles have a certain degree of autonomy, but the driver is required to intervene in the event of a problem), as well as supervised autonomous public-transport services (for example: shuttles operating in Paris – La Défense and Lyon) (Figure 2.2).
- In **2019**, driverless cars were tested in France. However, for the time being, fully autonomous vehicles (level 5) are not yet authorized to be on the open road, and are unlikely to be for a long time to come. It is important not to confuse the concepts: car, prototype, preproduction vehicle, production vehicle, and full-production vehicle.

Today, though numerous technologies are built into ordinary vehicles, which are capable of doing more than a standard ADAS (see Chapter 3), the legislation does not allow a vehicle to partly or entirely substitute for human decision-making while driving. The driver's action must always supersede the machine's predictive analysis.

Figure 2.2 Example of an autonomous shuttle.

Thus, with the official regulations as they currently stand in relation to autonomous driving, there is no legal way to surpass level-3 autonomy (or in rare cases such as valet parking, level 4) and most use remains at lower levels.

Example from the Île-de-France region

The Île-de-France regional transport authorities and Île-de-France Mobilités (IDFM) have invested to transform infrastructure in order to increase testing, and finance the installation of dedicated lanes with embedded information systems, on the region's highways (in partnership with the State, which owns the infrastructure) and on certain roads managed by local authorities, to allow fleets of between 20 and 40 autonomous vehicles to circulate.

The work was due to take place in 2020–2021 to allow *in-situ* experiments to be conducted (Figure 2.3). Trials of autonomous shuttle services have already been run, and the Olympic and Paralympic Games to be held in Paris in 2024 are intended to showcase France's expertise in this field.

2.2.5 The USA and autonomous vehicles

A major connected problem

Two nations that are major players in the automotive industry and autonomous driving – China and the USA – do not subscribe to the UN regulations. Industrial actors in China and the USA work with their own complex sets of standardization rules (see below), but are not bound by the laborious discussions at the UN, and it takes time to find a compromise that is acceptable to dozens of States. In fact, industrialists in these nations are advancing more swiftly, but have not yet rolled anything out on a large scale, and there are still many ripples from the announcements that have been made.

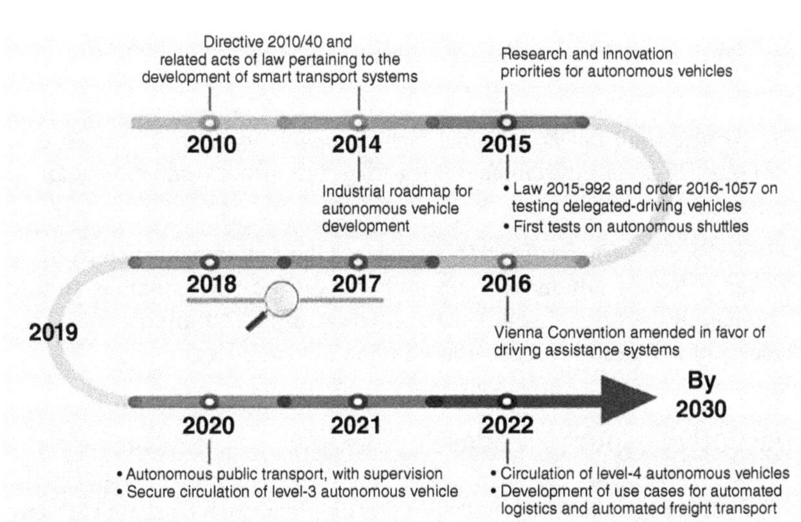

Figure 2.3 Projected schedule for "autonomous vehicles" expertise in France.

As usual, there are financial considerations underlying the technical concerns. If and when the USA announce their first standards for autonomous vehicles, they will attempt to impose them on all automakers, even beyond their borders, including in Europe. However, at the time of writing in 2021, it is estimated that both the Chinese and the Americans will need another 3–5 years to create a full legal framework that is shared in all the country's states or provinces, and then implement standards for more advanced autonomous-driving functions.

In the USA, in all matters relating to "autonomous vehicles" (types, driving, tests, development milestones, etc.), reference must be made to the Department of Motor Vehicles (DMV) and its local iterations, state by state. Indeed, each state is free to introduce its own regulations and bills of law.

In view of the industrial actors and the potential "early adopters" in these states, California and New York are highly active. However, in light of the rising commercial demand for driverless vehicles (taxis, last-mile logistics, vehicles for the disabled, etc.), many other states are now beginning to take up the challenge.

For details of the regulations, readers are recommended to refer to two main texts:

- California Vehicle Code (VEH)
  DIVISION 16.6 – Autonomous Vehicles [38750]

This document defines the express terms – for example:

- (a) "*Autonomous mode*" *is the status of vehicle operation where technology that is a combination of hardware and software, remote and/or on-board, performs the dynamic driving task, with or without a natural person actively supervising the autonomous technology's performance of the dynamic driving task. An autonomous vehicle is operating or driving in autonomous mode when it is operated or driven with the autonomous technology engaged.*
  - etc.

and:

- Adopted Regulatory Text
  Title 13, Division 1, Chapter 1 Article 3.7 – Testing of Autonomous Vehicles

Brief extract

  - *A motor vehicle shall not be operated in autonomous mode on public roads in California except as permitted under Vehicle Code section 38750 and the regulations in this article.*

Additionally, in 2018, the California DMV established the Autonomous Vehicle Tester (AVT) program, whereby automakers are authorized to test autonomous vehicles without a human in the driving seat. For information, as at **27 January 2021**, there were six holders of test permits for driverless autonomous vehicles.

2.3 Legislative aspect – Codes

For a representative example of the regulatory side of things, we look again at France, at the "*Code de la route*" (Highway Code), whose content is very closely linked to the text of the Vienna Convention. On this subject, let us recall three simple, classic and day-to-day examples that are representative of the French Highway Code[2] as in force in early 2020, and discuss whether a "level-5 autonomous vehicle" can conform to it.

2.3.1 Article R415-5

This article, which is incorrectly known as "priority to the right" is as follows:

> "When two drivers approach an intersection on different roads, the driver coming from the left is required to give way to the other driver, unless otherwise specified by this document.
>
> Any failure to abide by the rules of priority set by this article is punishable by the fine set for class-4 offenses. Any driver found guilty of such an offense will also have their license suspended for a maximum of three years. That suspension may be limited to driving for purposes other than professional activity.
>
> Such an offense will automatically incur four points taken off the driver's license".

Issues relating to level-5 technology

To begin with, when dealing with level 5, the term "driver" must be globally replaced by "vehicle", as there is no longer a driver to speak of.

- How is it possible to enter a square on the edge of a city, or a massive circular stretch of road such as the *Étoile* in Paris, at 6 p.m.? It becomes completely impossible, and the vehicle must "take liberties" with the Highway Code to join the flow of traffic.
- What happens if the vehicle coming from the left, which is only "required to give way," does not?
- The vehicle will be given a fine, and go to take out its wallet to pay it... and then remember that it does not have a wallet!
- It is impounded for three years... but it is released every day in order to go to work!
- Whose license is stripped of the four points? The vehicle does not *have* a license!
- Many similar questions arise.

In short, the legislation on penalties must evolve and, of course, we must break away from rather humorous remarks such as those above, but they do adequately demonstrate how far we still need to go.

2.3.2 Article R110-2 – on stopping

Definition of stopping: "when a vehicle remains stationary on a road for a moment, for as long as it takes to allow passengers to enter or exit, the vehicle to be loaded or unloaded, with the driver remaining at the controls or nearby so as to be able to move the vehicle if necessary..."

Question relating to level-5 vehicles

How, and on what basis, does the vehicle judge "as long as it takes" and interpret "if necessary"?

2.3.3 Article R417-10 – on stopping and parking

The article reads as follows: "Any stationary or parked vehicle must be positioned so as to cause the least possible obstruction to circulation. A stationary or parked vehicle is considered to obstruct public circulation if it is:

- On the sidewalk, in the case of mopeds or similar;
- In spaces reserved for the stopping or parking of public passenger transport vehicles, taxis, official carpooling vehicles or vehicles assigned to a public service. However, the police authorities may define certain hours during which parking is authorized;
- Between the road edge and a solid line, when there is not sufficient space remaining between the line and the vehicle to allow another vehicle to pass without crossing or driving on the line;
- In parking spaces where the vehicle is blocking another vehicle from getting into or out of a parking space;
- On bridges, in underpasses, tunnels, and under overpasses, unless otherwise provided for by the police authorities;
- On emergency stop lines, unless in case of an absolute necessity;
- On a public road especially designated by decree of the police authorities;
- In front of the drivable entranceways to local buildings;
- Double parked, except bicycles, two-wheeled mopeds, and motorcycles without a sidecar;
- In front of electric vehicle charging stations;
- In loading bays, reserved for delivery vehicles. However, the police authorities may decide on certain hours when parking is permitted;
- In meeting places, outside of a designated parking spot;
- In pedestrian zones, with the exception of bicycles using designated cycle parking infrastructure;
- Above signposted access to underground installations.

Any obstructive stopping or parking under the terms of this article is punishable with a fine as defined by Class-2 offenses. When the driver or owner of the vehicle is not present, or when asked by officers, refuses to move the vehicle so as no longer to cause an obstruction, the vehicle may be immobilized and impounded."

Questions relating to level-5 vehicles

The quote above may be astonishingly long, but imagine the headache of all these articles on prohibited stopping and parking, which a hapless level-5 (driverless) vehicle's onboard artificial intelligence (AI) must take into account, when offering valet service (a level-4 service). A passenger could come back two hours after leaving the vehicle to find it still circling the neighborhood in search of an appropriate space, having been unable to solve the problem.

Of course, all the above points are pure fantasy at present, as in order to comply with the Vienna Convention, in a so-called "autonomous" vehicle, the driver must be able to oversee the vehicle's functions and resume control at any moment (up to level 4). Thus, there will always be a driver (or, at the very least, someone) who is responsible, and can therefore be punished in the event of an accident. These conditions become a legal constraint in the context of projects to develop level-5 autonomous vehicles on the open road. Consequently, from the standpoint of the safety of the traffic system, the conditions imposed are wise in that it is difficult to conceive of an artificial intelligence system that is sufficiently reliable to substitute a driver in every way – and which is perfectly cognizant of the ins and outs of the Highway Code. It should also be noted that in the aeronautics industry, both a pilot and copilot must be in the cockpit, despite the fact that autopilot is available and is widely used.

In addition, it is important to take account of other, more delicate notions. For example: often, before a pedestrian crosses the road, they exchange glances with the driver of an oncoming vehicle – the pedestrian wants to be sure that the driver has seen him and intends to stop. Sometimes, the rights of way as set in law are not necessarily respected, be it out of habit, necessity, or courtesy. Indeed, a pedestrian may wave to a driver who has already been waiting for a long time to let the driver go, or the other way around. In this type of situation, the Highway Code offers no viable solution. In practice, other rules fill in the gaps where matters are not covered by the law, which is not designed to govern in fine detail – other rules such as custom, usage, and other social mores. These rules, and the way in which they are implemented, vary from one country to another, and even from one locality to another. Merely rendering the Highway Code as computer code is not enough, in this situation, to ensure the vehicles' autonomy. What would artificial intelligence do to resolve problems such as this?

As autonomous vehicles refine their artificial intelligence through coming into contact with practical situations and coexisting with vehicles already on the road, the solutions found and repeatedly employed by the machines will constitute new habits, in which case the computer code simply "begins to create". It remains to be seen how the AI's habits coexist with human ones. This is why the integration of truly autonomous vehicles into society currently remains a veritable challenge: the need to find a common language that is shared by all road users – in France, in Europe, but also, ultimately, all over the world.

2.4 Normative aspects

Though autonomous vehicles represent a market that is still emerging, even at this early stage, we need to develop norms to structure that market, or else allow the ecosystem to develop by itself. To facilitate the emergence of autonomous vehicles, two approaches can be envisaged to establish norms:

- Firstly, encouraging innovation by leaving the field open to proposed initiatives;
- Secondly, taking the view that a lack of standardization could lead to a fragmented ecosystem.

From a normative standpoint, this leads to two competing hypotheses:

- Either we should act in advance, to guide the development of the market, or
- Once the market has established itself, we should spring-clean the excessive number of proprietary solutions that have emerged and impose the most fitting standards on the whole market.

It should be noted that the current market in applications for autonomous vehicles straddles the boundary between these two hypotheses, because the true industrial market is getting going, and that there is still time to structure these applications with conformity to norms. We shall now briefly introduce the main standardization bodies that deal with autonomous and connected vehicles.

2.4.1 ISO, CEN and IEC, CENELEC

Note, firstly, that the CEN (*Comité européen de normalisation*, European Committee for Standardization) has struck agreements with the ISO (International Standardization Organization) and, secondly, the CENELEC (*Comité européen de normalisation en électronique et en électrotechnique*, European Committee for Standardization in Electronics and Electrotechnics) has done likewise with the IEC (International Electrotechnical Commission). All four organizations exercise the advantages of international standards for harmonizing trade and markets. To complete this preliminary structural section, in the CEN and ISO, France is represented by the national standardization body AFNOR, which, by delegation, is represented by the BNA (*Bureau de normalisation de l'automobile*, Automotive Standardization Bureau).

2.4.2 BNA

The BNA's mission is to:

- Study the requirements for standardization in the areas of road vehicles, emergency systems and two-wheeled vehicles (whether motorized or not);
- Establish a yearly schedule for standardization in these domains;
- Provide assistance to the standardization bodies in these fields – notably helping to draft norms; define the positions of French players, which will be championed within the European and international bodies (the CEN and ISO); appoint French experts to serve at the CEN and ISO; train and deploy experts; hold the secretariat for ISO/TC22 "road vehicles" and one of its subcommittees, CEN/TC301 "road vehicles" and numerous working groups belonging to the different committees followed by the BNA.

Thus, in relation to standardization, the BNA is the body to go to in France. There are, of course, equivalent organizations in other countries.

2.4.3 ETSI

For the connected part of a connected (and autonomous) vehicle, for many years the ETSI (European Telecommunications Standards Institute) has been developing standards in ITS (Intelligent Transport Systems) and all aspects relating to telecommunications with the connected vehicle (V2X solutions, etc., which we shall examine in Section 4.2) and with road infrastructure (traffic lights, road surface coatings, etc.).

2.4.4 ASCQUER

To complete this section on the norms and standards in place, we must also mention the road infrastructure, without which many functions of autonomous vehicles (e.g. ADASs) could not exist, or would not work as well as they do (for example: good guiding lines on the road surface, clear signage, traffic lights of known colors, etc.). In this domain, in France, the ASCQUER (*Association pour la certification et la qualification des équipements de la route*, Road Infrastructure Validation Association) has the objective of ensuring the network infrastructure (signs, traffic lights, message panels, white strips, etc.) is consistent (for example: broken lines on the road of the same size, defined colors, etc., so as to be absolutely sure of the lane guidance; signage with consistent iconography to indicate danger, slowing speeds, etc.).

This association helps implement procedures to assess and certify compliance, in order to ensure product quality in relation to manufacture, usage, and installation. It supports standardization efforts at national, European, and international levels to the development of innovation and to the drafting of regulations. In addition, under the banner of AFNOR Certification, it assigns usage rights for NF standardization and carries out the assessment procedures and compliance checks of CE certification. The ASCQUER participates in standardization, technical approval, and compliance certification programs.

2.5 Legal aspects

A number of accidents causing bodily harm, involving autonomous vehicles, occurred during the first few years of their use. This underlines the fact that the question of where responsibility lies in relation to users and operators of these "driverless" (or nearly driverless) cars was far from clear. The following sections briefly set out the multi-layered legal aspects surrounding autonomous vehicles and the technical consequences of these legal considerations when designing a vehicle.

2.5.1 International aspects

The Vienna Convention, defining the international regulations on road traffic, stipulates that "Every driver shall at all times be able to control his vehicle," but since 2016, it has allowed for automated systems "provided [...] they can be controlled or deactivated by the driver." In the USA, in September 2016, the House of Representatives passed a federal law to make the rollout of autonomous vehicles easier by preventing individual States from imposing overly restrictive regulations (the Self-Drive Act, H.R. 3388). For example, Arizona has introduced greater incentives to test autonomous vehicles on its roads to compete with California, which has stricter controls on such testing.

2.5.2 Example: French national aspects

Legislative framework tailored to connected autonomous vehicles

To give readers a concrete idea of existing legislative frameworks that are tailored to connected autonomous vehicles, let us consider France as a case study, which closely mirrors the actions taken at European level.

Recap and additional information:

August 2015: the law on energy transition for green growth authorized the testing of autonomous cars on public roads ("vehicles with partially or totally delegated driving") (L. no. 2015-992, 7 August 2015, art. 37, JO 18 August).

August 2016: the order implementing the above law stated that the driving of a vehicle with partially or totally delegated driving on a public road was to be subject to the issuance of an authorization, to ensure the experiment would be conducted safely (Ord. no. 2016-1057, 3 August 2016).

March 2018: the decree set the conditions for issue of such authorization (Dec. no. 2018-211, 28 March 2018).

May 2018: wishing to develop urban transport in the coming years (MaaS, Mobility as a Service), France presented its strategic guidelines for public action in the development of autonomous and connected vehicles, aiming to have level-3 autonomous vehicles (conditional autonomy) on the road by 2020 and level-4 autonomous vehicles (high autonomy) by 2022. The draft law authorized the government to pass orders to construct a suitable legislative framework. The purpose of this measure was to enable the government to take legal actions to regulate the use of connected autonomous vehicles on public roads. It aimed to do so by specifying the conditions for such vehicles to be used and, in particular, the applicable liability system; the August 2016 order was limited to circulation of these vehicles for test purposes.

September 2018: article 43 of the draft "PACTE" law, relating to the growth and transformation of enterprises, was examined and amendments passed in committee. After an initial phase (2014–2017) to validate the technology in the real world, the rollout of this strategy aimed to bring about a wider variety of use cases and test how the vehicles perform when driven by non-expert drivers, and even when driverless. A number of important points emerged:

- **Evolution of the concept of a driver outside of the vehicle (e.g. for valet service)**: the concept of a driver is crucial in relation to road law and for determining the relationship to events on the road. The ability to designate a person outside of the vehicle, who can at any moment resume control of it, as a driver, is essential. A range of important amendments have been adopted, with the aim of better defining what or who a "driver" is and the role of a driver:
  - The concept of a "driver outside the vehicle." To satisfy the stipulations of the Vienna Convention, this must be a human driver who is in charge of overseeing the experiment, rather than the driving of the vehicle itself, but must be able to take over control if need be;
  - Resumption of control from outside the vehicle helps to manage incidents and situations where the vehicle urgently needs to stop or to move;
  - The external driver must be able to take control of the vehicle in order to ensure the safety of its occupants, but also of other road users.
- **Public transport lanes**: an amendment has been passed, extending the types of lanes in which autonomous vehicles can travel:
  - The traffic authority can authorize testing in the reserved public transport lanes. The objective is to allow the local traffic authorities to judge whether such authorization is appropriate, without disrupting the existing traffic in these lanes.
- **Attempt to define a liability system**: in relation to the driver's liability, the amendments look at the sequence of the driver's actions over time, whether inside or outside the vehicle (levels 3 and 4, for example):

  - This ensures that the delegated driving system keeps the driver informed of its status in real time. This information is important in defining the driver's liability;
  - When the vehicle gives the driver auditory and visual prompts to resume control, it will take the human a certain amount of time to do so. The acceptable response time must be clearly defined from the outset. Only after this window has passed may the human driver be held liable – in other words, they are liable once they have had sufficient time to resume control.

On the other hand, many amendments regarding civil liability have been voted down – notably two of them, which proposed better developments to define the transfer of responsibility between the drivers inside and outside the vehicle. After restating that "the driver can only be held liable once she/he has resumed control of the vehicle," it was stressed that "if the automated system is deactivated, then the person inside the vehicle shall be deemed liable. If the system is activated and that person has been alerted in good time, they shall not be held liable."

Thus, the debate about civil liability and insurance for autonomous vehicles will not be solved by the PACTE law. Article 43 speaks only in the context of testing. The subject of the insurability of autonomous vehicles in use cases of commercial rollout will be addressed by the LOM (*loi d'orientation sur les mobilités*, Mobility Guideline Law). In the PACTE bill, the representatives notably specified the system of liability that applies in the event of an accident:

- Greater flexibility of the framework to authorize autonomous vehicle testing, as these vehicles are, to a greater or lesser extent, able to operate without human intervention;
- The possibility of conducting tests without a driver inside the vehicle (currently, the legislation specifies that there must be an operator on board);
- The text specifies the system for determining civil and criminal liability, notably stipulating that it will be the authorization holders – not the drivers – who will be held liable should an accident occur in autonomous mode;
- While the original draft bill stated that the driver should be held liable once more as soon as the system requested him/her to take back control, two amendments have been adopted, specifying that in the event of an accident, the driver can only be held liable after a certain reaction time to allow him/her to resume control on prompting. That time window may be specific to each test, but must be defined, and the drivers must be made aware of it;
- Tests are "absolutely necessary" because "countries which lead the way will gain a great deal in industrial terms, and in terms of innovation." In addition, we can clearly see the fundamental problems that this poses in terms of liability, which must be defined as precisely as possible;
- Improved public information about the circulation of a delegated-driving vehicle. The actors in districts where autonomous vehicles are to be tested in real-world conditions must be informed in advance of their rollout. The aim of providing such information is to increase citizens' awareness and also to educate users, allowing them to acclimatize to the presence of autonomous vehicles in their environment. The means of disseminating the information can be determined by regulations. This amendment also specifies that the test must serve to assess the safety and acceptability of these vehicles, so the data collected must lend themselves to such an assessment.

2019: the Transport Minister and Environment Minister introduced the LOM, which aimed, firstly, to allow regional authorities to make their own decisions and, secondly, to propose alternatives to motor vehicles. Autonomous and connected vehicles are among the proposed measures to encourage the development of new transport solutions. These vehicles must drive "in accordance with the Vienna Convention of 8 November 1968, by 2020 or 2022." Thus, the legislative framework created by the law allows, by 2022, for vehicles on France's roads "whose driving functions have been delegated, in part or in full, to an automated driving system." In addition, the adaptation of transport infrastructures to the needs of new technologies, as well as to testing and pilot programs, "is a crucial step for networks' future performance and the attractiveness of France." Finally, transport will continue to be part of investment programs targeting innovation in order to encourage industrial developments such as autonomous vehicles, as well as alternative forms of fuel and energy.

For information, in France:

- Renault is intending to commercially release fifteen semi-autonomous models by 2022.
- Stellantis (PSA+FIAT Chrysler) is hoping, by around 2022–2023, to commercially release the first models that will allow drivers to take their eyes off the road. On 27 March 2019, the CEO of PSA (Carlos Tavares) announced that the group would not seek to go beyond level 3 in relation to autonomous vehicles. Thus, for the time being, they are not expecting to develop functions characteristic of level-4 autonomy (delegating certain tasks to the vehicle's system), and certainly not to make fully autonomous (level-5) vehicles. With a smile, Carlos Tavares explained the reasons behind this decision: "In view of the additional cost of the technology, the cost of the car becomes such that someone who can afford it is no longer behind the wheel, but in the back seat..." Stellantis is not the only company to have acknowledged that it would be difficult to surpass level 3 within a reasonable time period, and impossible to achieve by 2022, in light of the regulatory and legal obstacles, and the reticence of insurers. The year 2030 is often cited as a target to achieve total autonomy of private vehicles. Of course, this does not preclude the possibility of bringing autonomous shuttles (level 5) into service on restricted or private circuits (e.g. ski resorts or university campi). Readers may refer to the foreword and preface of this book, quoting remarks from certain prestigious engineers, who state that level-4 and level-5 autonomous vehicles will not be truly industrially reliable and viable until around 2040 in the best-case scenario, and that they will not make an impact on road conditions until 100% of vehicles on the road worldwide are autonomous or, at the very least, interconnected. In light of their announcement, we see the real differences in points of view between the directors of automakers, mechanical or computer engineers, thinkers, and policy-makers. The fully autonomous car is no longer a priority.

Criminal liability in the event of an accident

The PACTE law, adopted in October 2018, clarifies the system for determining criminal liability in the event of an accident during autonomous vehicle testing. It contains a clause that exonerates the driver of an autonomous vehicle from criminal liability for offenses committed when the driving delegation system, which has been activated in accordance with its conditions of use, is operating (article 43 of the PACTE law – see the amendment in the previous section). It is hoped that the legislative framework will be adapted so as not to hamper tests, which are unavoidable waypoints on the path of "safe" development of autonomous and connected vehicles. However, it is far from certain.

In a ruling issued in November 2018, the Council of State held that the regulatory framework should be governed by the Vienna Convention on road traffic (November 1968). It drew the Government's attention to the fact that policy decisions to be enacted by order "can only enter into force once the Convention on road traffic has been revised [...] Only that convention can allow for the use of highly automated vehicles on public roads." The Convention was, indeed, amended in 2016, but in terms that are subject to interpretation, and which do not allow the use of fully autonomous vehicles – i.e. without a "driver/supervisor" on board.

Legal and regulatory impacts connected to the introduction of autonomous vehicles

Let us now look at the legal and regulatory impacts that may come about as a result of the introduction of autonomous vehicles.

Determination of legal liability

In the event of an accident, who is responsible? At present, if criminal liability is to be assigned, it lies with the driver, because we are dealing with autonomous vehicles that do still have a driver (up to level 4). This is absolutely not the same thing as a "driverless car with no steering wheel". However, the company that obtained the authorization to place the car on the road may be held civilly liable.

Historic example of an accident In February 2016, while the autonomous "Google Car" had covered nearly 1 500 000 km in autonomous driving mode (and later a stint with UBER) – it was involved in an accident... and about time, too! The autonomous vehicle was found to be at fault. It was traveling slowly (at 3 km/h) when it collided with a bus traveling at 25 km/h in a parallel lane. Ultimately, the accident only caused minor damage to the autonomous car's wheel and fender, but the very fact that it occurred sparked a lively debate, around the world, about autonomous driving, safety, and the reliability of driverless cars.

Comments

1. Up until that point, collisions involving autonomous vehicles had all been caused by human error (either the driver present in the vehicle who wrongly intervened or by another road user).
2. The first fatal accident involving this type of vehicle was a collision with a truck at over 120 km/h. Ultimately, the inquiry cleared the automated system, finding that the driver's behavior was not appropriate to allow him to take over the control, even after several requests from the system.

Analysis of the circumstances of the accident Even at this stage, it is interesting to study and dissect the different stages of the accident, so as to understand how and why the autonomous vehicle's algorithm misinterpreted the situation, because it allows us to draw certain interpretations.

Let us return to the example in the above paragraph. The accident took place in the heart of a city in California. Driving in the right-hand lane of a boulevard, the autonomous car detected the presence of an obstacle (sandbags, for roadworks) in its lane, and was forced to turn into the left-hand lane. However, at that moment, a short distance behind, there was a bus. The autonomous vehicle did indeed detect the presence of the bus, but as the autonomous vehicle knew the Highway Code like the back of its metaphorical hand, it applied what it took to be the priority rules, turning into the left lane in the belief that the oncoming vehicle on the left would have to slow down or stop to allow it to pass. Unfortunately, that was not the case. Thus, the front left wing of the car collided with the front right side of the bus, damaging the fender and several sensors on the autonomous vehicle.

The benefit of dissecting this incident lies in the realization of the fact that the algorithm ought to have tried to anticipate – and, in particular, allow for – human behavior. It may be considered that had the algorithm been less advanced, it would simply have stopped the vehicle and waited for a better opportunity to turn into the left lane on the boulevard. It was not a detection error that caused this accident (the car's sensors correctly registered the bus's position and speed), but an interpretation error. This behavior is anchored in a constructive logic whose aim is to give autonomous vehicles a more "human" conduct and artificial intelligence (however, that is not enough if it also needs to take account of motorists who fail to respect the Highway Code, cut in, and motorcyclists and cyclists, ambulances, etc., driving erratically and acrobatically!).

For the time being, level-5 cars are not available to the general public, but we shall see liability shift away from drivers (who will no longer exist in level-5 cars) toward automakers or operators. The choice of decision-making algorithms must be fair and transparent. This fundamental dilemma must not be concealed from the public, because it is a true democratic challenge, the rules of which must be decided upon by Parliament after a nationwide debate. However, between the starting point (today) and the destination ("tomorrow"), there will be a lengthy transition phase, and the process will be very interesting to observe. Many actors (e.g. governments, insurers, etc.) call for in-vehicle recorders to be installed to determine whether the driver or the onboard computer was in control at the time of an accident, but such mini-spy cameras are not always welcomed by the public. Nor are they acceptable to the CNIL, or under the GDPR!

Similarly, with the eventual arrival of 100% autonomous vehicles, insurance as we know it, where the owner of the vehicle is insured, is likely to vanish, but will certainly be replaced by something else. Others believe it will increase sharply (see Section 2.6 for a discussion of insurance).

2.6 Insurance-related considerations

Up until now, the model of car insurance has remained largely the same for decades. However, a host of new connectivity techniques and technologies (such as ADASs and most forms of autonomy) are reshaping the future, in terms of "who pays, for what, and when." At present, it is not possible to categorically state who will be liable for vehicles that drive themselves (either partially or totally) and, apart from a few comments from automakers committing to accepting a portion of liability, nobody wants to answer these questions precisely. What is certain, though, is that the technology is coming and, sooner or later, the issue will have to be addressed.

Let us now briefly examine the role of insurance and insurers, and the questions of liability in the rollout of autonomous vehicles.

2.6.1 Civil liability

In regard to civil liability, in France, Law no. 85-677 of 5 July 1985 establishes a liability system that enables victims (having suffered bodily or material harm) to be indemnified quickly and certainly. The insurance system based on the obligation for insurance to cover that liability appears to be able to operate as it currently stands in cases of automation, including full automation. Thus, the present legislation need not be a barrier to the development of automated vehicles. This needs to be modulated in each individual country, on the basis of the local laws.

In relation to autonomous vehicles, there is no clear definition of who, or what, the "driver" is; the vehicle owner is presumed to be its keeper. This exclusive liability system is paired with an insurance system based on the obligation to have insurance to cover civil liability. This national device, which offers particularly effective protection, ensures the victims receive compensation from the insurer of the vehicle at fault (except in the event of inexcusable or intentional misconduct). Subsequently, the insurance company may pursue reimbursement from the person responsible. The existing legislation ensures that, in all circumstances, the party at fault remains solvent.

This two-part system appears to be applicable to automated vehicles, even when no driver is present. The absence of a driver, or lack of control by a driver, makes no difference, for the application of the liability system or for the obligation to insure against civil liability. The Victim compensation would still be based on the concept that the automated vehicle is at fault, irrespective of whether a human driver is present or at fault. Once that compensation has been paid, incidents can be examined on a case-by-case basis to establish where the liability lies (with the automaker, OEM, software supplier, other vehicles, infrastructure, etc.). Notably, it will be a matter of determining whether there were defects or failures in the automated system, and thus whether liability lies with the automaker, the designer, the software supplier, or any other party involved in producing that automated system.

2.6.2 Criminal liability

In criminal law, an offense is held to have been committed by the responsible party who is identified. In addition, by virtue of the principles of legality of offenses, penalties, and adaptation of penalties to fit the offense, only the person identified in the offense may be held liable. Consequently, recognition of a liable party other than the driver (or the holder of the vehicle registration) requires the criminal liability laws to be amended. With respect to the driver's criminal liability, as driving tasks continue to be automated in vehicles, it is important to distinguish between two situations:

- Cases where the driver must remain at the controls of the vehicle (up to and including level 4), using certain automated driving functions, and where they must, at all times, be able to resume control. In these cases, no reform of the law would seem necessary. However, it would be helpful to examine how to deal with highly automated systems, where the driver can rely entirely on the system for long periods of time, which raises the question of responsibility, notably in terms of taking back control, to address the fact that the driver is always held criminally liable even though they are relying on the system practically all the time;
- Cases of level-5 vehicles, where the driver is not required to keep their eyes on the road (or where there may not even be a driver). Here, also, it would be wise to examine the possibility of creating exceptions to the existing legislation or regulations, implementing specific rules in the Highway Code to govern driverless (or potentially driverless) vehicles (notably, stipulating that such-and-such an article of the Highway Code is, or is not, applicable).

With the aim of allowing the rollout of highly automated vehicles by 2022 and future developments beyond, the system for determining liability needs to be adapted, to reflect the changing respective roles of the human driver and the automated system, for the corresponding use cases. Such adaptations will need to be tabled, in parallel with the corresponding changes to the Highway Code, in a working group, notably including the Ministries of Justice, Transport and Industry in each country. In the meantime, to facilitate the development of tests that are exempt from the Highway Code, transport guidelines will provide a liability system that fits with the particulars of each test – specifically, the commitment of the test manager, who holds the authorization, to ensure it is run safely. Magistrates, lawyers, and police officers will be trained on the issues of liability arising because of the development and rollout of autonomous vehicles.

2.6.3 Who pays in the event of an accident?

This is another puzzle to be solved: who is to pay in the event of an accident – particularly during the lengthy transition period between conventional and autonomous vehicles, when the two coexist on the road? Should the financial burden fall on the automaker, the OEM, the technicians who developed the software, the insurance companies – or the vehicle owners?

According to certain experts, for semi-autonomous vehicles, it is the driver who is liable, except when the vehicle is parking itself, in which case the OEM will be responsible. However, there could be an insurance fund, which is paid into by all insurers to finance compensation in the event of an accident caused by an autonomous vehicle. Other thinkers imagine that automakers and OEMs will take out insurance with huge corporations, with policies that cover their entire fleet of self-driving vehicles. Of course, the cost of the insurance would be partly passed on to the customers, but those customers would no longer need to worry about their own policies directly, as the liability for any collision would fall on the automaker.

Whichever system is chosen, the solution has not yet been found, and until it is, fully autonomous vehicles are not likely to be seen on our streets.

2.6.4 Other problems to be addressed

To conclude this section, let us point to a few more examples of problems that need to be dealt with, concerning the handling of insurance in the case of:

- Theft/diversion of autonomous vehicles (levels 3 or 4) when they are autonomously searching for parking spaces, by hacking their geolocation systems;
- Level-5 autonomous vehicles that are hacked and used for smash-and-grab robberies; etc.

2.7 Moral and ethical aspects surrounding autonomous vehicles

All over the world, in differing contexts with their own cultures and sensitivities, thousands upon thousands of workers in the automotive industry (mathematicians, electricians, software engineers, vehicle testing specialists, etc.) are working exclusively on developing a level-5, driverless, autonomous vehicle. However, this presents us with another, highly complex, issue to solve, because a level-5 vehicle will have major influences on the onboard electronics (e.g. the choice of datarate, networks, algorithms, etc.) and software solutions (e.g. in relation to AI solutions and management – see Section 4.3). Only a few years ago, the driver alone was master of the vehicle and was solely responsible for moral and ethical conduct. The consequences varied from one country to the next and from one civilization to another. Now, in relation to autonomous vehicles, we must consider the fact that, when it rolls off the production line, the vehicle will have its *own* sense of "morality" and "ethics" – the sense with which its maker has imbued it, following recommendations, national laws, European or global laws, or other guidance, and will be aware of that sense when it is bought.

2.7.1 Dilemmas to be solved

The *trolley problem* was set forth in 1967 by Philippa Foot. In 2015, researchers from France's CNRS, MIT, Harvard, and the University of California developed a website that gives access to a simulator, called *Moral Machine*. It allows users to judge ethical situations connected to the use of autonomous vehicles. This simulator creates, and points out, well-known examples of ethical and moral dilemmas that may be faced by artificial intelligence, designed to make driving decisions in place of a normal "biped."

Consider the example in Figure 2.4 (one example among hundreds). Faced with the choice between running over a pedestrian who is crossing the road or sacrificing its passengers' life or safety, which option would the computer driving the vehicle choose?

It is no simple matter. We ourselves cannot know in advance what we would actually do in such a risk situation, because often we respond instinctively at the last moment – our real-world behavior has very little to do with theoretical ethics as studied by academics. In the heat of the moment, when we see an obstacle, we do not consciously ask ourselves "Should I plow into the tree or run over that pedestrian?" – we "hit the brakes and hope for the best..." often thinking mainly of ourselves, selfishly! There is no way to program such a decision into a human being!

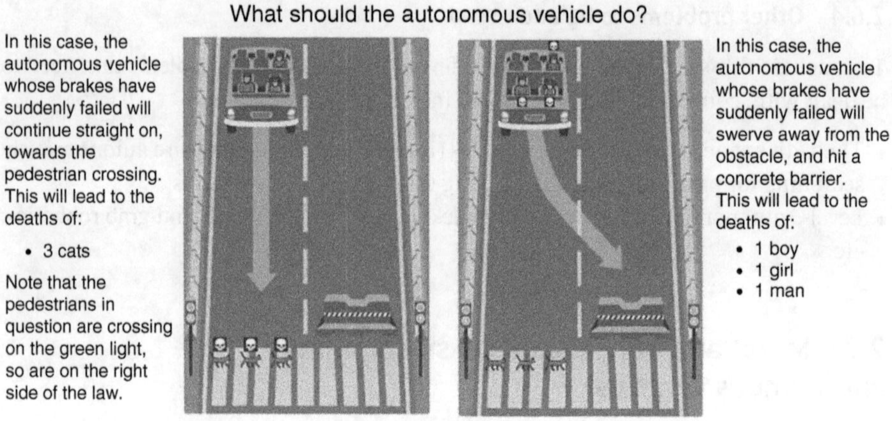

Figure 2.4 What would the vehicle's AI choose?

2.7.2 A solution to the problem

From a technical standpoint, it seems premature to reflect as deeply as we already have in this chapter. The true, achievable technical objective is to enable the vehicle to detect an obstacle as early as possible, using its equipment that is capable of scanning its environment at over a hundred meters (200–300 m – see Section 3.2), in all types of weather conditions (rain, snow, fog, fallen leaves, etc.), so the vehicle does not have to make a specific "ethical" or "moral" choice. Such preventive systems should provide redundant and complementary information about the vehicle's environment. Once captured, this information can be processed by the local AI algorithms. However, the question remains of what to do in the case of very short-range detection, and management of ethics in the (hypothetical) example that a child runs out into the road from between two parked cars, chasing a ball. Currently, the only solution is to hit the emergency brake! Today, one of the most significant challenges is to simulate and validate these systems (see Chapter 6) to demonstrate, before production starts, that the vehicles will achieve between 100 and 1000 times a lower probability of accidents than with a human driver. This naturally leads us into a more technically detailed discussion of the AI systems in autonomous vehicles (see Section 4.3).

Firstly, though, let us look at the issues of safety and cybersecurity.

2.8 Security

We turn now to matters of security, which are crucial in any discussion about autonomous vehicles and must be implemented when, in the real world, the vehicles are connected to one another, or to networks such as the Internet. To begin with, we shall select one of the many "official" definitions of security: in this case, the definition given by the ETSI:

- "Security is the capacity to prevent fraud, and the possibility to safeguard information, its integrity and confidentiality."
- Please note: "security" must not be confused with "safety"!

Now let us look at the core of the subject. The debate about security and cybersecurity, which we shall touch on here, is vast and represents one of the most major concerns in the world of autonomous and/or connected vehicles. Indeed, by virtue of the principle of similar applications, these vehicles are closely interlinked to the issues of:

- Humans' physical safety;
- Security against hacking and other forms of cyberattack;
- "Personal data" belonging to individuals – biometric, behavioral, geolocalized, etc., and therefore sensitive, needing to be handled with care (see Section 2.10.5 – GDPR) – and with high "end-to-end security" in application.

In 2016, a joint inquiry conducted by the IEEE, Agile-Io, and Eclipse IoT noted application developers' concerns about the way in which they construct their projects. End-to-end security topped the list. Alas, today, security remains a major problem. It will continue to be a prominent issue in the coming years and will remain difficult to solve, doubtless requiring new initiatives from the world of industry.

2.8.1 Weak links

In view of our lengthy experience in the field of security,[3] we were certain 20 years ago that security would become a major issue. Even today, data security in vehicles is often inadequately considered, if at all. How many vehicles are truly secured to prevent, or at least limit, this new potential battleground for hackers? Everywhere, this subject is on the agenda, not because the industry has suddenly become aware of it, but mainly for fear of media reprisals against industrialists in the event of problems and the fall-out for the companies' image and cost.

Let us briefly take a look at a (non-exhaustive) list of the background problems of typical weak links and "holes in the chainmail" of security in the elementary chain of connected vehicles, which allow hackers a great deal of room for maneuver and attack:

- There is little or no true security in the field, because the issue was not considered at the outset;
- The issue of security for autonomous vehicles has never truly been examined in depth;
- The philosophy of security has not been built in and the existing security measures are not designed to be integrated when the vehicles enter into production;
- There is no consensus as to how security should be implemented in vehicles;
- There are connected vehicles on the market that "cannot be secured;"
- Nothing, or very little, is in place to allow a lambda user to easily download an update or security patches.

Observation

We can observe the following facts, which, fortunately, are in the process of changing:

- Today, it is becoming clear that we cannot go backwards and that it is extraordinarily complicated to attempt to implement security retroactively;
- For large-scale consumer products, industrialists, wishing to limit their production costs, avoid/will avoid adding security;

- Passwords disclosed at the time of purchase (if any) or created by the user are changed only very infrequently;
- There are structural weaknesses in the networks;
- The web, network services, and mobile interface are insecure;
- The Cloud interface has little or no security;
- Authentication and authorization procedures are insufficient;
- There is a lack of uniformity in the encryption methods used for data transport;
- There are many questions still to be answered in relation to confidentiality;
- The security setup is insufficient;
- There are security issues in the software/firmware that are embedded in the components;
- There is an overall lack of physical safety, etc.

Ultimately, hackers are willing to exploit any vulnerability they find anywhere along the chain. They seek out and pinpoint points of weakness, or weak links, which they can use for financial gain. For a hacker, it is more advantageous to harvest/capture data from a connected vehicle that are stored in the Cloud than to hack the vehicle itself. Consequently, increasingly, it is data storage and big data servers that are being targeted.

To address the issue of data protection, it is crucially important to know where those data go on their journey: over the Internet, by means of which successive chains, and where are they stored in the Cloud? Indeed, in which Cloud? In which country? Who places them there? Etc.

Having made these observations, we can now take a look at some possible solutions.

Possible solutions

To remedy these weak links, there are a number of potential solutions, bearing in mind the fact that in order to implement security:

- It is important to know your enemy (doing so will point to potential solutions, but of course this comes at a cost);
- It is crucial to define the security targets. To do so, we must:
  - Know and understand the requirements;
  - Assess the risks and consequences (designer and user);
  - Know how to respond in the event of a problem;
  - Know how to communicate with clients/users;
  - Determine the price we are willing to pay.

Choosing a security target

To begin with, we must establish the "security target" we wish to achieve. This refers to two important groups of parameters:

- The parameters we wish to secure (the branches of the target);
- The levels of security we wish to attain for each such branch (the levels on the dividing radii of the target; with this in mind, we can also plot a spider graph that shows the surface covered by the security target).

Figure 2.5 shows an illustrative, highly simplified example of a security target. It may seem simple, but it still takes a great deal of time and effort to produce!

Next, we need to know whether and how the security objective can be achieved and at what cost. We must constantly bear in mind the following simple and pragmatic questions:

- Are the gains worth the investment?
- Does the chain as a whole offer end-to-end security?
- Is there still a weak link in the chain and, if so, where?

2.8.2 Levels of security applicable in vehicles

When we speak of "security" in autonomous vehicles, the concept of the level of security quickly comes up. Let us briefly discuss some such possible levels.

No security

This, of course, is the bottom-most tier, but it is true that numerous simple vehicle applications, or applications for older vehicles, simply do not need security, because the information being transmitted is not sensitive, in the worst-case scenario, could be altered without causing a catastrophe.

With security

Security for the messages transmitted and received by vehicles typically entails encrypting the data. However, even with the same principles and mathematical theories for encryption (AES, RSA< ECC, etc.), the levels of security can range from nothing at all to very high indeed. Let us explain.

"Paper house" security

This phrase refers to a very strong lock, fitted to a traditional Japanese house whose walls are made of paper. No matter how strong the lock, anyone wishing to gain entry

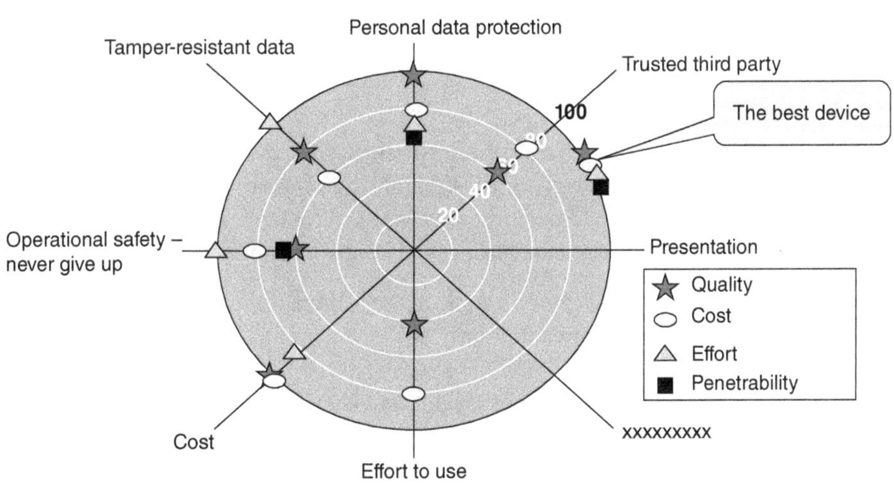

Figure 2.5 Simplified, illustrative example of a security target.

can simply pass through the walls effortlessly! Unfortunately, this is indicative of what many companies have done, implementing software codes on cryptographic units and believing that this is enough. Any hacker worth their salt will make short work of breaking the codes, using very simple tools. Thus, it is a simple and cheap approach, which is not fit for purpose.

"Sandcastle" security

Sandcastle security includes a microcontroller with a built-in "soft" cryptographic unit, which has a certain amount of protection and security at the hardware level. Of course, it is better than paper house security, which has no value at all, but under no circumstances can sandcastle security defend against all types of attack. Such units would help deter ordinary people from attacking the system, but certainly would not represent a hurdle to a talented hacker. Hence its name: it is a castle, but one made of sand!

"Stronghold" security with Secure Elements

Stronghold security is top of the range. Drawing inspiration from the medieval methods and techniques used to build true strongholds, we build a veritable fortress around our cryptographic unit, located at the very bottom of stacked levels of hardware, with multiple drawbridges, floodable moats, puzzle mazes, dungeons, oubliettes, near impregnable keeps – everything we need to truly dissuade the poorest and the best of attackers! Strategies range from attack deterrence to defense, etc., to full-on self-destruction and scorched earth policy, with the trigger phrase: "You shall never take me! I will perish if I must, but I will never reveal my secrets!"

In the wake of these lyrical flights of fancy, let us return to the technical detail. This construction defines a veritable *Secure Element*, which is the only truly secure solution (for years, all bank cards have been built around a base such as this one). The embedded solutions in level-3 and level-4 autonomous vehicles are gradually increasing to this level of security, which, in the near future, will become a necessity as a result of the type of data transported, which will be increasingly sensitive, personal, biometric, behavioral, fragile, and critical.

Now that you are aware of the issues involved in designing and implementing security in your vehicle, it is up to you to choose your camp. Should things go wrong in the future, at least you cannot say that you were not warned.[4]

2.8.3 Cryptography

In the context of autonomous and/or connected vehicles, the main purpose of using cryptography is to be able to provide guarantees, mainly in relation to security. Cryptography – encryption and decryption of data to protect the information – is one of the cornerstones of security and is implemented using both hardware and software. It facilitates:

- Identification, which helps clear up some of the unknowns, but is not sufficient;
- Authentication, which ensures that only authorized vehicles are connected to the network;
- Integrity, which checks that the message has not been altered between dispatch and receipt, and has not been interfered with by unauthorized persons;
- Confidentiality, which prevents an unauthorized user or device from obtaining sensitive information, whilst ensuring that the data are duly received by the target user;
- Non-repudiation in sending of messages between the links in the chain, which ultimately relates to the notion of a transaction and a contract between two actors.

In cryptography, various types of algorithms may be employed: symmetrical cryptosystems (e.g. AES), asymmetrical algorithms (such as Rivest-Shamir-Adelman [RSA], elliptical curve cryptography [ECC], and homomorphic functions), and Blockchain and quantum cryptography techniques, which are currently emerging and may be in phase with the production of level-5 vehicles by 2035–2040.

Conditions for the public adoption of a secure vehicle

The conditions for the uptake of secured vehicles by the general public include:

- The desired functions and applications, whether they replace old ones or are entirely new;
- An affordable price tag;
- Assured security at all levels (end to end);
- Assured privacy, in all aspects;
- If possible, interoperability with other systems, etc.; and
- Of course, satisfactory performance in terms of health, ethics, standardization, regulation, societal and environmental issues, etc.

In order to bring this about, it is necessary for:

- Security to be assured for the vehicle itself, using an appropriate cryptographic unit and, if possible, a true Secure Element;
- The entire chain – vehicle → network → processing → Cloud – to be tamper resistant from end to end;
- All problems of privacy management and other problems concerning personal data, in relation to privacy regulations, to be resolved (which should win praise from journals and consumer associations);
- Privacy-by-design to be implemented, preferably;
- Operational safety to have potential fallback positions to guarantee a minimum level of safe operation (otherwise, brand image will suffer greatly in the press or by word of mouth);
- Everything to be properly put together, on a technical level, to deal with long-distance communication, low/high datarate and low energy consumption; and
- Of course, the product to be "inexpensive," so that the general public can afford it.

To reduce costs, the silicon crystals in the integrated circuits must be small and inexpensive, so the security mechanism must be microscopic and not occupy more space than the circuit's actual primary function (see Sections 4.5.3 CAN FD and 5.3.8 Internet Switches).

2.8.4 Vulnerabilities and attacks on the vehicle chain

There are threats at all levels of the architectures of connected vehicles. Everywhere, there are potential vulnerabilities that could be exploited for malevolent purposes (cracking weak passwords, malware such as viruses, Trojan horses, etc.). To ensure the appropriate level of security for vehicles and their infrastructures (networks), it is necessary to conduct risk analyses and implement appropriate protection measures. Obviously, the levels of safeguards implemented must correspond to the tolerated risk levels and the likelihood of their occurrence.

2.8.5 Applications to standard vehicles and autonomous vehicles

To dot a few "i"s, connected vehicles often carry a great deal of personal data, in the regulatory sense of the term (biometric data, heart rate, stress level, location, etc.), and treat the individual in question with confidentiality and security. The levels of security guaranteed must generally be defined in all software packages all along the length of the chain, by the design of the different electronic cards, choices, and applications of the integrated circuits. Ultimately, they depend heavily on the potential access routes that an attacker could exploit to damage the different elements mentioned above.

2.9 Cybersecurity aspects

For the past decade or so, cybersecurity has been an important topical issue in relation to the electronics and networks in autonomous vehicles. Although there has been ample press coverage of these risks, the hacking of autonomous vehicles is still a new phenomenon. There is still a great deal of fear and misunderstanding about the subject, because not all the barriers to development have yet been fully overcome. The coexistence with conventional cars, the reliability of the electronics and the cameras, and their security against potential hackers are improving, but caution must still be exercised. Proof of this can be found in the attempts, some years ago, to seize control remotely, and in the history of accidents in autonomous and/or connected vehicles.

From a technical point of view, we shall now briefly examine these topics and demonstrate that, individually, they do not present too much cause for alarm.

2.9.1 General

Why might cars be hacked? The reasons are simple:

- Basic axiom 1: in cryptography, if something has been done, it can be undone;
- Basic axiom 2: in principle, anything can be hacked. It may take time, effort, and money, but it can be done;
- The burgeoning number of sensors, software suites, and connections in a vehicle leads to new security issues that need to be addressed;
- An average vehicle contains over 150 million lines of code, a great many ECUs (computers), and a large number of wireless connections to internal and external communication channels;
- The more functions and networks a vehicle has, the higher the potential for it to be hacked or attacked;
- Potential attack vectors include cellphone connections, Bluetooth, Wi-Fi, etc.;
- Many vehicles are connected to the Internet (even if it does not have advanced functions to stream music or provide traffic updates to the navigation system, the vehicle can be Internet-connected for the e-call automatic collision reporting system);
- In view of the complex embedded systems and logistics, the financial investment necessary to plug all possible "security holes" for millions of new and existing vehicles means it is practically impossible to protect vehicles (see Figure 2.6 for an example of attack points).

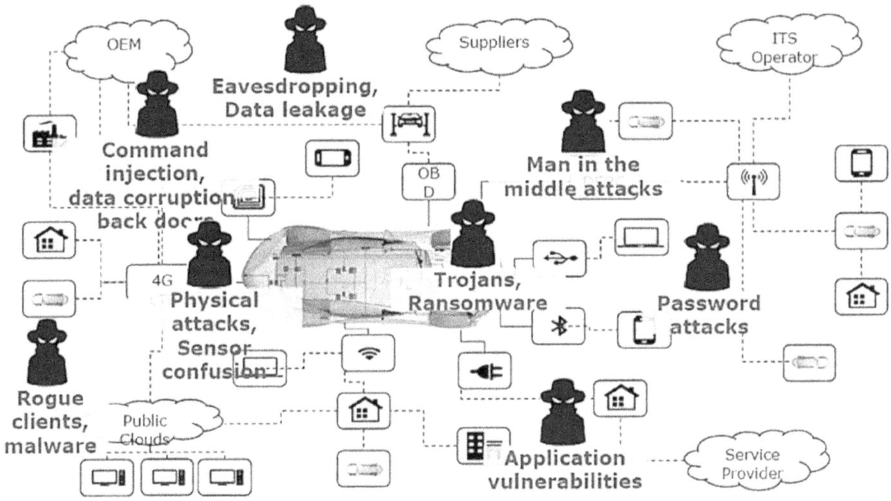

Figure 2.6 Example of the numerous attack points at which a connected vehicle is vulnerable.

2.9.2 For private vehicles

Unless it becomes truly lucrative for a hacker to invade the systems of a private vehicle (e.g. a top-of-the-range vehicle, a collector's vehicle, etc.), this type of hacking is likely to remain little more than an academic exercise. On a personal level, there is no need to panic. It is unlikely that hackers will attack private vehicles. Of course, it may always happen, and if it does, the hackers will probably aim to attack millions of vehicles at once, rather than just a few vehicles, one by one... so we are talking about fleets of vehicles.

2.9.3 The problem with fleets of vehicles

In many regards, it is a more attractive financial proposition to hack entire fleets of vehicles (e.g. robot-taxis, transporters, etc.) than individual cars, because it is easier to indiscriminately hack a large number of similar vehicles than to try to find a specific vehicle belonging to one person. To learn how to hack the vehicles may take between six months and a year. Once a hacker has found a way to hack a type of vehicle, it is often easy to hack the whole fleet. Once they have found a way to hack one of a brand's vehicles, they will soon work out how to hack almost all of that company's new vehicles. Besides, it is easier to hack all the vehicles through the Internet than to find one driven by a particular person. In short, alas, hacking entire fleets offers far more potential financial rewards. This will become a particular problem as autonomous vehicles become more and more widespread, presenting the danger that someone will attempt the "hack of the century" on a fleet! For example, imagine that one morning, a fleet of robot-taxis refused to start, with their dashboards simply displaying a telephone number to call in order to pay money (automobile ransomware) if we want them to start.

2.9.4 What is to be done?

In the knowledge that there are many potential attack vectors, and that nothing is perfect, what can we do to assuage the fear of hacking?

- Firstly, check that the vehicle software is up to date by checking with the provider. With this in mind, numerous automakers have already drawn up lists of the vulnerabilities in their vehicles' software;
- Rectify these vulnerabilities with *over-the-air* updates using the cellular network;
- Systematically repair these vulnerabilities when the vehicles are brought in for regular servicing;
- In addition to cellular modems (which are necessary and mandatory in order to save lives using automatic collision notification [ACN]), some of the risks for the vehicles could be dealt with earlier on in the supply chain.

Automakers recognize the dangers of fleet hacking and build massive security into the vehicles, including at the level of a global security architecture, with protection by encrypting communications and networks, both in and out of the vehicles. In addition, they are working hand in hand with network operators who provide the vehicles' connectivity, constructing security nerve centers in which analysts can hunt down security threats across whole fleets, and other system-level protections. Also, it must be borne in mind that everything needs to be constantly kept up to date: the job of security is never finished, because the adversaries are highly crafty, and unless you manage to outwit them and stay one step ahead, they will have won.

2.9.5 Cybersecurity and insurance

Another important point is how cybersecurity in autonomous vehicles will impact our insurance premiums.

The technology, with its accompanying need for cybersecurity, brings another change that could have a lasting impact on motor insurance. If the vehicle's known or projected security systems are "circumvented" (by hacking), who will pay for the resulting damage? Will insurance companies foot the bill, or require consumers to pay an additional premium for cyberprotection? This could well be the trigger for insurers to offer "absolute" coverage, including protection against cyberattacks, therefore covering situations in which the vehicle has been hacked or compromised, similar to a vehicle only being covered if a breakdown has caused an accident. It all depends on how the vehicles' security is implemented and their vulnerability when they are hacked.

Who is to blame?

Cyberthreats are among the potential problems that are beyond the driver's control. In the event of cyberattacks, the following question arises: "If you are not driving the vehicle, are you responsible, as its owner?" If the security system does not work, or is unable to respond, to what extent is the automaker responsible? The higher the degree of autonomy a vehicle has, the more difficult it becomes to determine where the fault lies in the event of hacking. In addition, auto insurers will need to assess a mixture of fully and partly autonomous vehicles, and those that are still driven by humans.

Connectivity could help clear up confusion created by this mixture of vehicles. In the past, there were only a limited number of points that needed to be examined in order to truly understand and assess the risk. Now, in view of the advent of autonomous and connected vehicles, there are an infinite number of data points to be examined. This ought to have more of a positive impact than a negative one. For example, in the event of an accident, the data can tell us whether the light was green or not, and who was at fault. The data can greatly help to extrapolate the situation, and determine whether the vehicle behaved as it was supposed to, even in certain litigious or critical cases.

Cost

A significant number of advanced security functions have emerged as the result of ADAS technologies. As usual, these functions were first made available in top-of-the-range vehicles, before being built into cheaper models. This has an impact on the theory of insurance costs, because:

- If, as is predicted, autonomous vehicles helped to reduce the number of accidents, then insurance costs should go down;
- Also, certain collisions should be less serious;
- These new safety functions will help ensure people survive accidents;
- If things become less expensive to produce, they should not be expensive to repair or to insure.

The likely reality is this:

- In light of the complexities of the systems, repair costs will increase;
- Consumers may not be able to rely on lower insurance premiums.

This does not mean the insurance premiums will go up. However, as always, there will be a balance that needs to be struck.

2.10 Societal aspects – GDPR

In the context of this book, why speak about regulations on individual and societal freedoms?

- The first answer to this question is trivial: because it is the law, and like it or not, we must be familiar with it and comply with it;
- The second is more concrete: in the world of autonomous and connected vehicles, there is a huge amount of information/personal data processed and uploaded in relation to individuals (e.g. biometric, behavioral data, etc.), and it is wise to treat these “personal data” with all the necessary precautions.

To some readers, the coming paragraphs may seem a little off topic, but the matter really is at the (electronic) heart of autonomous and connected vehicles, and must not be ignored.

2.10.1 Regulations on individual and societal freedoms

As before, let us begin with a little history, context, and general culture. Looking back over the past 30 years, we can summarize the history of individual freedoms, data freedoms, and privacy:

- **January 1978**: France saw the publication of the *Loi informatique et libertés* (Law on information technology and freedoms) and the founding of the CNIL (*Commission nationale de l'informatique et des libertés*, National Data Protection Commission);
- **December 2008**: mandate 8436 was issued by the CEN, the ETSI, and the CENELEC. It recommended analyzing the gap between the standards/norms existing in RFID, data protection, and privacy protection;
- **May 2011**: the subsequent analysis highlighted that there were no existing standards relating to a privacy impact assessment (PIA) mechanism and public information;
- **January 2012**: the KoM (kick-off meeting) took place, promising the publication of a standard within a maximum of two years;
- **July 2014**:
  - Two milestone norms: EN 16 570 (signage and public awareness) and EN 16 571 PIA (process for RFID applications) were published,
  - EN 16 571 passed the Registration Authority;
- **January 2015**: "Mandate 530" was published. It was a request to the European standardization bodies regarding privacy and personal data protection;
- **May 2018**: the GDPR (*General Data Protection Regulation*), published in May 2016, came into force.

Let us recap.

2.10.2 France's Loi *informatique et libertés*

For example, in France, enacted in January 1978, article 1 of Law no. 78-17 (with its successive amendments) on information technology, files, and freedoms states that "Information technology should serve each and every citizen. It needs to be developed through international cooperation. It must not be allowed to threaten human identity, human rights, privacy, or individual or public freedoms." This says all that needs to be said in relation to applications.

The CNIL

In France, the purpose of the CNIL is to provide professionals with guidance in complying with regulations relating to information technology, and assist private individuals in keeping control of their personal data and exercising their rights. The CNIL also analyzes the impact of technological innovations and emerging usages on privacy and freedoms. It works in close collaboration with its European and international counterparts to develop a harmonious regulatory framework.

The CNIL and connected vehicles

In 2016, the CNIL launched projects notably pertaining to the personal data of a vehicle driver and their interaction with the environment and road infrastructure. For this purpose, the organization brought together actors from an enormously wide variety of backgrounds (the automotive industry, insurance, telecoms, automobile, and public authorities) and developed the framework that would serve as a "compliance toolkit" – a compliance pack specific to connected vehicles. It would also encourage a new regulatory vision, which championed a positive, "privacy-by-design", approach. This toolkit comprises a set of recommendations to help ensure conformity with privacy regulations from the vehicle design stage. It is imperative that automakers come together and provide drivers with solid guarantees in this area.

"Connected vehicles" compliance pack

To safeguard automobile users' personal data and foster innovation-based ecosystems, this pack offers guidelines as to responsible data use in the next generations of vehicles. The goal is to integrate personal data protection from the product design stage, ensuring the process is transparent and the data can be checked by the subjects themselves. This approach is designed to build users' trust in these technologies, thus ensuring their continued development. It is the expression of a regulation that is both adaptive and coordinated.

Working method

The working method advocated by the pack is based on three scenarios:

- **"IN → IN"**: the data collected in the vehicle remain in the vehicle and are not transmitted to a service provider:
  - Example: an eco-driving solution that processes the data directly on board the vehicle, to display eco-driving recommendations to the driver in real time on the onboard computer;
- **"IN → OUT"**: the data collected in the vehicle are transmitted to somewhere else in order to provide a service to the person in question:
  - Example: a pay-as-you-drive insurance policy;
- **"IN → OUT → IN"**: the data collected in the vehicle are transmitted out of it in order to trigger an automatic action in the vehicle:
  - Example: Infotrafic – a dynamic solution that computes a new route in the wake of a road accident.

Compliance "toolkit"

To ensure it is effective, this toolkit must periodically be revised. Remember that all data that can be attached to an identified or identifiable natural person (potentially identifiable, notably by the vehicle license plate or serial number) are personal data that are protected by the *Loi informatique et libertés* and the GDPR:

- Example: data on journeys taken, the state-of-repair of parts, dates of technical checkups, number of miles covered, or driving style are personal data that can be traced back to a specific natural person.

The pack raises awareness among economic actors in the automotive sector as to the principles of transparency and fairness in relation to data collection. As a minimum, the data subjects in question must be informed that their data are being collected and, preferably, consent should be obtained.

If certain configuration procedures are adhered to every time the vehicle starts up, there is a risk of making the driving experience less pleasurable. Thus, the rules applicable to processing need to be determined on a case-by-case basis, notably in relation to the chosen scenario, the nature of the data being collected, and the users' legitimate expectations. Privacy by design must be encouraged as far as possible. One such approach is to establish easily adjustable dashboards so that users can keep an eye on their data. In addition, actors are encouraged to adopt the IN → IN scenario, whereby the data are processed locally in the vehicle, without being transmitted to an external service provider. This approach offers a good level of reassurance in terms of users' privacy and also means that the data controllers are subject to less stringent data protection requirements.

Remember the following five key points on data protection:

- Purpose: define the objectives and what the data will be used for;
- Relevance: ensure that the data are relevant. Only those data that are necessary to achieve the objectives may be harvested;
- Data conservation: data may be kept for only a limited period of time and must then be erased;
- Rights: the data subject must have been informed and retain a certain number of rights over their data: the right of access, the right to rectify data held about them, and the right to withdraw consent to their use;
- Security: the data must be kept secure and confidential.

2.10.4 Mandate 436

In 2008, the European Commission in Brussels issued Mandate 436,[5] which offered a detailed description of the issues of privacy, individual freedoms, and societal aspects relating to RFID, including NFC, the Internet of Things (IoT), and connected vehicles, which are only a few specific (if major) branches.

2.10.5 GDPR

In 2014, the European G29 passed the General Data Protection Regulation, commonly referred to as the GDPR. This text, integrating aspects relating to personal data and privacy by design, was published in the OJEU in May 2016, and came into force on 24 May 2018. It replaces the former Data Protection Directive.

The GDPR, which is around 150 pages, is free to access. Readers are advised in the strongest possible terms to download it from the web and read it carefully, because a considerable amount of data relating to natural persons, identified either directly or indirectly, will be sent back by autonomous and/or connected vehicles. Sooner or later, these will be considered personal data. The GDPR offers a sufficient level of protection for personal data processing within EU Member States, and establishes new recognized rights for data subjects in terms of protecting their personal data. In principle, it does not set out specific norms, because it defines the requirements that must be satisfied in order to properly process data, and also the regulations that apply when data are abused. At an extremely high level, it also sets forth the potential penalties for non-compliance.

Thus, the GDPR is the shared data protection law which is to be followed in all EU Member States. A string of some 40 articles authorizes Member States to enact measures at national level to adapt the GDPR to local requirements.

Let us now take a brief tour to look at the different types of data that need to be protected in autonomous and connected vehicles.

Personal data

What exactly is implied by the label "personal data" and what is the legal framework that applies in Europe? The following laws and regulations answer these questions:

- France's *Loi informatique et libertés* (personal data protection), in force since 1978;
- Mandate 436 and PIA for applications in RFID, IoT, and therefore vehicles;
- The GDPR; and
- Finally, privacy by design.

Definition of personal data

Any information pertaining to a natural person, who is identified or may be identified, directly or indirectly, by reference to an identification number or to one or more unique elements (see article 2 of France's *Loi informatique et libertés*) constitutes "personal data". If such data are to be collected, the so-called "data users" are obliged to inform the "data subject" and obtain their consent.

What kinds of personal data are we dealing with?

The main categories of at-risk, sensitive personal data that are currently collected or there are plans to collect, notably in projects to create autonomous and stroke or connected vehicles, cover the following domains:

- Biometric data (bio [living] + metric [measurement]): the contours or shapes of the hand, the fingers, fingerprints, vein patterns, temperatures, face shapes, iris patterns, etc.:
  - Example: when driving, in the context of preventative activities, diagnostics, heart rate, blood pressure, transpiration, muscle contractions, brain waves, etc.
- Behavioral data, driving habits, braking style, acceleration style, posture in the car seat, etc.:
  - Example (see Figure 2.7): the anti-stress car seat from Faurecia (Active Wellness 2.0). This car seat, which is capable of detecting fatigue or stress in its occupant and reacting to relieve that state, contains specially designed sensors that are able to detect the heart rate and respiratory rate of all the drivers and/or passengers, as well as other biometric medical data. In response, the car seat provides a specific type of massage, or air circulation, to revive a tired occupant. It has an automated seat adjustment function, based on the driver and front passenger's body morphologies, allowing the vehicle to measure their specifics when they enter and are riding in the vehicle, using a camera. Electromechanical devices adjust the length of the headrest, height and inclination of the seat, and the angle of the backrest to match the occupant's morphology. It also tailors the triggering of the seat and dashboard airbags, and the airbag lights on the dashboard, using the communication protocol CAN.

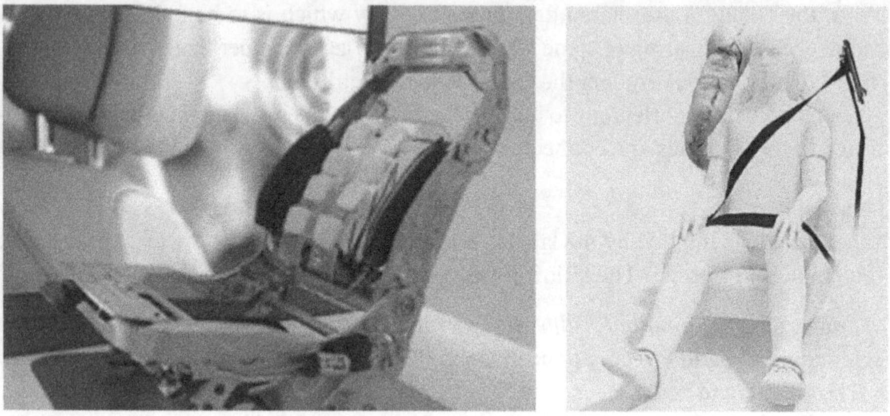

Figure 2.7 Example of a smart biometric and behavioral seat (*source*: Faurecia).

- Data on geolocation, mobility, route, monitoring the user's movements during a range of activities (by means of a connected thing, a mobile app, etc.) or monitoring an employee's vehicle during the course of their duties (using GPS/GSM data);
- Personal data collected within a company: employees' trade-union membership, absence, exposure to professional risks, occupational health risk factors, car journeys, etc.

Autonomous and/or connected vehicles and personal data

Autonomous and/or connected vehicles that, for whatever reason, have an identification system, a range of data sensors, and the capacity to process this information are part of the family of connected things. Given that the autonomous and/or connected vehicles are as close as possible to the people, most in phase with them and their biometric and behavioral data, the data are often captured, measured, and put to use as personal data. Therefore, the companies involved (automakers, OEMs, operators, etc.) will accumulate vast quantities of such information about the physical users of this type of vehicle. The major issue, then, is to ensure that natural persons, who are potential customers, have trust in the market of smart vehicles, which are sometimes viewed as intrusive, tracking their movements and journeys, storing their biometric, behavioral, and private information – in short, their sensitive health and wellbeing data – and potentially being able to transmit it to anyone, anywhere in the world. Thus, we must anticipate and ensure that autonomous vehicles are compliant in relation to the various aspects of personal data protection. In addition, most autonomous and/or connected vehicles are based on RF communications or similar techniques (RFID, NFC, BLE, Wi-Fi, GSM, etc.) using HF, UHF, and SHF communications. Naturally, the information with which they deal is personal data.

Having digested this most meaty of starters, let us now move on to the main course: the details of the GDPR.

Implementation of the GDPR in the automotive industry

To correctly implement the GDPR in the industry of autonomous and/or connected vehicles, we must be in compliance with the regulations. Thus, we must have, or appoint, a DPO and a DC:

- The role of the DPO (data protection officer) is to ensure that their organization is compliant with the data protection regulations;
- The DC (data controller) is jointly responsible and appoints the people in charge of enforcing the data protection regulations. In principle, the DC is the person, public authority, service, or organization that determines the means and ends of the data processing.

Major companies (notably, big manufacturers and the "big four" – Google, Apple, Facebook, and Amazon: GAFA – some of which produce or sell vehicles) and other organizations that handle vast quantities of personal data must respect the scope and implications of the GDPR. In France, any personal data breaches must be reported to the CNIL and to the data owners. The text also stipulates the amount of the financial sanctions that can be incurred as a result of a breach (€20 million or 4% of the company's total annual turnover, so for automakers, for example, that would represent millions of euros). The burden of proof falls on the data controller, who must keep the regulation up to date and conduct impact assessments. Note, finally, that the right to be forgotten is expressly mentioned. In short, this regulation sets out a great many data protection measures with which businesses, both big and small, must comply. If this all sounds like double Dutch, then let us discuss its applications to vehicles, specifically.

Standard example of vehicles: this is a vehicle that, in addition to driving you around, measures a great many things: your fatigue, your stress levels, your level of aggression at the wheel, and your heart rate, measured using an ECG (these biometric and behavioral data are duly treated as personal data). The vehicle sends all these data, via an app, to your mobile phone. The phone, in turn, sends the data up to a Cloud (which Cloud? Who knows?!). We do not know the geographic location of that Cloud or the use that is made of the data (for example, they might be sold on to an insurer, or – joy of joys – to a funeral provider – this has indeed happened in the past!). The data then return to your mobile phone by an unknown route, etc. In short, there is a lengthy and complicated chain of communication that needs to be examined at every step.

Formalities that must be observed by businesses processing personal data

The DCs of any automotive companies processing personal data must produce solid activity logs (up to date, of course) on processing, to prove in the event of an inspection by the authorities (*a posteriori*) that they are compliant and respect the regulations. In addition, it is clear from the regulations that the two concepts – privacy by design and privacy by default – require the companies to implement technical measures and forms of organization appropriate for the issues and rights of the people involved, from the moment the products are configured. Of course, this should always have been the case. Finally, in the sector of novel technologies, such as autonomous and/or connected vehicles, prior to any processing that could compromise the rights and freedoms of the data subject, a privacy impact assessment must be conducted.

PIAs

Sooner or later, it is necessary to quantify all the risks relating to the impact of your autonomous and connected vehicle on users' privacy. At that point, a formal *privacy impact assessment* (PIA) report must be drawn up, in accordance with the European norm. The norm gives a detailed description of the procedure and methodology to be followed. Copies must be kept and produced if questions are asked.

A privacy impact risk is a scenario describing an event that is undesirable to the company or to the data subjects. Depending on the sensitivity of the data and the risk of privacy impact from their processing, the data controller must:

- Conduct a full PIA;
- Produce a PIA report, in order to understand the life cycle of the data in relation to their nature and format, the purposes and contributions of the processing of those data, for the company or for the people to whom they relate.

The DC must then rank the undesirable events in order of seriousness and of the true likelihood they will occur. With all the risks having been objectively identified, any threats that are likely to cause such an event must be anticipated by the DC and the DPO.

There follow a number of examples of high-level risks:

- Unauthorized access to the data by a third party;
- Unintended modification of the collected data;
- Loss or erasure of the collected data.

The impact assessment determines whether the protective measures the company is taking are sufficient in relation to the identified risks and whether the residual risks are acceptable. With that in mind, the data controller will be able to validate or invalidate the PIA. Finally, the report must be made available to the authorities. If it is published or disseminated by the company, this may help win consumers' trust.

Note that privacy-related factors must be assessed (PIAs and similar) before any new applications are launched for each vehicle.

Privacy by X

The concepts of data protection are divided into two paradigms: *privacy by design* and *privacy by default.*

Privacy by design The ultimate goal is not to wait until the end of a project to think about a PIA, but instead to think about it at the very outset and keep it in mind throughout the project. This is known as *privacy by design*. The security chain relating to privacy must be examined, verified, verifiable, and guaranteed, from end to end. Example: a vehicle that, as often happens, measures a range of personal data, such as the posture of the driver or passengers.

Privacy by default *Privacy by default* is the corollary of *privacy by design*. It refers to the data controller ensuring the highest possible level of personal data for the subjects. It entails:

- Carrying out a PIA at the pre-design stage;
- Harvesting personal data only for specific purposes and being able to demonstrate how they are relevant to achieve a particular goal;

- Ensuring that the default settings are appropriate to protect the user's privacy;
- Ensuring the data are kept confidential from the very start;
- Informing the user of the risks and of the option to change the settings themselves.

2.11 Aspects relating to health recommendations

For many years, there has been a profusion of electromagnetic and radio frequency sources in our world, which represent a concern for both human health and the environment. These subjects are always topical, from a scientific, political, and public-relations standpoint – on the one hand, these new products are greatly appreciated by the general population, but, on the other, they arouse mistrust, notably because of the electromagnetic waves that they generate. The numerous technologies used in connected vehicles (V2X, V2I, Wi-Fi, etc.) are likely to increase the levels of exposure for the general population or users (by means of new mobile equipment or by encouraging new forms of behavior). They bring up a range of questions (relating to their biological and clinical effects, epidemiology, regulations, usages, levels, etc.) and a similarly broad range of concerns – notably relating to the possible impact on human health. It is therefore necessary to address health concerns in relation to human exposure to electromagnetic fields.

Certain entities, associations, etc., while they have no actual power to impose regulations on State administrations, are highly expert and thus are well placed to make recommendations to the authorities as to certain values or criteria to be observed in their particular fields. It is then up to the authorities to accept, recommend, adopt, impose (etc.) these values, by law or by decree.

One such organization, operating worldwide, is the International Commission on Non-Ionizing Radiation Protection (ICNIRP). The World Health Organization (WHO) is another.

ICNIRP

The ICNIRP issues recommendations on maximum acceptable levels. One of the most representative parameters is the specific absorption rate (SAR), expressed in W/kg, for signals up to a frequency of 300 GHz.

2.12 Aspects connected to environmental regulations

The majority of commercially available autonomous vehicles will be electronic and use large batteries. Thus, questions about their environmental impact and the recycling of their parts (often electrical or electronic) at the end of their lives are inevitably raised time and again. People often view these issues as being barriers to their use. It is also important to note the regulations in force. There follow a number of points on which to reflect.

2.12.1 Recycling

Besides the fact that the presence of electronics makes matters more difficult and that it is possible to remove the purely electrical/electronic parts, notably the batteries, sensors, and ECUs/microprocessors to retrieve the copper, silicon from the integrated circuits, antennas and rare metals, these problems are tricky to solve. However, to encourage recycling, it is important to focus on eco-design, which, unfortunately, has not yet been well developed by vehicle companies. In addition, new technologies will not necessarily simplify the problem, because the hardware used will be increasingly complex, as sensors and different components are integrated with other hardware. Thus, it will be impossible to separate them and recycle them using current technologies. In the long term, they will increasingly become a major source of waste.

Here are a few official elements to light the way.

2.12.2 Electronic waste processing

Electrical and electronic equipment (EEE) often contains substances or components that are environmentally hazardous, but there is significant potential to recycle the materials from which they are made (ferrous and non-ferrous metals, rare metals, plastics, etc.). For example, in France, the Ministry of Ecological Transition is in charge of regulating waste electrical and electronic equipment (WEEE). In view of these environmental issues, a channel has been put in place for the management (collection and recycling) of such waste, based on the principle of the manufacturer's continuing responsibility. This greatly limits the range of applications and usages (even for professional purposes) of vehicles (also see Section 2.1.4).

2.13 Aspects connected to public acceptability

Before permanently removing steering wheels, accelerator, and brake pedals from level-5 vehicles driving on the open road, it is sensible to conduct in-depth studies of the opinions of future users. For transport (shuttles, etc.) and other niche uses for fleets of vehicles, it may almost make perfect sense to use autonomous vehicles. However, what about the individual vehicle user, a member of the general public (the consumer of mass-produced vehicles)? In terms of changing attitudes, changing ways of life, and new types of transport, it will be a different story. What do we do about the personal aspect, individuality, etc.? Are we to become sheep, meekly obedient to level-5 autonomous vehicles?

2.13.1 Human factors

The issue with delegated driving vehicles, as it relates to human factors, is not simple to grasp for a variety of reasons.

Concerns

The concerns about such vehicles fall into a number of categories:

- **Technical concerns**: this category, of course, includes all doubts about the technology, pertaining to the electronics (malfunctions, breakdowns, unexpected reboots, the potential threat of hacking, etc.) and software bugs associated with such systems;

- **Concerns about the implications for the pleasure of driving**: many drivers today take pleasure in driving and wish to continue to enjoy the feeling of being in total control of their vehicle. They would certainly not welcome an automated system that would completely take away that joy. A number of people who drive for the experience – for the feeling – would even be prepared to turn to motorcycles if fully autonomous cars were to come on to the market... and it is more difficult to design a truly autonomous motorbike (although that has not stopped Yamaha and BMW from trying)! Obviously, these drivers do not experience the same joy every morning when they encounter the same traffic jam in the same place on their way to work, and would be happy simply to be transported without worrying about such things – they could read their newspaper in peace, round off their night or make some calls, etc. In parallel, they would be delighted to retain the pleasure of driving when they go off on vacation, but would encounter the same traffic jams on the highways on days when huge numbers of people are traveling. In conclusion, perhaps we need a partially autonomous car – a level-4 vehicle with a steering wheel, in which the autonomous functions can be switched off. The concept cars developed by Renault Symbioz (Figure 2.8a) and PSA 508 e-Legend Concept have a retractable steering wheel (Figure 2.8b and c) so these are level-4 cars that are almost level-5. Do we need to create a level-4½ or 4¾ car? How would we cope with a true level-5 car that has no steering wheel or pedals?
- **Concerns relating to the resumption of human control and safety recommendations**: numerous studies have been conducted by automakers, OEMs, and universities on the subject of the human driver resuming control, and the time it

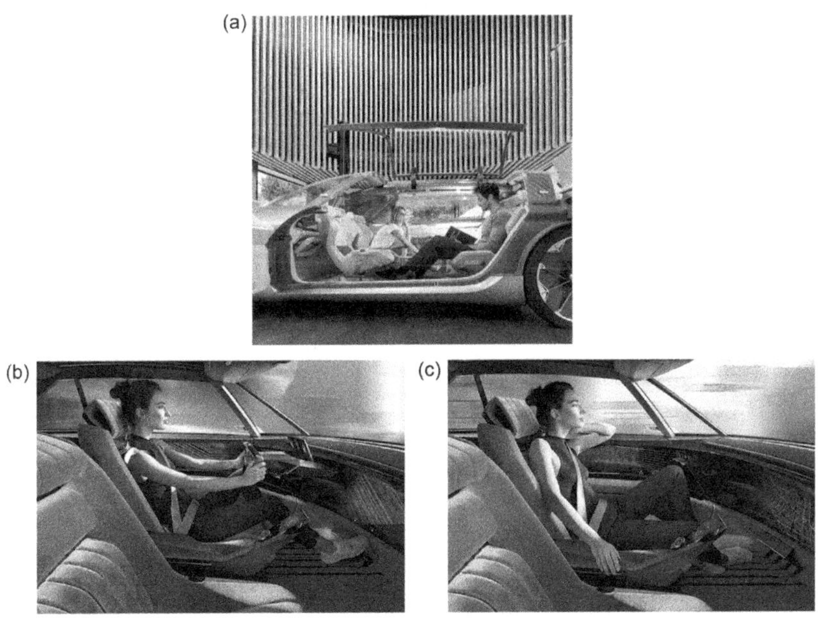

Figure 2.8 (a) Renault Symbioz, (b) and (c) Peugeot e-Legend Concept. *Source*: Renault and Peugeot.

would take in level-3 and level-4 vehicles, in the event of any problem with the autonomous system (response time, panic, appropriate management of the resumption of control, writing of recommendations, information for the safety of passengers and of people outside the vehicle, etc.).

Financial acceptability and drivers' needs

Similarly, while the hardware costs of vehicles may go down, the technology has a price tag. Thus, even ten years from now, "autopiloted" autonomous cars could still cost €5000–€10 000 more than a "normal" vehicle. In addition, we must consider the acceptability of autopilot functions:

- In **2012**, a study conducted by JD Power reported that only 20% of drivers would be willing to buy such a vehicle if it represented an additional €2000 cost;
- In **2015**, according to the Observatoire Cetelem, 55% of Europeans said they would be interested in an autonomous vehicle. This represented an increase, but still not enough to be able to speak of real excitement about the prospect;
- In **March 2017**, in a report published by the American Automobile Association, 54% of Americans would not feel comfortable sharing the road with driverless cars;
- In **September 2017**, a study conducted in Germany showed that only 26% of Germans would consider getting into an autonomous vehicle at that point and only 18% would consider driving one;
- In **2018**, while safety issues were prominent, 45% of respondents also pointed to a "lack of clarity as to liability;"
- In **2019**, having seen reports of the first accidents (even minor ones), road users were a long way from being won over by the idea of autonomous vehicles;
- In **2020–2021,** the world has been rather too busy dealing with COVID-19 to make much progress in the field of autonomous vehicles.

2.14 Aspects relating to battery technologies in electric and autonomous vehicles

Why speak of battery technology in a book that is fundamentally about autonomous vehicles and networks? As we shall show in the coming sections, there is a major problem that has yet to be solved: the electricity consumption of the part of the vehicle dedicated to autonomy (processors, integrated circuits, networks, datarates, etc.). This needs to be determined in order to calculate its true autonomous range – the distance it can really cover, as opposed to the commercial figures published on the basis of specific tests (remember the problems that certain automakers had with CO_2 emissions, running special programs for the official tests).

2.14.1 Electric and autonomous... but over what kind of distance?

By the time they actually enter into mass production (which should be around 2035), the level-5 autonomous vehicles on the open roads will very likely use powerful electric motors rather than combustion engines. Thus, anything that consumes the battery (and there will be a huge number of things) will reduce the vehicle's range. As we shall see in Sections 3.2 and 3.3, the cameras, radars, lidars, ADASs, their data fusion systems, artificial intelligence, etc. will need (very) high-datarate networks. As such, they will consume a certain amount of energy and detract from the stated range by the automakers. For simplicity's sake, let us consider a very concrete example:

Example: on a snowy winter's night, when I start up my car, I put on the heating, the windshield wipers, the fog lights, and the radio (or the infotainment system) to get additional information about the route. In principle, all the ADASs are activated to help prevent collisions and keep the vehicle in the correct lane... the radars, lidars and other sensors are all working, etc. All of this inevitably consumes energy from the battery. Of course, it is to be hoped that less energy is expended on this than on actually moving the vehicle, but all these factors draw upon the battery and, in so doing, reduce the true functional range of the vehicle, potentially quite significantly.

Important additional remarks: the NEDC (*new European driving cycle*) consumption test or the WLTP (*worldwide harmonized light vehicles test procedure*), which has been in force since 2017, are still only laboratory tests. The experience of drivers of electric vehicles has produced a general formula: the 80–50 rule, which holds that in the best possible real-world driving conditions, we can only expect an actual range of around 80% of what the NEDC says; meanwhile, in the worst possible conditions (e.g. in winter or on the highway), we can expect around 50% of the NEDC stated range. It is important, therefore, to pay attention to the stated figures in terms of range.

In view of all of the above, the following few paragraphs are by no means "off topic" in relation to the range of "electric" (or other) autonomous vehicles.

2.14.2 Functional principle of an electric battery

Remember that a battery is an electrochemical device that stores energy, primarily in chemical form, and releases it, on demand, as electricity. There are a wide variety of batteries, but all are composed of two electrodes (positive and negative), immersed in an electrolyte that exchanges ions to maintain electrical neutrality. The chemical nature of the redox couples determines the potential of the battery, expressed in volts.

Main parameters of a battery

There are numerous parameters by which a battery may be qualified. Chief among these are:

- Capacity (in Ah/kg): the quantity of electrical charge Q (in coulomb, which is to say in ampere-seconds) that a battery can store per unit mass. Capacity is measured in ampere-hours per kilogram (Ah/kg) and is a key parameter;
- Quantity of energy (in Wh/kg): the quantity of energy stored per unit mass is obtained by multiplying the capacity (as stated above) by the electrical potential at the battery terminals. Thus, it is expressed in watt-hours per kilogram (Wh/kg), which is, in fact, an energy density by mass (joules/kg).
- Life: to quantify a battery's lifespan, it is tested by a user performing a large number of charge/discharge cycles. Batteries are generally required to withstand over 2000 cycles. The rate at which energy is stored and restored is also taken into account.

Safety

Finally, we must ensure that the batteries are safe to use, considering the following parameters:

- Charge time (full recharge or quick top-up);
- Place and type of charging (at charging points, by energy harvesting, at the user's home, at the office, etc. and the hacking opportunities);
- The battery's operational temperature and cooling method;
- The true range we can expect from the vehicle with different types of use and in real-world driving conditions, as opposed to figures that have been partly skewed;
- The dates when different battery technologies are introduced.

Example

A standard car, of average weight, driven at a reasonable speed, in order to overcome the forces due to air resistance, road resistance, and the mass of the vehicle (parameters that are completely independent of the type of propulsion chosen) needs around 15 kWh of energy to cover 100 km. The watt hour is a measure of energy, which can also be measured in Joules: 1 Wh = 3600 J (one hour = 3600 seconds) and 1 kWh = 1 000 Wh = 3 600 000 joules.

Also, in the case of fully electric propulsion:

- The energy yield of an electric motor is around 85%;
- Therefore, under the same conditions as before, the motor must be supplied with 15/0.85 = 18 kWh to cover a distance of 100 km;
- Thus, with reasonable driving, to achieve the range of around 700 km to which users of standard vehicles are accustomed, the electricity tank (i.e. the battery) must have a minimum capacity of 7 × 18 kWh, which, with a small remainder, equates to around 140 kWh;
- Using lithium-ion batteries (a technology that is well established at this point), with an energy density of 200 Wh/kg, the total weight of the battery will be approximately 140 kWh/200 Wh/kg = 700 kg. This, of course, is rather too heavy.

For example, Figures 2.9a, b, and c show the battery zone in a Tesla Model S, for which the total vehicle weight (including the front and rear motors, bodywork, cabin, battery, etc.) is quoted as around 1900 kg.

Obviously, if you have a burning desire for hard acceleration, for speed in the depths of winter, and want to keep the same range of distance, you are free to do so, but you will need to either increase the battery's total capacity, and therefore its weight, or find a different battery technology that offers a better energy density by mass.

The total consumption (dissipated power) by the "autonomy entity" (sensors, data fusion unit, AI, actuators, etc.) is between a few hundred watts and 1 kW. Thus, the energy consumed for this function per hour is between a few hundred Wh and 1 kWh.

2.14.3 Present and future battery technologies

Lithium-ion battery technology

Almost all electric vehicles today use lithium-ion batteries, because automakers and OEMs can produce and test modules/packs of individual cells of lithium-ion batteries and can also provide after-sales service. At the time of writing, lithium-ion batteries

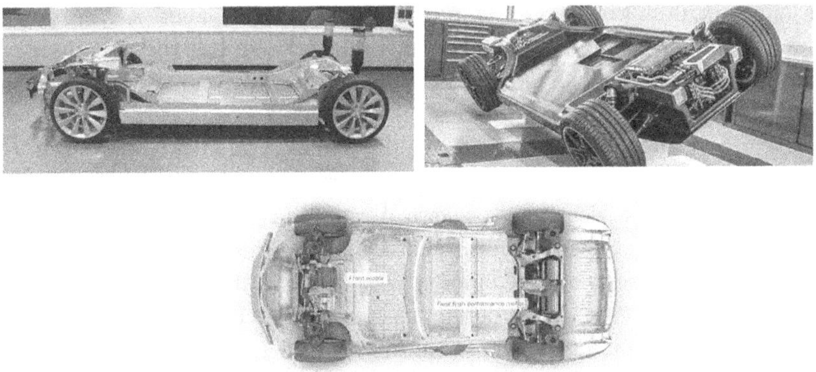

Figure 2.9a, b, and c Example of the battery in a Tesla Model S.

appear to offer the best compromise for electrification of power transmissions: they strike a reasonable balance between reliability, safety, life cycle, stored energy density, and thermal performance. Architecturally speaking, all lithium-ion cells are essentially the same, as they all have an ionic liquid conductor and an insulating sheet that serves as a separator, preventing the positive and negative electrodes from short-circuiting the system.

In addition, in a pack based on serial or parallel connections between thousands of individual cells, it is possible to increase the voltage and the pack's capacity. Currently, one of the main problems with lithium-ion batteries is their performance in relation to temperature:

- They are capable of withstanding very low temperatures (−20°C or lower);
- Conversely, heat can create problems, so these batteries need to be cooled with air or liquid, or, potentially, in time, phase-changing substances. If lithium-ion batteries become too hot, the phenomenon of thermal runaway may be triggered: the battery's chemical components are caught in a chain reaction that creates more and more heat; that heat accelerates breakdown and causes the release of both heat and gas. When this happens, batteries can catch fire or even sometimes explode. Although there are safety measures built into lithium-ion accumulators, further measures need to be put in place to improve safety, firstly in relation to the chemical component and secondly in relation to the technology, creating multiple mechanisms (redundant on multiple levels) to alleviate the risks.

Another important point is their degradation and power loss:

- Lithium-ion batteries degrade over time, meaning that, with the same charge capacity in the battery pack, the range is reduced. They also have an expiration date, which is based on numerous factors. Unfortunately, though, it is difficult to accurately predict when this will happen, and after how long they will no longer be able to hold a charge, before they have to be replaced.
  - Example: it should be noted that even with a significant loss of capacity (for example, in a Nissan Leaf or a Tesla Model S), it is still possible for the vehicle to run for another 80 or 90 km. This statistic should be viewed in combination with the fact that around 80% of the population travel less than 60 km per day.

- In the long term, this also raises the issue of how to recycle the batteries. What do we do with the batteries when they are taken out of the vehicle? Can they be reused or recycled? Certain automakers (such as Toyota and Renault) have already begun thinking about recycling their batteries and giving them a second life.
  - Example: the "*Advanced Battery Storage*" project aims to build multiple stationary electricity storage sites with batteries from electric vehicles, to foster the integration of renewable energies into the electrical grid. Primarily, such a system would be based on reclaimed electric vehicle batteries, combined in containers. However, a stock of new batteries would be stored, with a view to "future standard exchanges as part of after-sales care". Thus, customers could bring a used battery to the storage site and swap it for a new one.

Solid electrolyte battery technology

Up to this point, electric vehicles have suffered from a great many ills: reduced range, excessive charging time, insufficient infrastructure, prohibitive cost, the risk of explosion, etc. There has been a huge amount of progress made in terms of the distance that electric vehicles are able to cover on a single charge, but the main concern is still the amount of time that it takes to recharge, even despite the fact that fast charging stations can dispense enough charge rapidly to cover a crucial few miles (by comparison, it takes only a few minutes to completely fill the tank in a combustion-engine vehicle). Nevertheless, these constraints are still present, meaning that electric cars cannot (or cannot easily) make their mark as an alternative to combustion technologies. Certain companies believe that improvements can be made to lithium-ion batteries with new technology, which replaces the liquid electrolyte with a solid-state electrolyte (a glass plate) (example: the patents Fisker holds for a graphene solid-state battery). This technique reduces the risk of explosion and promises better performances:

- Better battery efficiency, and therefore greater capacity;
- Higher energy density, which has risen from 200 Wh/kg for a conventional lithium-ion battery to nearly 800 Wh/kg, or even 1000 Wh/kg, representing a four- to fivefold increase;
- Range of around 600 to 700 km which is equivalent to that offered by a full tank of gas in a standard vehicle;
- Faster charging, with the charge time reduced to just a few minutes;
- Significantly increased lifespan, as these batteries can be charged many more times than can traditional ones, which become worn out after a few years;
- Significantly broader temperature range: from –20 to 100°C, compared to 15–35°C with today's technology.

At present, the solid-state battery appears to be the most promising technology for the coming years and is expected to break into the market by around 2022. Unfortunately, it will still be some time (2023) before we are able to achieve an 800 km range from around a minute's charging. For its part, Toyota is already promising to bring out a model based on this technology by 2025.

Example: the deluxe sports car EMotion, from Henrik Fisker's company, has a range of over 640 km, but also takes less than 10 minutes to receive a top-up charge that will take it another 200 km, and Fisker have pledged to reduce this charge time still further.

In addition, the Japanese have announced a collaboration to bring this technology to market by 2025. Toyota, Nissan, and Honda have joined Panasonic and GS Yuasa (two battery manufacturers), aiming to achieve a range of 525 km by 2025 and 800 km by 2030. In Europe, Saft has set up another alliance with Manz, Solvay, and Siemens to stem the hegemony of Asian corporations, speed up solid-state battery research, and be capable of putting forward a viable industrial solution for high-capacity evolutive blocks by 2025. If the solid-state battery lives up to its promises, it will be as competitive, in terms of cost and performance, as combustion technologies, and could also neutralize the threat posed by hydrogen technology (see the next section) – at least on small vehicles, because, for a 40-ton truck, it is probably not sensible to carry 8 tons of electric batteries; in this case, hydrogen is probably a better option.

As indicated above, one of the major issues with this technology surrounds the recharging of batteries and how long it takes. EDF's objective is clear: by 2022, to become the European leader in supplying energy to electric vehicles and EV charge points, supplying power to 600 000 electric vehicles (see Figure 2.10).

Such a scheme inevitably requires massive investment. In February 2019, a joint French and German EV financing plan was drawn up. Also included in the plan is the founding of two battery production factories: one in France and the other in Germany.

Fuel-cell technology

Another solution is to electrify vehicles using fuel-cell technology, based on compressed hydrogen gas.

Functional principle

A fuel cell is a system that generates electrical voltage locally by oxidizing a reductive fuel (such as hydrogen) on one electrode and reducing an oxidizing agent such as oxygen from air. Thus, a hydrogen fuel-cell vehicle is an electric vehicle carrying its own electric power station, in which hydrogen stored in tanks is sent to the fuel cell at the same time as a flow of air in order to provide oxygen. When the two substances come together, an electrochemical reaction takes place that produces electricity, and the

Figure 2.10 Example of a charging station at a highway rest stop.

only emission into the air is water vapor! The electricity generated is then converted to DC and fed into an electric motor, which drives the wheels while a buffer battery either recovers energy when the vehicle brakes and slows or provides assistance to the fuel cell during sharp acceleration. Fuel cell systems offer certain advantages in comparison to simple battery systems for electric vehicles.

Example of performance (Hyundai Nexo):

- A tank containing 6.4 kg of compressed hydrogen gives vehicles a range of around 600 km and flexibility of use that is very similar to conventional fueled vehicles;
- The electric motors are almost identical for vehicles with standard batteries and vehicles with hydrogen fuel cells;
- Overall, they offer the same benefits as electric vehicles, with a very quick response and instant torque delivery;
- In cold weather, electric battery vehicles draw power from the battery very quickly to warm up the cabin;
- The cooling fluid surrounding the stacks that convert hydrogen into electricity is heated and can then be used to heat the vehicle interior;
- For the same range of distance covered, the weight of hydrogen consumed is less than the weight of the battery, which helps reduce the overall vehicle weight;
- Fuel cell vehicles are safe, because the hydrogen is stored in carbon fibre tanks that can withstand huge levels of pressure (700 bars) and, in the event of a fire, the hydrogen is rapidly evacuated from the tank. In particular, that ejection is risk-free, as hydrogen gas disperses very quickly;
- The vehicle can be refilled in the same way as an electric vehicle is recharged. Instead of an electric plug, it is connected to a hydrogen nozzle;
- Filling time is identical to that for conventional gasoline-fueled vehicles (it takes only five minutes to fill the tanks with 6.4 kg of hydrogen);
- At present, hydrogen is still pricy. In 2021, the price sat at around €10/kg, so the fuel bill is approximately equivalent to that of a vehicle running on gasoline. For example, the Nexo has a stated NEDC range of 600 km, coming at a cost of around €60;
- Today, the main difficulty lies in refilling the vehicle with hydrogen, as the infrastructural network is still at the embryonic stage. Access is often restricted to private users (most such vehicles are used by local authorities or private companies). Thus, in order for hydrogen vehicles to sell, a network of service stations would have to be set up.

Final useful remarks, with feet firmly on the ground

To conclude this section, after the flight of fancy to look at futuristic vehicles, let us now come back down to Earth. Vehicle electrification is a highly important trend in the automotive market, and particularly in the autonomous vehicle market. It is predicted that by 2030, around half of this market will be "electrified," with 13% of vehicles being fully electric. Combustion vehicles are likely to account for 52% in 2030, so we must carefully monitor the global geopolitical situation, keeping a particularly close eye on the availability and supply chains of the metals that are crucial to the manufacture of electric vehicles.

According to the experts of the World Materials Forum (WMF), the BRGM (*Bureau des recherches géologiques et minières*, Geological and Mineral Research Bureau), and the consultancy firms McKinsey and CRU Consulting, the supply of certain materials could become problematic, leading to an uncertain future for electric vehicles and energy storage. The most notable consequence will be the soaring demand for certain metals used in batteries – primarily cobalt and nickel. Until 2030, raw materials are not expected to be a problem in relation to vehicle electrification.

Based on criteria ranging from the estimation of known reserves to the possibility of substituting or recycling a metal, and also taking political risks into consideration, the analyses and studies performed by the WMF and BRGM classify elements according to their risk level:

- Cobalt, classed as "red," tops the list of "critical" metals. The global demand for cobalt, which is an essential ingredient in batteries, could overwhelm supply between 2025 and 2030! In addition, according to the study, half of the supply comes from the Democratic Republic of the Congo, which presents a major political risk. Other sources could potentially be exploited, such as waste products from copper mines that contain cobalt. Some twenty years ago, successful tests were conducted in Uganda;
- With tungsten, in addition to the dependence on China, new uses are emerging, and developing rapidly in metallurgy. It can be used to make extremely high-performance alloys for aeronautics and for 3D printing;
- Tin also ranks highly among the important metals. In the case of tin, the problem is down to underinvestment in the development of mines;
- Such difficulties have already been encountered with zinc;
- Vanadium is also subject to uncertainty. It is estimated that there are reserves for up to 250 years and mining infrastructure is sufficient;
- Finally, three rare earth elements are classed as "red": dysprosium, neodymium, and praseodymium.

Thus, uncertainty persists as to whether one technology or another will win out in the automotive sector in the long term. We therefore need to pay close attention to a number of metals that are likely to gain different levels of prominence, depending on the technologies that are adopted. Of course, certain materials are absolutely critical, but, above all, there is uncertainty as to the future choices of propulsion type and energy generation, which is crucially important. Electric vehicle development will require €10 billion or $10 billion of investment in infrastructure for every million new vehicles brought to market. The industry is perfectly prepared to accept a shift, as long as it is an organized shift and the players have time to adapt.

2.15 Other aspects

To conclude this first part of the book, there follows a new list of terms, which will be defined in their respective sections.

2.15.1 Tests

A test is an operation designed to make certain of the characteristics or qualities of something, or a trial or action undertaken with a view to producing or obtaining something, without being certain of the result.

Manufacturer tests

For an automaker, autonomous vehicle testing is one of the most difficult stages to pass. For example, we must determine whether a particular autonomous vehicle model is secure, define what security means in this specific context, quantify it, measure it, validate it, etc., and then disseminate that information to the general public. Is it possible to gage reliability by the distance that an autonomous vehicle has traveled during the course of its development? These are only a few of the huge number of factors that need to be tested.

Part of the answers to these questions is generally based on simple statistics, often leading to conclusions such as: "fully autonomous vehicles must have traveled hundreds of millions of kilometers in test driving, in all types of traffic conditions – sometimes even hundreds of *billions* of kilometers – to prove their reliability." Unfortunately, declarations such as this are known to be inadmissible, because the statistical argument sometimes leads to huge margins of error.

Example: results and estimation of test failure rate

The argument is generally presented as a problem of estimating the failure rate observed for a particular test (e.g. the number of accidents of such-and-such a type involving autonomous vehicles) in comparison to the known current failure rate – the rate of such accidents involving human drivers. The accidents are modeled as distinct, independent, and random events with a statistically constant failure rate. The failure rate of fatal accidents can be calculated by dividing the number of fatal accidents by the vehicle kilometers traveled (VKT). This method is simple, but the calculations use the wrong measurements: instead of dividing the number of accidents causing death by the VKT, they divide the number of deaths by VKT. This leads to an inflated failure rate for human drivers, because a single accident may cause multiple deaths and the number of deaths per accident may depend on a great many factors other than the driver's reliability.

Test centers

Every automaker has its own private test centers and circuits, closely guarded from prying eyes, often in the middle of the countryside, with walled enclosures, etc., where all prototypes, pre-production vehicles, etc., are tested (often under cover).

2.15.2 Trials

A test is a trial involving a task that must be accomplished, identically, for all subjects in rigorous application conditions, with a precise method for assessing success or failure. The term may also refer to the trialing of a product or device to check how it performs in all circumstances, which demonstrates or measures something. Alternatively, it may denote a standardized and benchmarked examination, used to assess the capabilities of the given product.

As is the case with any vehicle, the next phase (certification) for autonomous vehicles is preceded by many months/years of proprietary testing and refinement, when they are compared, contrasted, and tested in a complex environment: congested roads, difficult weather conditions, with different vehicle types present, the need to make decisions quickly, etc. As a result of these tests, new issues to be addressed will emerge;

so too will new methodologies, scenarios, and test equipment. These scenarios help to fine-tune the autonomous vehicles so that they conform to all safety requirements (technical regulations on automobiles) and the Highway Code (various laws and the Vienna Convention). The hope is to develop standardized and benchmarked tests by which to assess the capabilities of the autonomous and/or connected vehicle.

What exactly is the situation on the ground for autonomous vehicles, in terms of experiments, tests, or the legal framework? At the very least, we can say that there is still a great deal of work to be done.

2.15.3 Validation

Validation refers to the action of approval or the fact of being approved, or confirmation from the competent authority of the legal validity of an act or operation that has been carried out.

Validation and simulation of autonomous vehicles

Once the broad outlines have been laid out, given that it is simply not possible to have a vehicle drive for millions of kilometers during the preliminary phases, we must establish rules regarding simulations for the validation of an autonomous vehicle in relation to two criteria: usage aspects and technical aspects.

- Usage: validation of usage aspects might include, for example, determining whether a user can use their phone whilst the driving is delegated to the automated system, with a level-4 autonomous vehicle, etc.
- Technical: technical validation would include defining whether it is necessary to install a system to monitor the physical presence of hands on the wheel and take action when the driver does not respond to repeated requests to take over control of the driving once more. Note, in passing, that there is no such thing as zero risk.
  - This question has previously been examined, in the case of the fatal accident involving a Tesla with its autopilot engaged. According to the NHTSA, prior to the collision, the driver had only had his hands on the steering wheel for a total of 25 seconds during a 37-minute journey, and had ignored numerous alerts during that time.

Discussions are ongoing between the authorities and automakers on all of these issues.

A step in the right direction, perhaps

Level-2 vehicles have long since been validated and are worthy of approval. Their functions are limited: the vehicles merely take care of the longitudinal and lateral positions – in other words, pressing the pedals and turning the steering wheel. However, between level 2, which is already in place, and level 4, which is expected to be brought out in 2022, is there really space for a level-3 vehicle? The reality is a little more complicated than it might seem. Discussions are taking place at a European level, notably driven forward by Germany, but, as yet, there is nothing concrete in place. Indeed, for the moment, it is a usage regulation, with imprecisely defined parameters, which would "authorize [and this is at the heart of the debate] the driver, under certain technical conditions which have yet to be defined, to carry out what are known as sub-tasks, such as looking at their screen to check their e-mails, or watching a short video."

Example: it may be validation for a function that is "*eyes off* subject to the driver's responsibility," but "we do not know what the solution will converge upon." In any case, it still will not be a true level-4 function.

For their part, certain automakers have announced that they will begin working with veritable test fleets of *eyes off/hands off* autonomous vehicles from 2022.

2.15.4 Homologation

Homologation is the action of approving something or the approval of an act by a legal or administrative authority, which must be gained before the act can be carried out. The term may also apply to official recognition of performance by a league or federation, in accordance with established criteria.

The testing and homologation stages are essential and mandatory precursors to the registration and commercialization of any new model. The aim of this procedure is to verify that the vehicle respects the technical regulations governing automobiles in relation to safety, steering, and braking, but also such things as signage and lighting, etc. Only with this figurative green light will a vehicle be able to drive on open roads or public roads, and autonomous vehicles are no exception to this rule.

One point that must be made is specific to the European Union: since 1990, the homologation of a vehicle by one Member State, by a national technical bureau (such as UTAC in France), is valid for the whole of the EU.

According to certain experts: "the regulations for homologation of level-3 and level-4 autonomous vehicles are a thing of madness! These driving technologies will be available by the end of 2021, and the homologation regulations should also be ready."

Homologation, then, is another story entirely, because if the tests exist, a judicial or administrative authority must be established to issue homologations. Indeed, if prototype, experimental, pre-production autonomous vehicles are driving on the roads, then all well and good. However, they are not production-level vehicles, but vehicles that have only been approved for use in a specific experimental context. We shall need to wait for the associated regulation to evolve before we can approve mainstream production models to circulate on the open road. Today, any new vehicle, to gain homologation in a European country, must conform to some fifty regulations, defined by the United Nations Economic Commission for Europe (UNECE). For level-3 and level-4 autonomous vehicles, a further ten tests are planned. Automakers are required to carry out thousands of tests and/or simulations to ensure that their vehicles will correctly respond to hundreds of dangerous situations: emergency braking to avoid hitting a pedestrian, driving straight on, taking a bend, in the daytime, at night, in fog, etc. On this topic, UTAC CERAM and automakers are working on a standard list of the most common accident-causing scenarios, which could ultimately lead to a draft ISO standard in this domain.

- For example: the autonomous driving system must be able to assess the driver for itself: are they sitting in the driving seat? Do they have their hands on the wheel? Are they paying attention to the road?

The same is true of homologation of software. Indeed, artificial intelligence systems will pilot autonomous vehicles, and algorithms will allow the vehicle to learn on its own and become a little bit better every day (for example: at Tesla, which can automatically and remotely install software updates that alter the vehicle's functions). This poses a series of questions:

- How can we give final approval to a computer program that can and will constantly evolve?
- Which protocols will protect the software from cyberattacks?
- How can we be sure that the owner of an autonomous vehicle will not independently install a computer "patch" or device that could compromise the vehicle's safety? Recently, the NHTSA has banned a magnetic device that fooled the "Autopilot" system in Tesla vehicles into thinking that the driver still had their hands on the steering wheel. Such a ruse negates the benefits of level-3 and level-4 vehicles!

At present, automakers have authorization to test prototypes of level-4 autonomous vehicles as defined by the SAE standard and are scheduled to complete the testing of such vehicles by 2022. Once more, remember that at this level, it is a question of full delegation of driving ("the machine is in charge of all operations, all driving tasks, and therefore relieves the user – as, when the driving is delegated to the system, we can no longer speak of a driver as such – of any need to remain vigilant"). However, in accordance with the Vienna Convention, there must always be an additional level of security in place, in the form of a human presence ("an expert driver will always be behind the wheel, ready to intervene"). In any case, the planned deadlines are fast approaching and testing programs are focusing on level-4 autonomous vehicles (in which the driver delegates the driving to the autonomous system in a given geographic area and in certain conditions) that can be used by lambda users by 2022–2023. Note, however, that these vehicles will not be able to be commercialized (and thus receive homologation) unless the regulation changes by that time, which currently seems unlikely.

In parallel, two regulations will come into force, setting the standards for three new functions:

- Valet parking;
- Highway assistance;
- Autonomous highway driving.

These regulations define requirements in terms of security, emergency maneuvers, driver takeover time, driver monitoring, HMI, etc. It should be noted that they could make the safety audit more stringent and necessitate the addition of a black box in vehicles, as is carried in airplanes. They will also take account of the conclusions on accidents that have occurred over recent months involving semi-autonomous vehicles, with the aim of improving user safety. Thus, vehicles can only be commercialized after certification when the norms and regulations in force allow – but when will these modifications happen? For the time being, there is no sure way to answer that question. At present, the various stakeholders are communicating amongst themselves, and the texts are unlikely to be finalized just yet. However, in view of competition between countries (e.g. the USA, China, etc.) and between automakers, the pressure is on to speed up the definition of certification procedures, though the fact that there are still technical issues that have yet to be resolved.

UTAC CERAM

France's UTAC CERAM (*Union technique de l'automobile, du motocycle et du cycle – Centre d'essais et de recherche appliqué à la mobilité*) is a private and independent

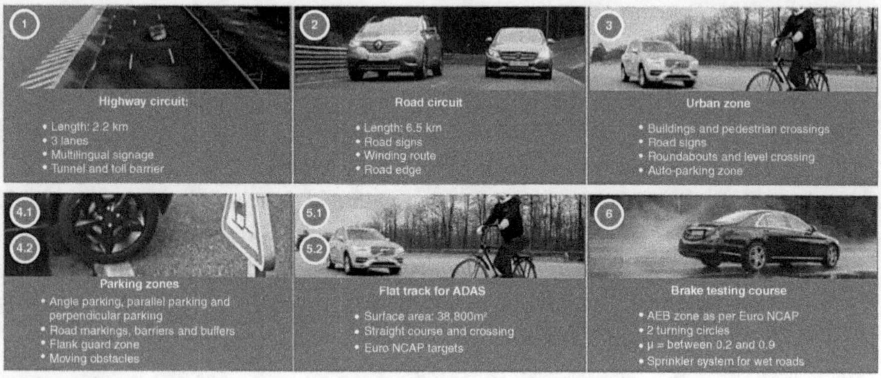

Figure 2.11 Examples of equipment on the UTAC CERAM tracks. *Source*: https://utacceram.com/teqmo.

group that provides services in all areas of regulations and approval, ISO 17025-accredited tests and technical assessments (relating to the environment, security, endurance and reliability, certifications, driver training) and also serves two official roles for technical testing (OTC) and standardization (BNA). UTAC CERAM is partner to the French State authorities and is the body in charge of testing vehicles' and their equipment's compliance with the regulations. It operates at two test centers (the autodrome in Linas-Montlhéry [Essonne] and Mortefontaine [Oise]) and at clients' own premises, but also outside of France. Its Teqmo technology center in Linas-Montlhéry boasts 12 km of test tracks associated with laboratories (to test matters relating to the environment, security, endurance; see Figure 2.11). It serves automakers, OEMs, software developers, telecoms operators, infrastructure suppliers, startups, and more.

2.15.5 Certification

Certification is a written assurance – or a procedure of authentication of an act – or a testament given by a national or international body, asserting that a device complies with the standards in force.

2.16 Projected schedule for autonomous vehicles

To conclude this lengthy series, the table below, as at May 2021 (Figure 2.12), presents a summary of the projected schedule for the development of autonomous vehicles (of all levels) and connected vehicles.

This completes the torturous tour through the regulations, norms and standards. It was a necessary evil, which will ultimately do the reader good.

We have now come to the end of Part One of this book. In it, we have acquired a broad view of the wide range of aspects that need to be borne in mind when making the step from a vague idea to the concrete realization of an autonomous and/or connected vehicle.

		2020	2022	2025	2027	2030-35	2040
Level of autonomy		L3 approved	L4 approved	L 4	L5	L5	Gasoline no longer used
Regulations	LOM	PACTE		Yes	Yes	Yes	OK
	L4		OK		Closed routes	Open roads	OK
Insurance			OK for L3		OK for L5	OK for L5	OK
Smart road infrastructure					Yes	No	OK
Connected		V2x	LTE/5G	LTE/5G	5G	5G	OK
Range (distance)			Electric 700	700		800	OK
GDPR		OK					
			L3 on the road	L4 on the road		L5 Open road Mass production	

Figure 2.12 Projected schedule for the development of autonomous vehicles.

Notes

1 The authors extend their heartfelt thanks to Me Gaëlle Kermorgant – a barrister in Paris, a specialist in the automotive field, and along with the authors, a co-founder of the working group "GDPR Associates" – for her valuable cooperation, help, and involvement in constructing this lengthy chapter.

2 www.legifrance.gouv.fr/affichCode.do?idSectionTA=LEGISCTA000006159602&cidTexte=LEGITEXT000006074228.

3 "Secure Connected Things" by Dominique Paret and Jean-Paul Huon, ISTE Ltd., 2017.

4 "Secure Connected Things" by Dominique Paret and Jean-Paul Huon, ISTE Ltd., 2017.

5 M 436 – EN, *Standardisation Mandate to the European standardisation organisations CEN, CENELEC, and ETSI in the field of information and communication technologies applied to radio frequency identification (RFID) and systems.*

3

DAS, ADAS, HADAS, and AVs – L3, L4, L5!

Double Dutch? Ancient Egyptian hieroglyphics? Not quite – these are terms relating to modern vehicles! Decoded, the chapter title reads: "From simple Driver Assistance Systems (DAS) to Advanced Driver Assistance Systems (ADAS), Highly Automated Driver Assistance Systems (HADAS), and level-3, level-4, and level-5 Autonomous Vehicles!" Certainly, this is a great deal of ground to cover.

In this chapter, we shall trace the years-long journey from simple functions and the earliest driver assistance systems to the advent of fully autonomous vehicles. However, before getting down to the meat of the matter, let us list and describe the functions that an autonomous vehicle should/must be able to perform, unassisted.

3.1 Functions to be provided

To begin with, we need to know what an autonomous vehicle has to be able to "see" and what it has to be able to "do."

The answer is very simple: everything that a human being can see and do with the benefit of conscious discernment, and more...! Plus everything that humans cannot do (e.g. have eyes everywhere – even in the back of their heads)! Let us take a look at a few examples, among many others.

In front, to the side, and behind

Figure 3.1 represents the outline of what is presented in this chapter, and summarized in its numerous tables.

Figure 3.2 offers a fuller view of the problem, by more explicitly imaging the angles of view that these machines must have.

Figure 3.3 presents an overview of the technical details of what each of these sectors must, as standard, be able to cope with in terms of distance, angle of view, speed of action, and/or processing, minimum latency time, and appropriate technology.

- For information, this table includes approximate values of the raw data streams that these functions generate for each element (excluding any treatment to refine the received data).

Autonomous and Connected Vehicles: Network Architectures from Legacy Networks to Automotive Ethernet, First Edition. Dominique Paret and Hassina Rebaine.

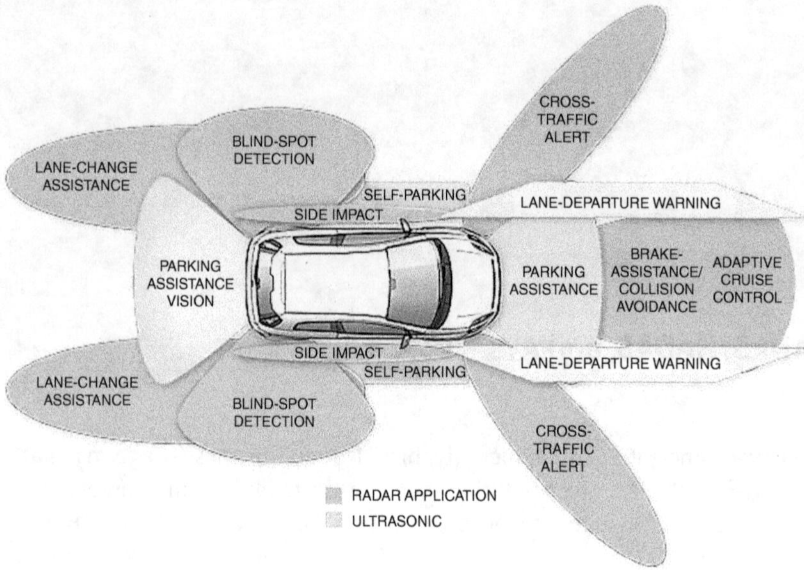

Figure 3.1 Examples of "types of vision" that an autonomous vehicle must have.

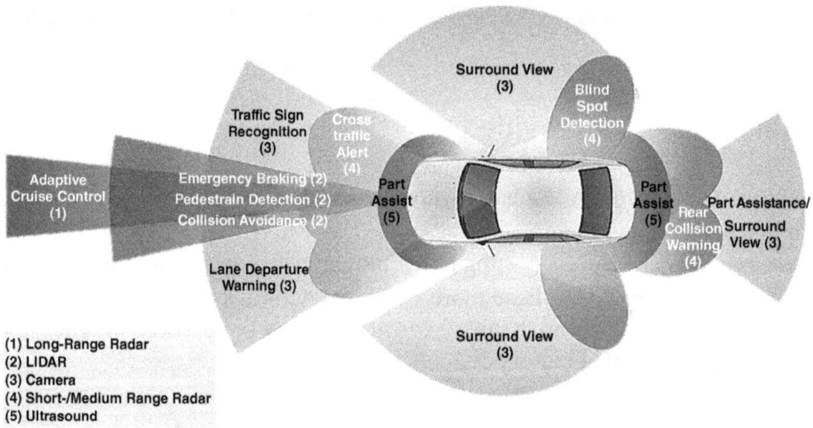

Figure 3.2 Examples of "angles of view" that an autonomous vehicle must have.

The usage of the technologies indicated above offers some idea of the number of each type of device required, indicated in Figure 3.4.

There are multiple possibilities as to how to create such a system, depending on the requirements and the available budget. Figure 3.5 offers a snapshot, taken at a given time, of two examples that will be referred to many times throughout this book.

Distance

Obviously, distance is a parameter that immediately springs to mind, but how are we to define the value of that distance?

	Distance		Angle of view	Speed	Latency	Datarate	Technology
Units			°	ms	ms	Mbit/s	
Forward	Far	Adapt cruising speed in relation to other vehicles and maintain safe distances	30				Lidar
	Middle	Brake when traffic in front slows suddenly	50–60				Radar
	Near	Parking	180				Radar Sonar
Front lateral	Far	– Avoid side-on impact – Recognize road signs – Alert driver to lane departure	120				Radar
	Middle and near	– Parking – Blind corners – Blind spots	90				Camera
Lateral	Near	Assist with parking maneuvers	90				Camera
	Middle	– Monitor – Blind corners – Blind spots	90				Camera
Rear lateral	Far	Avoid side-on impact	90–120				Radar Sonar
	Middle	– Parking – Blind corners – Blind spots	90				– Camera Radar
Rearward	Far	Monitor situation to rear	90				Camera
	Middle	Lane keeping	90				– Camera Radar
	Near	Parking assistance	180				– Camera Sonar

Figure 3.3 Overview of technical details of the functions required.

- The vehicle must be able to see at very short range (1 m or less). For example: automated parking in a very tight spot.
- It must also be able to see far, to detect potential hazards, and alert the driver to take the appropriate action, or activate the ADAS in partially automated vehicles, or respond fully autonomously in level-5 vehicles.

In conclusion, often the distances in question range from 0 to 200 or 300 m.

Technologies	Applications	Number of devices	Datarate per device (Mbit/s)	Angle of view
Long-distance radar		5	50	80
Short/medium-distance radar		4	45	180
Lidar	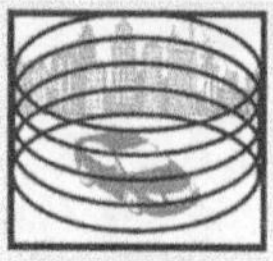	1	100	360
Cameras		5	100	120
Ultrasound sensors		15	30	360

Figure 3.4 Overview of technologies required to provide the technical functions.

Renault Espace (in 2017)	Valéo on VW vehicle (in 2017)
3 lidars, or long-range laser scanners: 2 front; 1 rear. 5 radars: 1 long-range front radar; 4 medium-range angled radars. 7 digital cameras: 3 with different focal lengths (short/medium/long range) at the top of the windshield; 4 cameras with 180° view (short range) beneath wing mirrors and on license plates. A belt of 20 short-range ultrasound sensors.	6 lidars around the vehicle: 2 on each side; 1 front; 1 rear. 6 radars: 2 front; 4 rear. 5 digital cameras: 1 front; 1 behind the windshield; 2 rear; 1 on the steering wheel to monitor the driver.

Figure 3.5 Examples of possible systems.

Far In order to manage safety, we need to be able to see in the distance, and brake if need be. Let us look at the problem from that angle, and calculate the true braking distance and stopping distance that a vehicle needs.

Reaction time To begin with, we must look at the reaction time: the time that elapses between the moment the human driver, or automated system, perceives or is alerted to a danger, and the moment the brake is actually pressed or activated.

- The reaction time of an alert driver is around 1 second (of course, if the driver is tired, has been drinking, or is under the influence of drugs, they will need longer to react).
- The reaction time of an electronic system may be considerably less than this (around 500 ms) – see the discussion below.

Reaction distance As a corollary to the reaction time, the reaction distance is the number of meters traveled between perception of the hazard, or alerting of the driver thereto, and the physical pressing (or activation) of the brake. Roughly speaking, this distance is equal to $(v/10) \times 3$.

For example: at 100 km/h, reaction distance = $(100 \text{ km/h} /10) \times 3 = 30$ m.

Braking distance The braking distance is the number of meters traveled between the pressing or activation of the brake and the vehicle actually coming to a stop. In addition, there are other factors that may increase braking distance (e.g. a road surface that is in poor condition or wet, brakes in poor condition, or used tires). Generally, the following formulae are accurate:

- In dry conditions: $(v/10)^2/2$
  - Example: at 100 km/h → $(100/10)^2/2 = 50$ m
- In wet conditions: $((v/10)^2/2) \times 1.5$
  - Example: at 100 km/h → $((100/10)^2/2) \times 1.5 = 75$ m

Stopping distance The stopping distance is the sum total of the reaction distance and braking distance.

Example: if a driver is traveling at 120 km/h, his or her stopping distance on a dry road will be 108 m.

Figure 3.6 gives the distances (in m) associated with these parameters.

We shall show, later on, that in view of the computation time and processing time for the images captured by the cameras (see Section 3.2), at 120 km/h, visibility of at least $144 + 24 \approx 170$ m is required, and that, for safety's sake, designers must allow a distance of 250–300 m to react, and also to be able to distinguish details.

Near Let us now look at the case of near vision – typical examples here include parking maneuvers, detection of pedestrians crossing near to the vehicle, and so forth. In this case, the distance falls to a few centimeters, with high precision and excellent reproducibility, irrespective of adverse weather conditions (cold/hot, rain, snow, etc.).

Angles

As previously indicated in Figure 3.3, depending on the target applications, the angles of view required vary enormously:

Speed	Distance in meters (m)				
	Ideal conditions	Dry road		Wet road	
	Reaction distance	Braking distance	Stopping distance	Braking distance	Stopping distance
20 km/h	6	2	8	3	9
30 km/h	9	4.5	13.5	6.75	15.75
40 km/h	12	8	20	12	24
50 km/h	15	12.5	27.5	18.75	33.75
60 km/h	18	18	36	27	45
70 km/h	21	24.5	45.5	36.75	57.75
80 km/h	24	32	56	48	72
90 km/h	27	40.5	67.5	60.75	87.75
100 km/h	30	50	80	75	105
110 km/h	33	60.5	93.5	90.75	123.75
120 km/h	36	72	108	108	144

Figure 3.6 Braking and stopping distances.

- Small (from a few degrees to around 10°): such would be the case with a system that needs to look a long way in front of the vehicle (such as adaptative cruise control), but without being overly concerned with what is nearby to the side.
- Medium (approximately 50°): such would be the case with a system that needs to look fairly close to the vehicle (e.g. road sign recognition, lane monitoring, lane changes, blind spots, etc.).
- Wide (roughly 120 to 150°): such would be the case with a system that needs to look near to the vehicle (monitor other vehicles on the road and in the city, to the fore and to the rear, parking assistance, etc.).
- 360°: such would be the case with a system whose purpose is to give a complete view of the vehicle's surroundings. Sometimes, such 360° vision can be obtained using multiple systems (three or four), with a field of vision of around 120° each.

Time/rapidity

The reaction speeds required vary depending on the desired applications. Generally, they are:

- Medium for long-distance view systems;
- Ultra-rapid for medium- and short-distance view systems.

Latency

The latency times are also a function of the desired applications. Generally, they are:

- Short – for example, for lidars needing to see far, and therefore giving the system a little time to react;
- Extremely short – for example, for radar or sonar sensors that need to detect a vehicle or pedestrian at very close quarters when parking.

Datarates

The datarates are also dependent on the desired applications, as well as the technologies that are used. As we shall see later on, they are generally:

- Slow/medium for numerous sensors;
- Extremely fast for lidars and analog or uncompressed digital cameras.

Figure 3.7 offers examples of the order of magnitude of these values, depending on the intended applications.

Environment

Of course, the quality of these parameters must be appropriate to deal with the many different types of environment encountered in automobile applications, including:

- Rain, snow, fog, etc.;
- Sun, darkness, driving in tunnels and underpasses, sudden glare, etc.;
- Gusts of wind carrying dead leaves;
- Cyclists walking their bikes across the road, etc.

In short, the system must be able to cope with all possible conditions!

3.2 Sensors and the technologies they use

3.2.1 General

In order to satisfy the requirements of the functions and technologies described in this chapter, we need sensors, which are the basic devices. What are these sensors, though, and how do they work? To begin with, here is a brief recap of the definition: a sensor is a device that transforms the state of a physical variable into a usable value.

Domain	Description	End-to-End Latency Requirements	Bandwidth Requirements
Powertrain	Controls the components that generate power and transmit to the road	< 10 μs	Low
Chassis	Controls steering, brakes, suspension	< 10 μs	Low
Body and Comfort	Radio, A/C, window, seat, and light controls	< 10 ms	Low
Driver Assistance and Driver Safety	Controls systems designed to increase safety	< 250 μs or < 1 ms depending on the system	20–100 Mbps per camera
Human–Machine Interface	Controls displays and other interfaces that interface with the driver or passengers	< 10 ms	Varies by system, but is growing

Figure 3.7 Summary of technical characteristics of the connections required for different applications.

Autonomous vehicles require numerous sensors because, in time, the goal is for these systems to be able to operate without a driver, and therefore replace the driver's vision, hearing, sense of smell, etc. Importantly, such a system must be able to achieve results in parity with those of the human eye, as this is the organ that is most heavily involved in driving. Thus, visual perception is one of the main key points for applications in autonomous vehicles. With all the sensors, the system must be able to instantly reconstruct the scene taking place in front of, behind, and 360° around the vehicle. In the human eye, visual information is transmitted as an electrical impulse to the brain, which reconstructs the scene on the basis of the information received, relating to color, shape, motion, spatial location, and so forth. Thus, we need to do the same with the vehicle: this is what it must do by observing the world around it using sensors and building a model of its environment in the onboard computer.

Hence, all the sensors on board an autonomous vehicle must be capable of reproducing all human faculties needed for driving. Thus, in a manner of speaking, we see a reproduction of the human brain, with the computer recreating the areas of the brain involved in reaction, location, etc., but – and there is always a "but" – these new technologies, based on measurements and analyses thereof, have/will have limits. An improperly defined requirement could lead to measurements being misinterpreted, with the risk of disastrous results, including running down a pedestrian or two!

To briefly conclude this general discussion, it must be understood that the only way a system can perfectly reliably locate objects is by having multiple layers of redundancy in terms of optical sensors, cameras, radars, lidars, etc., in addition to odometers, inertial navigation systems, infrared sensors, ultrasound sensors, etc., and then a computer takes a certain amount of time (as little time as possible, naturally) to analyze all the data supplied by all these sensors, fuses those data, generates a map, applies predetermined rules of conduct, and finally manipulates and activates the steering, accelerator, and brakes!

3.2.2 Vision

Let us begin our discussion with sensors that are responsible for vision. It almost goes without saying that the system's vision must be excellent, both near and far, in daylight and in darkness, whatever the weather conditions (fog, snow, etc.), and cover a full 360° around the vehicle.

The "optical" sensors described below (cameras, radars, and lidars) yield information about the environment, allowing the system to precisely calculate the vehicle's location, in terms of position and orientation. The sensors often have "local intelligence" capable of segmenting the data and classifying moving objects or obstacles in the vehicle's environs. When speaking of such sensors, it is also important to distinguish the following points:

- Manufacturers of optical materials (lenses – objective lenses, prismatic lenses, etc.);
- Manufacturers of electronic components (semiconductors, integrated circuit boards, etc.);
- Manufacturers of subsystems (often OEMs) who assemble all the components into mechanical modules, and may include a little or a great deal of local intelligence in the cameras, radars, and lidars.

3.2.3 Cameras

Cameras are passive sensors (they receive information in the form of light).

Visible-light video camera

By definition, a visible-light camera is sensitive to a spectrum similar to that of the human eye (wavelengths between 400 and 1000 nm). A "visible-light camera" comprises one or more photoelectric sensors (using CMOS or CCD technology), which convert energy from received photons (with wavelengths in the visible spectrum) into analog electrical signals.

Principles underlying monochrome and color cameras

In general, the targets of photographic sensors in cameras are rectangular matrices of photodiodes (e.g. in 1280 × 1024 format in a CMOS). Each one of these photodiodes is a sensor, which constitutes a pixel (a picture element). They are sensitive only to the intensity of the light received, not to its color, so, fundamentally, the cameras are "monochromatic" or monochrome.

Monochrome cameras become "color" cameras by the addition of a *Bayer color filter array*. That is, an array of RGB (Red, Green, Blue) color filters is interposed between the incident light and the photodiode matrix. Each filter allows only one of the R, G, or B components of the incident light to pass through. As shown in Figure 3.8, the Bayer filter array is composed of 50% green filters, 25% red filters and 25% blue filters, imitating the physiology of the human eye. Indeed, the human retina uses cones in daylight, which are more sensitive to wavelengths corresponding to green. The signals thus collected can be interpreted to produce color images.

The output signal, which is fundamentally analog, represents the luminance of a monochrome image or a color image. Often, the analog luminance value of each pixel in the signal is directly digitized – i.e. converted into a digital signal of 8, 10, 12, 14, or 16 bits depending on the required precision or resolution. In addition, very often, in the camera, the signal can be pre-processed (being digitized, having gain added, etc.), by means either of a linear function or of a log function, to combat certain phenomena of glare (see the example in Figure 3.9A to D).

The formats used to transport the digital signals obtained are often USB2.0, USB3.0, Camlink, GIGE, Ethernet 100 Mbit/s, and similar. From an optical standpoint, the

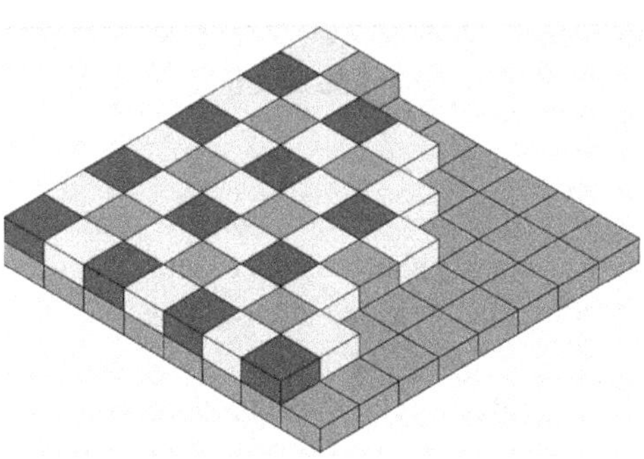

Figure 3.8 Bayer color filter array.

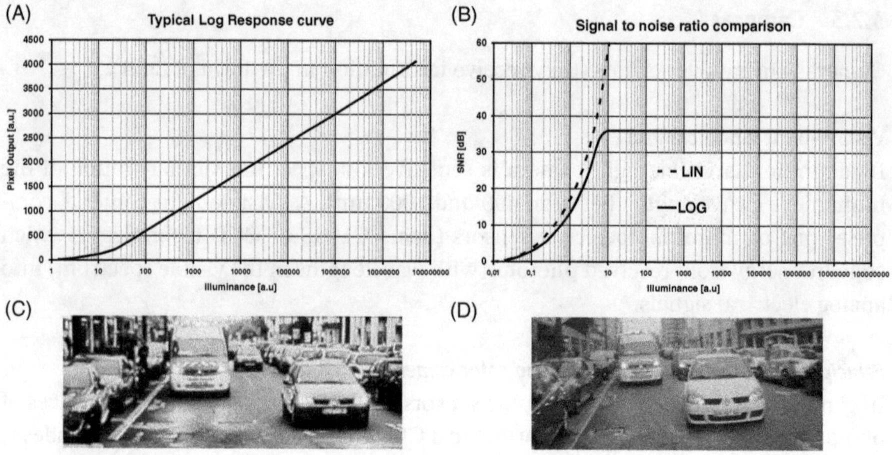

Figure 3.9 Examples of processing to combat glare.

cameras' angles of view vary depending on a range of parameters (lenses, focal lengths, sensor size, etc.). Note that there are optics that can deliver extremely wide angles of view (fisheye lenses can offer vision up to 180°).

Applied technologies

In today's market, there are two opposing theories and application technologies. As usual, the ratio of quality to functionality to price will, ultimately, make all the difference.

Stereoscopic cameras The technology in stereoscopic cameras approximates human vision: the "eyes" deliver overlapping images to the "brain", which processes them and determines aspects such as the field depth, peripheral movement, and the size of objects. In autonomous vehicles, stereoscopic cameras are used to cover the whole surrounding scene. In general, two cameras mounted on the outside of the vehicle are placed a few centimeters apart, to obtain a view of the surroundings in which the images overlap, due to an intentional slight parallax. The associated software creates a 3D world on the basis of a 2D image. Each camera has a field of view of around 50° and is accurate to a depth of field of only about 30 m. Very often, the positions and configurations of stereoscopic cameras differ from one autonomous vehicle model to another. However, as is the case with the rest of the vehicle, there are redundant technological solutions available as a backup, which would allow the vehicle to remain roadworthy even if the stereoscopic cameras malfunction.

Mono-cameras The use of a mono-camera also draws inspiration from the properties of human vision. To obtain depth perception over very short distances, human vision uses both eyes, making a stereoscopic camera system an appropriate solution. However, the interpretation and analysis of scenes typically encountered in driving rely on vision over much greater distances. All the parameters relating to depth perception over larger distances must be interpreted by only one eye. In this way, it is possible to identify shapes (vehicles, pedestrians, etc.), perspective, shadow, textures (strips on the road, text on road signs, etc.), and movement indicators that the human visual system uses to understand the visual universe surrounding the vehicle.

The discovery of this mechanism of vision has had a significant effect on the market – prior to this, it was believed that sensors such as stereo cameras and radars were simply superior. Thus, the majority of automakers have begun using mono-vision cameras as the main sensor/source of information for their *advanced driver assistance systems* (ADASs), which are capable of semi-autonomous driving. These cameras, which are often placed at the top of the windshields, behind or in the side rear-view mirrors, are designed to monitor the road ahead, detect traffic lights, interpret road signs, and, with the help of the onboard computer, detect moving objects, such as pedestrians or cyclists (e.g. pedestrians wearing white in sunny conditions, emerging onto the road from between two white vehicles, or cyclists dressed in black at night, emerging from between two dark-colored parked cars), and lane guidance on highways (though not in urban environments). They can also be used for rear-view and surround-view systems (the latter offering full 360° vision).

Advantages

Visible-light cameras offer the following benefits:

- They use reflected light;
- They work very well in daylight;
- They can see for an arbitrary distance if their field of view is narrow;
- Their orientation can be changed;
- They can easily distinguish color;
- They generally do not contain moving parts;
- They are better than other devices in terms of reading signs, because they can easily discern traffic lights, brake lights, indicators and other types of lights, read road signs, and keep track of other vehicles' trajectories, whilst also being on the lookout for pedestrians and other obstacles;
- It is possible to obtain very good resolution for faraway objects;
- They are capable of analysis over 3000 rows of pixels;
- In view of this high resolution and color distinction, in theory they are able to understand/interpret things in the scene that could not be detected by a lidar;
- Mono-cameras are small;
- They are not bulky, so a large number of them can be used;
- Even with a wide field of view, mono-cameras are relatively inexpensive.

Drawbacks

Of course, there are also a number of disadvantages to the use of cameras:

- At night, they need lighting, and must use the light emitted by vehicle headlights, among other sources, which may not always be sufficient;
- They must constantly readjust to cope with the variation of ambient lighting. Objects encountered on the road while driving are often obscured by, or cast, moving shadows, and may be lit from any direction – or not at all.

Examples of integrated circuits for cameras

The number of cameras and other imaging technologies used in vehicles is rapidly increasing. Without wishing to unduly publicize any one technology, we have chosen one significant example among many possible others: ON Semiconductor's AS0140/42 CMOS imaging circuits (see Figure 3.10):

- Quarter-inch 1-megapixel image;
- 45 fps frame rate at full resolution, or 60 fps with progressive scanning of 720 rows (720 p.);
- Consumes 530 mW in 30 images/s mode, in the high-dynamic range (HDR);
- Offers distortion correction and superposition of multiple colors;
- Analog (NTSC) and digital (Ethernet) interfaces;
- Adaptative local tone mapping (ALTM), to eliminate artifacts occurring during the image capture process;
- 93 dB dynamic range for high- and low-light applications;
- Multi-camera synchronization means it is easy to increase the number of cameras installed in vehicles;
- 30% space gain in relation to the conventional solutions using discrete sensors and processor components. This offers designers the freedom to implement camera solutions without impacting the vehicle's style or esthetic appeal;
- Automobile temperature range from −40°C to +105°C.

Examples of cameras

Figure 3.11 presents a table of the main features of some of the cameras available on the market (see Figure 3.12).

Figure 3.10 ON Semiconductor's AS0140/42 CMOS imaging circuits.

		Bosch MPC2	New Imaging Technologies - NIT MC1003-PGC	Realtek RTL9020A
Image format	Pixels	1280 × 960	768 × 576 1280 × 1024 2/3 inch	High-definition video
Pitch			6.8 μm	
Shutter			*Rolling or snapshot or differential*	
Digitization			8 or 14 bits Logarithmic response	Video compression up to full-HD resolution
Field of view	Horizontal	50° (nominal)		
	Vertical	28° (nominal)		
Resolution		25 pixels/°		
Frame rate		30 images/s	50–150 images/s - Frame scanning - Progressive scanning	
Exposure dynamic range		110 dB	140 dB	120 dB
Wavelength		400–750 nm		
Current consumption		< 5.0 W (0.35 + at 14 V)	4 W	Ultra-Low Power standby mode (35 μa)
Operating temperature		−40 to + 85°C (+105°C)	+20–70°C	
Video			In Y In RGB	AVB (audio video bridging)
Connectivity		2× CAN/CAN + Ethernet *Optional: FlexRay* *2× digital in/out, windscreen*	USB 2.0 and USB 3.0 interface GigE Vision standard	PHY and multi-port switches for simple 100BASE-T1 non-shielded twisted-pair cable

Figure 3.11 Main features of a number of cameras on the market.

Examples of camera systems

Most major OEMs (Valéo, Continental, Bosch, etc.) rarely provide automakers with simple cameras, but rather with "smart" cameras (Figure 3.13), which come with numerous inbuilt functions that either do the job of, or serve as a prelude to, ADAS. From their point of view, this offers a range of advantages:

- The need to transmit only information that has been filtered, preprocessed, processed and refined – therefore not a voluminous flow of data at a very high datarate,

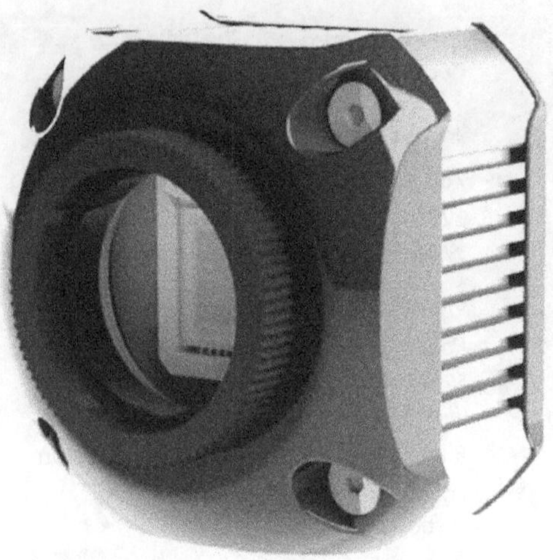

Figure 3.12 NIT MC1003-PGC.

Figure 3.13 Example of Valéo's smart camera.

transportable via Ethernet, but a flow that is often compatible with CAN/CAN FD networks;
- Above all, the ability to sell their customers more than simple cameras: they are selling a portion of their knowledge and intelligence, which offers added value.

Of course, that added value swells the OEMs' coffers – not those of the automakers, who are eagerly awaiting the dawn of Gigabit Ethernet, in order to be able to handle the raw video feed, at a high datarate, and process it (including data fusion) on the vehicle's ECU.

In addition, in numerous new vehicle models, the driver has a rather reduced direct physical view of their immediate environment (side windows and rear windshields that are increasingly small), outward vehicle shapes are dictated by aerodynamics, consumption and protection of pedestrians, making it practically impossible to perform maneuvers safely and accurately, and park. OEMs that offer multi-camera driver assistance systems, providing a perfect view of the vehicle's surroundings. Miniature cameras (in the front radiator grill, in/under the license plates, in the rear-view mirrors, etc.) record a 360° view of the surroundings and transmit the images to a smart control unit, which processes and painstakingly optimizes them, and then displays them. The driver sees a bird's-eye view of the vehicle, but also a three-dimensional image of the environment, and full visibility on the vehicle's color screen. Thus, the driver can drive and maneuver in safety, thanks to excellent visibility of the blind spots and all obstacles in the vehicle's vicinity.

Example Let us take an example showing multiple application scenarios based on a system using between one and four cameras and an ECU to:

- Check blind spots;
- Assist with changing lane on the highway;
- Assist with kerbside parking without damaging the tires and rims;
- Use a surround system offering a multitude of ways of seeing, rear view, mirror systems, etc.

We shall also look at evolving solutions combining four cameras with a horizontal field of view of >180°, thus being able to see all round the vehicle and create any type of display.

With a single camera

The module is made up of the system's optic and a color image sensor with a resolution of 1280 × 960 pixels. Thanks to its resolution, this sensor offers a wider aperture angle and a significantly increased range for object detection, now over 120 m. It is based on a dual-core ARM microcontroller, and the algorithm means that a simpler, less expensive optical system can be used. This variant of versatile camera allows automakers to integrate an extensive range of functions into the ADASs of their vehicles using just one sensor.

With two cameras

The version using two cameras offers a 360° 3D image to assist with parking maneuvers and slow-speed maneuvers. The two cameras are affixed as follows:

- One mounted in the front, in the radiator grill;
- The other mounted in the back, near to the license plate.

In view of its attractive cost, this system is particularly well suited to use in entry-level vehicles.

With four cameras

One in the front, one in the rear, and two in the wing mirrors.

This four-camera system is able to:

- Monitor the vehicle's surroundings;
- Recognize pedestrians;
- Alert the driver and/or even halt the vehicle in emergency situations.

The system is particularly suitable for city driving, because, amongst other things, it is capable of:

- Instantly recognizing pedestrians in the vicinity;
- Alerting the driver to vehicles cutting into their lane;
- Braking;
- Automatically parking, even without the driver actually being in the vehicle.

Estimations of datarates used for digital video in AVs

Let us linger over this point, because it is important to clarify, from the outset, exactly what we mean when we speak of digital datarates from cameras, radars, and lidars, because in certain real-time video applications (meaning there is no data compression), sooner or later, the raw data feed (i.e. unprocessed data) must be fed into a computer, which will merge the data from the various devices involved in autonomous driving of the vehicle.

Take a look at an example:

- The fastest man in the world can sprint 100 m in just under 10 seconds, which equates to 10 m/s = 36 km/h.
- To avoid problems such as *frame flicker* and *line flicker* in an image acquired by interlaced or progressive scanning for display to a human being, it is sensible to choose a refresh rate of approximately 30 to 50 refreshes per second (that is, every 20 ms) with progressive scanning.

When the first image, I1, is digitized, the computer quantifies the significant luminance differences from one pixel to the next (by a mathematical technique called differentiation), indicating the likely presence of objects, pedestrians, etc. The entire image is separated into blocks (e.g. measuring 8 × 8 pixels), which are labeled.

When it comes to the next image, I2 – that is, 20 ms later, or once the second image has been fully processed, i.e. 40 ms after the start of the first one – the system looks for those particular blocks and analyzes whether they have shifted along the *x* or *y* axis in the pixel matrix, and/or have grown or shrunk homothetically. From the new coordinates *x*" and *y*", it can deduce a vector V representing an element of the object's relative velocity in relation to the vehicle. The point of application of that vector is in image I1. It has a direction (right, left, up, down) and an intensity value (the gap between *x*' and *x*", and between *y*' and *y*"). Similarly, from the variation in the block's homothety, the system can determine whether the object is getting closer (the image is growing) or getting further away (the image is shrinking); and if its size remains constant, this means that it is traveling at the same speed as the vehicle itself. For the sake of certainty, the same operation is performed with image I3, at which point, all the necessary conclusions are drawn (braking, changing lane, changing direction, etc.). However, even when operating very quickly, the whole process has taken somewhere between 60 and 80 ms when driving at, for example, 3 × 36 km/h = 108 km/h, which equates to 30 m/s. Thus, in the space of 80 ms, the vehicle will already have traveled 2.4 m before doing anything at all! In addition, certain facts also need to be considered – human performance, mechanical capability, environmental performance (braking distances, etc.).

In view of the computation time taken to process the images, at 120 km/h on a wet road (see Section 3.2.1), a vehicle needs visibility of at least 2.4 + 108 = ~ 110 m, and, for safety's sake, the system needs to be designed to see over a distance of 200–300 m and be able to distinguish details. In addition, the target of the camera (or radar or lidar) must have a large number of pixels, and the necessary aperture of the beam of the field of view cannot be too wide. In this case, we can calculate the chord of the arc subtended by that angle of view, on the basis of the distance (see the example in Figure 3.14).

At a 300 m distance, with a beam whose aperture is 35°, giving a field of view that is 180 m wide, a camera target measuring 728 horizontal pixels will give a definition/object resolution of a minimum of 180/728 = 24 cm in width. A different target measuring 1280 pixels will give a resolution of 14 cm.

Let us proceed with this example to its conclusion. In doing so, we can introduce the concepts and values of datarates that we shall encounter in the coming chapters (see Figure 3.15).

This illustrates one of the origins of the Gigabit Ethernet.

Video format, H264 standard and AVC To recap, the video encoding standard "ITU-T H.264" and the "ISO/CEI MPEG-4 Part 10" is the product of a Joint Video Team (JVT) made up of the ITU-T Q.6/SG16 Video Coding Experts Group (VCEG) and the ISO/

	Value of subtended chord	
	Beam aperture	
Distance	50°	35°
At 300 m	250 m	180
At 100 m	85 m	60

Figure 3.14 Value of the chord of the subtended arc.

Parameters		Units	Example 1	Example 2	Example 3
Refresh rate	Frames per second	Fps in Hz	50	50	30
Frame scanning	Type		Progressive	Progressive	Progressive
Number of pixels	Horizontal		1280	728	728
	Vertical		1024	576	576
Pixel sampling		Bits	14	14	8
Raw data		Mbit/s	1280 × 1024 × 50 × 14 = approx. 920	728 × 576 × 50 × 14 = approx. 300	728 × 576 × 30 × 8 = approx. 101
Beam aperture		°	30	30	
Resolution at 300 m		cm	14	24	

Figure 3.15 Overview of the example and calculation of datarates in Mbit/s.

IEC Moving Picture Experts Group (MPEG). In the MPEG, the same technology is also known as AVC (Advanced Video Coding) – see Section 5.4. Figure 3.16 recaps the main relationships between a number of widely used image formats and profiles, and the raw datarates required for the associated digital video signals.

Thermal and infrared cameras

Thermal and infrared cameras are sensitive to wavelengths of light greater than those in the visible spectrum, between 400 and 700 nm. According to the black body model, every object emits light, the wavelength of which depends on the object's temperature. As the wavelength of infrared radiation is temperature dependent, the different types of infrared radiation are linked to the heat waves given off.

- Human beings, who are naturally warm, emit a spectrum in the far infrared (FIR), between 8000 and 14 000 nm. These wavelengths correspond to those generally used in so-called "thermal cameras."
- Another range of wavelengths is also used in driver assistance systems: near infrared (NIR), between 750 and 1400 nm. The benefit of using these wavelengths is that they can be directly detected by traditional CCD or CMOS sensors, unlike FIR, which requires special, and more costly, sensors. NIR cameras are used, in particular, for driver monitoring applications, because infrared lighting of the vehicle cabin (which is invisible to the driver) means that the driver can be monitored even in very low ambient light (e.g. at night). Infrared cameras are cameras that use and detect light in the near infrared, whose wavelength is between 900 and 1700 nm.

Frequently, the sensors in these cameras simply reflect an image of the intensity of radiation from the source's temperature (thus, they are "single-channel" sensors, in the same way that a monochromatic camera displays "black and white").

	Maximum datarate in Mbit/s per profile				**Definition in horizontal and vertical/frames per second**
Levels	***Baseline, extended* and *main profile***	***High***	***High* 10**	***High* 4:2:2 and 4:4:4**	
3	10 Mbit/s	12.5 Mbit/s	30 Mbit/s	40 Mbit/s	720 × 480/30.0 720 × 576/25.0
3.2	20 Mbit/s	25 Mbit/s	60 Mbit/s	80 Mbit/s	1280 × 720/60.0
4	20 Mbit/s	25 Mbit/s	60 Mbit/s	80 Mbit/s	2048 × 1024/30.0
4.2	50 Mbit/s	62.5 Mbit/s	150 Mbit/s	200 Mbit/s	2048 × 1088/60.0
5.2	240 Mbit/s	300 Mbit/s	720 Mbit/s	960 Mbit/s	4096 × 2 160/60.0

Figure 3.16 Main relationships between various image formats and profiles, and the raw datarates required for the associated digital video signals.

Advantages

Below is a non-exhaustive list of the benefits of using infrared sensors:

- Thermal images can be recorded at any time, day or night. In fact, thermal imaging is the most widely used technique for night vision.
- An image can be displayed on a color screen in "false color," by associating a preselected color to a certain thermal intensity received to make the image easier to read: each color in the image corresponds to a temperature. An ADAS can begin to process data on the basis of that image and/or map.
- Thermal and infrared cameras are defined by their spatial resolutions (for example, the smallest discernible object would be 15 × 15 μm on a 640 × 512 pixel matrix) and their thermal resolutions (the slightest perceptible temperature difference). These two types of resolution are not independent and, generally, cameras are characterized by the graph showing the evolution of the thermal resolution with spatial resolution (the MRTD curve – Minimum Resolvable Temperature Difference).
- Infrared sensors are highly effective and can replace or complement the human eye. Their role is similar to that of cones in the retina. By detecting the colors of the lines being followed and tracking lines on the road surface, such a system can allow an autonomous vehicle to keep in its lane. Furthermore, it is capable of perfect vision in light and in darkness (white lines, followed by broken lines with response times for overtaking, work with yellow or red painted, overlapping lines).
- Owing to their high sensitivity and high dynamics (typically 120 dB), thermal/infrared cameras can produce excellent images even in the darkest night.
- Once calibrated, some such systems can also measure temperatures remotely.

In view of these performances, they are perfectly suited to a wide variety of applications. For example, Forward-Looking Infrared (FLIR) thermal cameras are installed in certain vehicles to improve the driver's view, allowing them to "see" up to four times further than headlights alone.

Limitations

Unfortunately, even with a decent resolution:

- Thermal cameras are costly;
- They must also be installed on the exterior of the vehicle, because LWIR (Long-Wave Infrared) cannot easily pass through glass, especially when, for reasons of comfort inside the vehicle, the glass is heat-resistant.

  At present, they are not widely used.

Camera processing

The processing of the signals from a camera can be divided into two broad categories: machine vision and computer vision.

Machine vision

Machine vision generally refers to localized digital image analysis. This includes processing such as analysis of roadside features, motion parallax detection, and parallax measurement in stereo images to estimate distance. These techniques are fairly well established. Certain problems are more difficult to solve, but solutions are being worked on. One example is the detection and reading of signs when the panels have been graffitied.

Computer vision

Computer vision refers to a set of more complex problems, nearer to the human brain's processing capacity – things such as object recognition, for example. A human driver can identify a human being no matter what the circumstances, irrespective of lighting, etc., quickly identify that it is a human being, how far away the person is, and even discern which direction they are moving in and what they are doing. This technology is essential and is at the very heart of autonomous driving. Not only does this mean having a large number of high-performance cameras on the vehicle, but also having efficient processors that can analyze the video feed from those cameras. Sophisticated algorithms – whether proprietary or open source – need to process incoming information in real time, at rates of around a gigabyte per second.

Example of video transmission standard

Digital cameras generally offer far more functions than simply providing an image. Image processing and the addition of results into the image data feed, checking external hardware, and real-time application processing are commonplace tasks for cameras today. Therefore, the camera programming interface has become more complex, its purpose being to provide a generic programming interface for all sorts of cameras and other devices:

- For example: the software interface for GigE Vision® v2.0 standard, based on the standard UDP/IP protocol, allows devices to communicate their generic software functions using standard XML files. The combination of UPD/IP and XML definition means that:
  - Software from any provider can be used. Thus, the image processing hardware becomes interchangeable;
  - Images can be transmitted at high datarates (125 Mbit/s), including control data, and can be connected to an Ethernet network;
  - Low-cost cables can be used over distances of up to 100 m;
  - Non-streaming peripheral devices can be controlled;
  - Data can be transferred at 10 Gigabit Ethernet and link aggregation is possible;
  - Compressed images can be transmitted (in JPEG, JPEG 2000, and H.264 format);
  - Multi-camera systems can be synchronized with a high degree of accuracy;
  - Design cycles are shorter and development costs less.
- Another example: Mobileye is another contender in the market of smart cameras for applications in ADAS. In fact, the system is a video sensor that, after processing, sends the vehicle's communication bus data/information including obstacle detection and road markings. These data can help the system to control the vehicle (e.g. emergency braking, etc.). From the standpoint of the automaker, these smart sensors are extremely practical, because they have inbuilt image processing algorithms and output only the usable data for mapping the vehicle's surroundings.

3.2.4 Radar

As we shall see, radars and lidars use essentially the same principle to measure distance.

Radars and lidars are active sensors (they emit a signal and then receive information).

Functional principle

Radar (the term derived from *radio detection and ranging*) is a system that uses electromagnetic waves with millimetric wavelengths to detect the presence and determine the position and speed of objects. The waves sent out by the transmitter are reflected by the target (a phenomenon known as *backscattering*[1]) and the returning signals (called radar echoes) are recorded and analyzed by the receiver, which is often collocated with the transmitter (this system is known as a "monostatic" radar):

- The distance from the object is calculated on the basis of the measured *Time of Flight* (ToF) for the signal to reach it and return;
- The direction of the object is known because of the angular position of the antenna where the return signal was received;
- The object's speed is measured by quantifying the frequency difference between the transmitted and received signals, which is due to the Doppler effect.

In the automotive industry, radar sensors are used to detect the position of surrounding objects, whether near or far, and to avoid and protect pedestrians (see Section 3.3 for ADAS applications).

In automobiles, there are two main types of radars, whose frequency bands Ka and W are imposed by the regulations in force in different countries:

- *Short-range* radars, working at 24 GHz (in the Ka band), are generally used to detect potential collisions to the side of the vehicle. For example, they are used for *blind spot warning* (BSW) and warnings when changing lane. If a collision is inevitable, then, amongst other things, the vehicle can automatically apply the brakes, pretension the safety belts, etc.
- *Long-range* radars, working at 76 GHz (in the W band) are generally used for forward detection. For example, this allows the ACC (*automatic cruise control*) system to maintain a reasonable distance from the vehicle in front (this function is known as adaptive cruise control).

For all these short- and long-range applications, there is an increasing demand for the use of the 79 GHz frequency band (the W band, which has a bandwidth of 4 GHz, ranging from 77 to 81 GHz as per ETSI EN 302 264, is not subject to time constraints or other limits on its operation, and is suitable for use in the vicinity of radio-astronomy sites), which is crucial in order to perform numerous automobile safety functions. Systems working at that frequency and with a larger bandwidth (4 MHz) are able to deliver better spatial resolution, and so better results, because they are more reliable and more accurate, and they can distinguish more clearly between objects present on the road:

- These devices are primarily designed to improve comfort when driving and are essential for new functions such as pedestrian or object detection, automatic emergency braking in urban areas, adaptive gear regulation, collision warning, blind spot warning, lane-change assistance, rear cross traffic alert, and reverse parking assistance.
- Such systems also help improve safety to a certain degree, because they are now able to offer proactive safety functions. Examples include pre-collision systems, designed to attenuate collisions or detect vulnerable road users.
- Other benefits: the higher the operating frequency, the smaller the total size of the radar. The same technology can be employed for all applications, and the risk of interference is low because the emitted power required is low.

Advantages

Radar sensors offer the following advantages:

- Because of the properties of the electromagnetic energy that is authorized for use in the Ka and W bands, radars can measure over short, medium, and long distances (up to a little over 200 m).
- In view of the millimetric wavelengths used, radar is not badly affected by fog and light rain. Overall, radar is able to operate in any weather conditions, where other optical sensors would fail.
- With each analysis, the radar returns a time value, associated with a specific distance. However, by exploiting the Doppler effect, it is also able to directly measure the speed of moving objects and other vehicles.
- Radars are generally installed in the front and rear bumpers of autonomous vehicles.
- Vehicles use radar to detect an imminent impact, or even brake automatically to avoid a collision, or constantly adjust the acceleration and brakes. In essence, such a system is an adaptive speed regulator, which always takes account of vehicular traffic around you.

Drawbacks

Of course, there are certain limitations to the use of radar:

- High-resolution radars require a greater bandwidth.
- Standard radars have poor angular resolution (approximately 3° around its maximum radiation field).
- Consequently, they have difficulty in determining whether a target (such as an object or a vehicle) is or is not in your lane, or whether it is on a bridge or on the road in front of you.
- In principle, fixed objects (such as the ground, signposts, warning signs, road closure bollards, etc.) also reflect radar signals, as does a parked vehicle. Thus, it is difficult to determine whether the object is completely off the road or is parked on the side of the road.
- Most automobile radars simply ignore signals returned by stationary objects. This is why automatic cruise control systems have long struggled (functioning poorly, or not at all) in urban traffic and in stop-and-go mode.

Applications

Changing lane, whether quickly or slowly, is always risky, especially if a vehicle in the adjacent lane is traveling faster than you or is in your blind spot (the area behind the vehicle which is hidden from the wing mirror). A lane change assistance system, using radar, offers invaluable information, telling the driver whether there are vehicles in the blind spot or coming up quickly from behind. Generally, such a system uses small radars (not containing any moving parts), located on both sides of the rear of the vehicle (Figure 3.17). These radars scan the surroundings with a multitude of separate beams and are able to detect any vehicle (cars, HGVs, motorbikes, etc.) that is near to the rear of the vehicle. Such radars are able to operate in practically all weather conditions.

Oncoming traffic and vehicles you are overtaking are detected, but no alert is raised. The driver receives the information in the form of a smart signal, built into the wing mirror. Thus, the information is received immediately, and in a form that is easy to understand. In summary:

- Though ultrasound sonar technology (see Section 3.2.6) is more mature and less expensive than radar technology, vehicle designers who are concerned about esthetics and the vehicle's appearance do not wish to have too many visible sensor apertures on the outside. With future designs, mini/affordable radars could, and ultimately should, replace ultrasound sensors.
- Generally, radar complements other systems in the vehicle, such as inertial measurement units, gyroscopes, and rotary encoders, sending highly accurate data to the vehicle's central processing unit so it can make the best decisions as to how to minimize the risk of an accident.
- Radar is often married with ultrasound sonars. Both have a narrow field of view and let the vehicle know whether something (be it a pedestrian or another vehicle) is passing through their beams. The information received can serve, amongst other things, to correct the vehicle's course, apply the brakes or activate the seatbelt pretensioners.
- Radar is capable of detecting driving infractions, detecting a truck coming towards you or telling you what a vehicle is doing when it is hidden in front of a truck you are following.

Figure 3.17 Valéo's blind spot radar.

- Radar can inform blind spot alerts and alert the driver to an unintended lane departure (see Section 3.3: ADAS).
- Systems using radars sound an audible alarm to alert the driver to an imminent danger and, in certain situations, can assume control of the vehicle to avoid the problem. An example is an adaptive cruise control system with long-range detection, observing the vehicle in front, and speeding up or slowing down to maintain a safe distance.

3.2.5 Lidar

The word *lidar* derives from a portmanteau of the words *light* and *radar* (it is the shortened form of *light detection and ranging* or, occasionally, *laser illuminating detection and ranging*). Lidar was first used in the automobile industry in the wake of the earliest autonomous vehicle projects. The design of high-performing but affordable lidar sensors will indubitably be one of the key factors in making automated vehicles safe, because such sensors offer certain advantages in comparison to other detection technologies – notably radars and cameras.

Let us first outline how lidars work.

Operational principle

Lidar is unlike sonar, which uses sound waves (see Section 3.2.6), and radar, which uses millimetric and centimetric wavelengths (frequencies between 1 and 100 GHz – see Section 3.2.4).

Lidar technology is based on measurement by optical remote sensing (telemetry – determination of the distance to an object), using electromagnetic waves, the frequencies of which are beyond 10 THz, with wavelengths of between 250 nm and 10 000 nm. The system illuminates a target with a beam of coherent, polarized light (either light from the visible spectrum or invisible in the infrared or ultraviolet – typically, infrared is used in the automotive industry), almost always emitted by a laser. On the basis of the known propagation speed and an analysis of the backscattered light reflected by the target, it can measure the distance or determine other properties of the target.

In order to do this, a lidar (see Figure 3.18) emits short pulses of invisible infrared laser light in a certain direction, and an integrated sensor measures and records the time taken to receive reflected light (measuring the time of flight of the reflected pulse). As the wavelength of the emitted light is known, consequently so too is its velocity ($\lambda = v\,T$):

- On the basis of the received power, the system can deduce the luminosity of the target to a high degree of accuracy;
- By measuring the time of flight between the sending of the pulse and detection of its reflection, it can calculate the distance between the target and the source.

The receipt and transcription of the data contained in the infrared spectrum received only produces monochromatic images, displayed in grayscale. These can be colorized

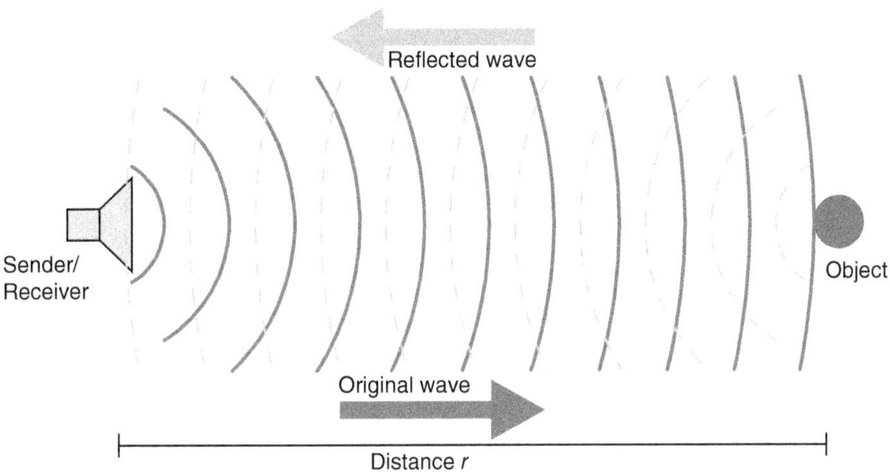

Figure 3.18 Operational principle behind a lidar.

as each designer sees fit. This principle applies to a majority of lidar applications in automobiles.

Another class of applications based on measuring the speed of a wave uses a laser, emitting in a spectrum at a very specific frequency. The method uses the Doppler–Fizeau effect, measuring the variation of the frequency of the reflected wave received on the basis of the target's motion and, from there, determining the object's speed. In the atmosphere and other diffuse media, many other parameters (concentrations of gases and specific particles, density, temperature, and so on) can be measured (for example, by studying clouds and fog) if we are able to isolate the effect of the various interactions between light and matter along the length of the laser beam.

Lidar is an excellent, and indispensable, tool for active remote sensing.

Echolocation technology

Generally speaking, as is the case with sonars and radars, lidar works on the basis of echolocation – the main difference between these methods lies in the reflective properties of the spectral domain of the electromagnetic waves used. Physically, a lidar is an opto-electronic system comprising a laser emitter, a receiver including a light collector (such as a telescope or another optical device), a photo-detector to turn the light into an electrical signal, and an electronic signal-processing system to extract the desired information (see Figure 3.19).

In the automotive industry, this type of sensor is highly effective in detecting obstacles, because after scanning in the x direction, it directly returns the distance information in the form of 3D point clouds. The advantage of lidar for an autonomous vehicle is that it can be used to measure distance in addition to a sensor such as a camera or radar (e.g. for ACC applications, to detect road markings and pedestrians).

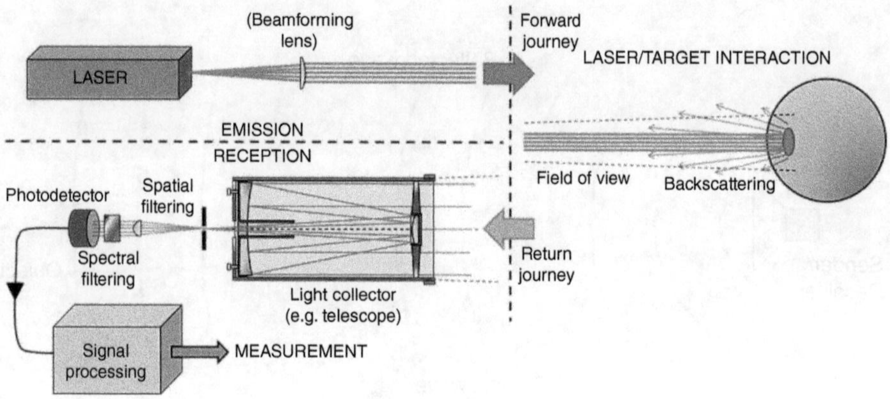

Figure 3.19 General diagram of a lidar system and the principle on which its measurements are based.

Comparison of various lidar technologies

In order to drive safely, an autonomous vehicle needs to "see" the world around it in 360° degrees and to process this information rapidly. The best way to achieve this is to design a system that sends out millions of light pulses each second, measures the time taken for those pulses to reflect off surrounding objects, and finally uses that information to construct a detailed 3D map.

The earliest applications designed specifically for enhanced and/or autonomous driving in vehicles date from around 2005. They were costly and did not conform to the reliability required by the automotive industry. For this reason, dozens of lidar manufacturers have emerged in recent years, with each one claiming, or attempting, to strike the right balance between the technicality, range of products, resolution, robustness, and cost of their solution.[2]

Currently, to achieve 360° vision, there are two main competing lidar technologies:

- The first, and oldest, is made up of a single, constantly rotating assembly that provides a circular 360° scan mechanically (on a mechanical level, this system is farily complex, and includes numerous moving parts, so the cost is high);
- The second, which emerged more recently, is a *solid state* system. It is static, with its own set of variants, so is intrinsically less mechanically complex and less expensive. However, it covers a smaller field of view (between 120 and 140°). Therefore, multiple units (between 3 and 6 lidars) need to be juxtaposed to achieve 360° vision and the data they produce need to be fused in order to cover the whole environment.

Now, let us look at a few representative concrete examples (of course, not an exhaustive list) of products on the market for these different solutions.

Mechanical scanning lidars

Operational principle

In order for the onboard computer to be able to reconstruct and present a 360° view of the environment, the lidar unit, installed on the roof of the vehicle so as to have free, unobstructed vision, is constantly rotating on itself (a *spinning ball*, rotating at

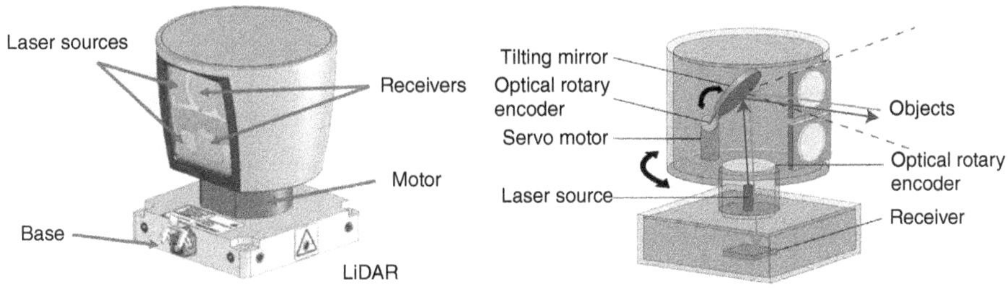

Figure 3.20 Example of elements in a spinning lidar (*source*: Velodyne).

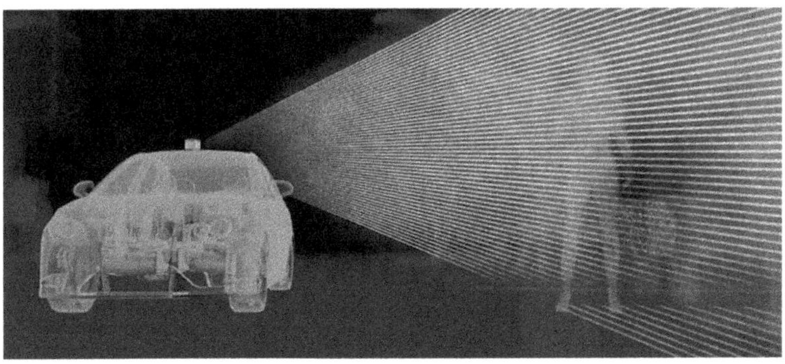

Figure 3.21 Example of scanning by a Velodyne Lidar 101.

a frequency between 5 and 20 Hz [or rotations per second], firing multiple laser beams to compile an image of the road) (see Figures 3.20 and 3.21).

Example: Velodyne Lidar For example, the HDL-32E, HDL-64E, and VLS-128 models from Velodyne respectively use sets of 32, 64, and 128 laser beams. These systems produce between 300 000 and 1 million data points per second (it should be noted that a minimum resolution of 300 000 data points per second is crucial for applications in autonomous vehicles). The quantity of information generated by a lidar can very quickly become voluminous, which could cause problems. Therefore, autonomous vehicles need to have powerful processors capable of handling all the information acquired in real time without phase changing.

In principle, because of the numerous moving parts, mechanical scanning lidars are less reliable and also more expensive. This is generally due to the fact that numerous optical components (lasers, sensors, etc.) are solid within a rotating case. In addition, for these applications, the systems must be able to function at any latitude, whether it is hot, cold, or damp.

A range of smaller lidars in smaller cylindrical casings (known as puck lidars) are easier to integrate into or onto a vehicle's bodywork, and are generally installed on the top of the vehicle, at the ends of the roof, in the bumpers, etc. (see Figure 3.22).

Figure 3.23 summarizes the functional properties of these lidars (also refer to Figure 3.24).

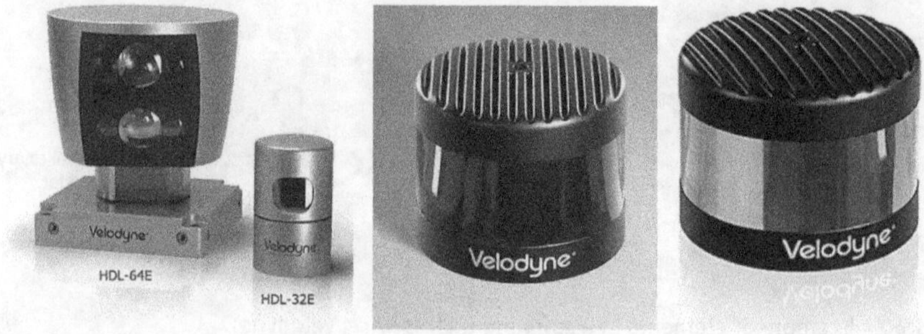

Figure 3.22 Standard Lidar and Puck VLS 128 models (*source*: Evan Ackerman/IEEE Spectrum – Velodyne Lidar).

Name	Puck	Puck Lite	Puck Hi-Res	HDL-32E	Ultra Puck	HDL-64E	VLS-128
Model	VLP-16	VLP-16-LW	VLP-16-Hi-Res	HDL-32E	VLP-32C	HDL-64E	VLS-128
Channels	16			32		64	128
Range (in m)	100				200	120	300
Accuracy	±3 cm			±2 cm		±2 cm	
Field of view (vertical)	+15° *to* −15°		+10° *to* −10°	+10.6° *to* −30.6°	+15° *to* − 25°	+2.0° *to* −24.9°	
Angular resolution (vertical)	2.0°		1.33°			0.4°	
Field of view (horizontal)	360°						
Angular resolution (horizontal)	0.1°–0.4°					0.08° – 0.35°	
Rotation rate	5–20 Hz						
Data points per second	300 000 (*single*) 600 000 (*dual*)			695 000 (*single*) 1 390 000 (*dual*)		1 300 000 (*single*) 2 200 000 (*dual*)	
Power	8 watts			12 watts		60 watts	
Weight	830 g	590 g	830 g	1.0 kg		12.7 kg	
Diameter	103 mm			85 mm		215 mm	
Height	72 mm			144 mm		283 mm	
Price	1000					75 000	

Figure 3.23 Technical features of Velodyne lidars.

Solid-state lidars

Achieving performances with lidars that simultaneously offer wide fields of view and the strict requirements of OEMS without depending on mechanical scanning is no easy task!

Figure 3.24 Example of an image from a Velodyne VLS-128 lidar (*source*: Velodyne Lidar).

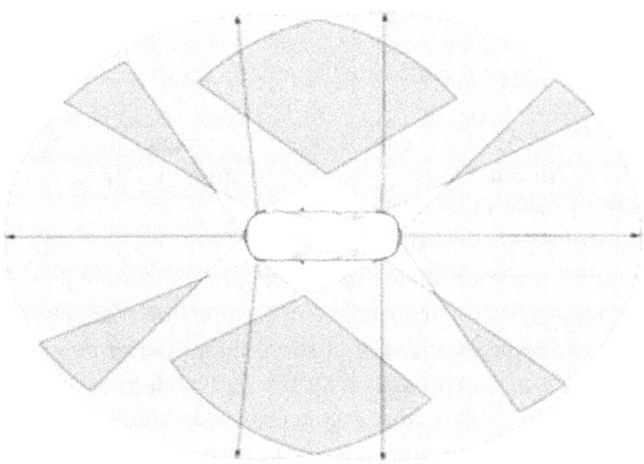

Figure 3.25 360° coverage using multiple lidars.

Operational principle

As before, the so-called "solid state" lidar also produces a series of laser pulses of short duration (between 3 and 20 ns). Of course, the shorter the laser pulses are, the greater will be the precision in different directions, which are used to measure the time between the emission of the laser pulse, its reflection by an object and its reception by the sensor. Given the constant speed of light propagation, the distance to the reflective object can be calculated by measuring the signal's ToF. Thus far, we have seen nothing new.

These lidars, which are, by construction, fixed, solid state, and roughly flat, therefore cannot cover a 360° field of view. Typically, their horizontal field of view extends only to 120–140°. One solution to this problem is to juxtapose multiple lidars (3, 4, 5, or 6) at different angles or points on the vehicle to cover the full 360° around the vehicle, ensure they operate in synchronicity, and then fuse their data in order to obtain usable results (see Figure 3.25). The results of that analysis can be used to produce a 2D or even a 3D representation of the environment that, in the automotive sector, is suitable for environmental perception in all types of traffic scenarios.

Of course, this presents us with the problem of bulk and esthetic appearance. Also, we shall eventually need to compare the cost of such a system with a spinning lidar. Indeed, if an autonomous vehicle needs to operate safely with an ADAS or a level-L4 or level-L5 system, it must be equipped with multiple sensors (one to the front and one to the rear, and possibly one at each corner of the vehicle – so between four and six lidars in total). All automakers must ponder the questions:

- What is the cost of four to six stationary lidars in comparison to that of one spinning lidar + data fusion + time?
- Should the lidars be located inside or outside the vehicle?
- Do they pose esthetic problems?
- etc.

Figure 3.26 offers an overview of the main parameters that must be considered when evaluating a lidar system.

Let us look at some examples of this type of solid-state lidars that are commercially available.

Example: LeddarTech lidar The semiconductor-based approach used by LeddarTech exploits existing optical technologies. It achieves good performances by optimizing the software to circumvent the limitations of mechanical scanners. Lidars based on this design are less expensive, easier to integrate into vehicles because they are smaller, and more reliable because of the fewer moving parts.

Semiconductor-based designs typically suffer from optical problems, due to the transmission and reception, which reduce the intensity of the system signal. This limits their range and their overall performance in comparison to mechanic scanners. The optimization of the performances of these lidars therefore needs to be noticeably better than that of mechanical scan lidars, which benefit from the raw power of their collimated lasers and a concentrated, highly targeted signal at their optical receiver. The technology used here uses digital signal processing capabilities, with high-performing acquisition algorithms and digital processing of optical signals. This considerably improves sensitivity, immunity to noise, and data extraction capacity. The result is a much cleaner signal, with very low levels of noise, which allows for rapid dissemination of accurate measurements, extending the effective range and the reliability of the measurements. The sensor also offers superior discrimination of lateral targets and the critical capacity to detect multiple objects simultaneously within its field of view. This technology is an integral part of new semiconductor-based lidar designs (see Figures 3.27 to 3.29).

These semiconductor-based solutions are designed for ADAS applications ranging from the simple (levels 1 and 2) to the highest levels of autonomous driving (levels 3–5).

Figure 3.30 gives examples of lidar scanners on the basis of their numbers of desired performances.

Parameter		Value	Comment
Celerity of light		= 300 000 km/s $= 3 \times 10^8$ m/s	
Duration of emitted pulse		= 10 ns	
Time of flight for detection at 300 m distance – i.e. a return journey of 600 m		$= 600 / (3 \times 10^8)$ $= 2 \times 10^{-6}$ = 2 μs	So as not to cause interference when receiving the reflected signals, the pulses must be emitted at least 2 μs apart
Refresh rate in *x* and *y*		50 Hz, so 20 ms	At 108 km/h, equal to 30 m/s,the vehicle will travel 0.6 m
Number of dots per image to be distributed in *x* and *y*		$= (20 \times 10^{-3})/(2 \times 10^{-6})$ = 10 000	
Choice of *x* × *y* matrix		$= 1250 \times 8 =$ 10 000	
Horizontal	**Desired field of view**	= 125°	
	Dots per image	= 1250	
	Resolution pf laser beam	= 125/1250 = 0.1°	
Vertical	**Rows per image**	= 10 000/1250 = 8	
	Resolution of laser beam	= 0.4°	
	Coverage of laser beam	= 8 rows × 0.4 = 3.2°	
Horizontal chord	125° at 300 m	= 530 m wide	
	0.25° at 300 m	= 1.3 m wide	
	0.25° at 150 m	= 0.65 m wide	
Vertical chord	8 rows × 0.4 = 3.2° at 300 m	= 16 m high	
Number of points analyzed per second per lidar		$1250 \times 8 \times 50$ = 500 000	
Digitization		on 14 bits	
Raw datarate for channel		= 7 Mbit/s	
Number of lidars used to cover 360°		= 4	
Total number of points per second for 360°vision		$= 7\,000\,000 \times 4$ = 28 Mbit/s	

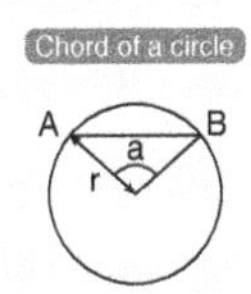

Length of chord [AB] $= 2 \times r \times \sin\left(\frac{a}{2}\right)$

Figure 3.26 Parameters to be taken into account when evaluating a lidar system.

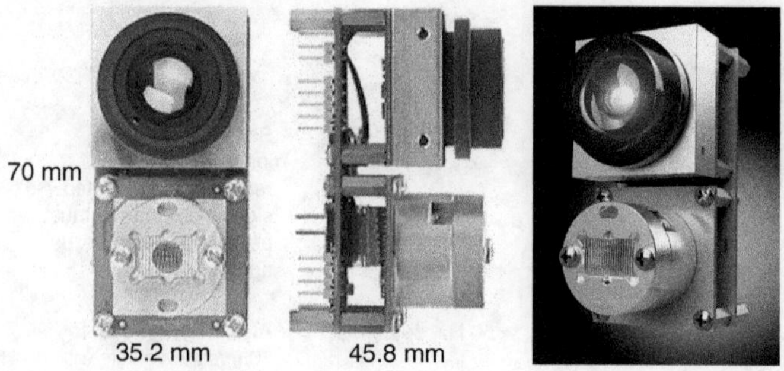

Figure 3.27 Examples of the "solid-state" lidars made by LeddarTech.

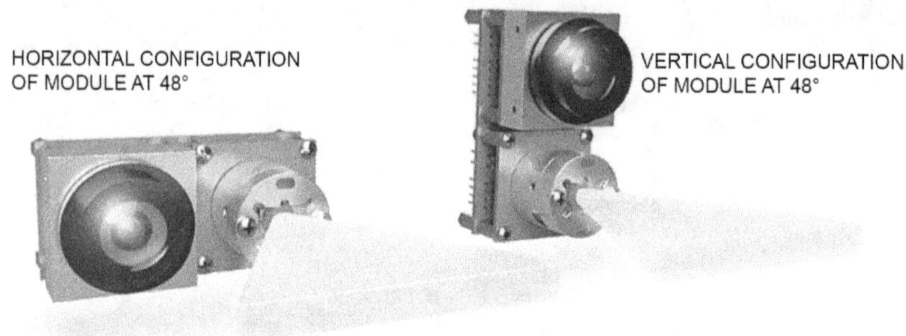

Figure 3.28 Beam distribution in the LeddarVu lidars from Leddar Tech.

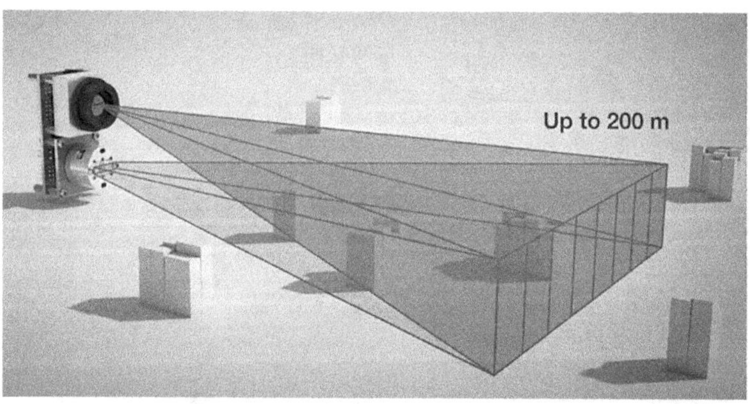

Figure 3.29 Bundles of lidar beams (*source*: leddartech.com).

- Performances (see Figure 3.31):
  - Detection range of up to 200 m for pedestrians and over 300 m for vehicles;
  - Resolutions up to 512 × 64;
  - Angles of view of 120° (horizontal) × 20° (vertical);

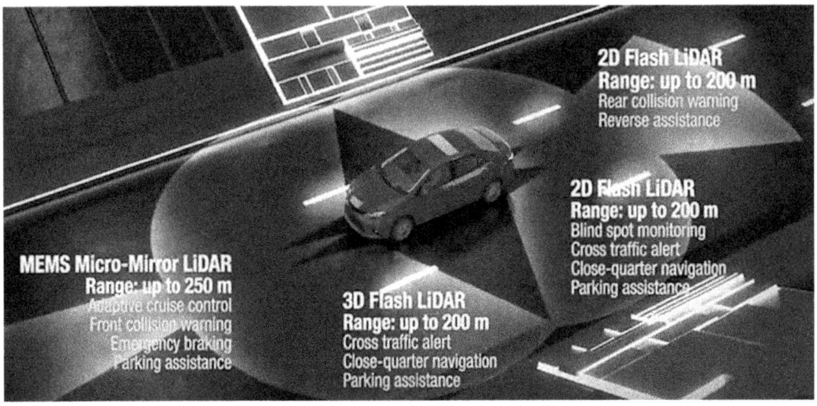

Figure 3.30 Example of a combination of lidars for an autonomous vehicle (*source*: Leddar Tech).

Technology	Time of flight, Output of distance and echo pulse width						
Infrared Laser	905 nm – (Eye Safety – Laser Class 1) IEC 60825-1:2014						
Horizontal	Field-of-View	20°	20°	48°	48°	100°	100°
	Angular resolution	0.25°					
Vertical	Field-of-View	0.3°	3°	0.3°	3°	0.3°	3°
	Angular resolution	0.8°					
	Layers – Multi-layer	8					
Data Refresh		Up to 100 Hz					
Distance Resolution	Exactitude	4 cm					
Axial Accuracy		10 mm					
Typical Range		185 m	121 m	118 m	85 m	61 m	34 m
Power Consumption		<2 W					
DATA I/O		SPI, USB, CAN, serial (UART/RS-485)					
Operational temperature		-40 to + 85°C					
Package Size		(70 × 35.2 × 67.5 mm) (70 × 35.2 × 45.8 mm) (73 × 40 × 65 mm)					

Figure 3.31 Technical features of commercial lidars.

- Rapid data acquisition, at a rate of up to 100 Hz;
- Multiple beam configurations for an optimal field of view;
- Compact and lightweight – between 110 and 125 grams;
- Highly robust – with no moving parts;
- Designed for applications both inside and outside the vehicle.

Example: Velarray lidar from Velodyne In late 2016, Velodyne announced the launch of Velarray semiconductor sensors, which could be used to make more compact lidars than their spinning predecessors, at a lower cost, and able to combine performance and reliability, with reduced bulk (125 × 50 × 55 mm). These lidars can easily be incorporated into the front, on the sides, and in the angles and corners of vehicles, to deliver the advanced safety expected of autonomous vehicles. These solid-state sensors also reduce the risk of malfunction, as they contain fewer moving parts and use a fixed set of lasers and receivers, rather than the spinning matrix found in earlier versions of the sensors (see Figure 3.32).

This sensor does not create a 360° image of its environment, as do other Velodyne spinning sensors. Instead, it covers an arc of the field of view 120° on the horizontal and 35° in the vertical, with a range of approximately 200 m, even for objects with poor reflectiveness.

Example: ScaLa "smart" lidar from AutonomouStuff The wide-angle lidar ScaLa v3.0 made by AutonomouStuff also includes a system to fuse data from multiple sensors to obtain a 360° view and track objects. The object classification the system offers (see Section 4.3) is suitable for ADAS solutions. As Figure 3.33 shows, this compact lidar can easily be integrated into a vehicle design.

- Technical features

The main technical features are shown in Figure 3.34.

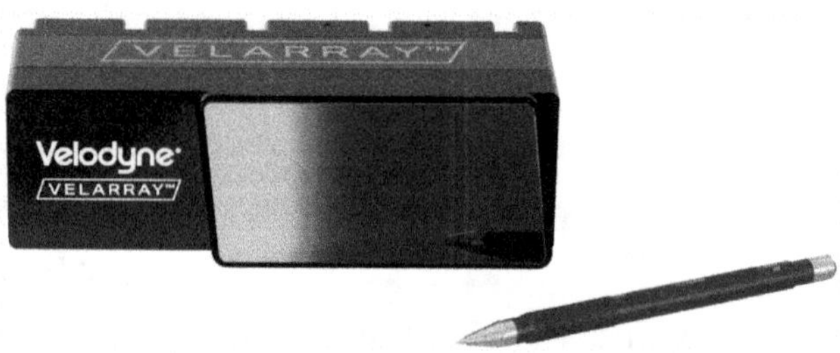

Figure 3.32 Example of a "solid-state" lidar made by Velodyne.

Figure 3.33 Example of the integration of the ibeo ScaLa lidar sensor from Valéo.

Features	
Principle	Time of flight, Output of distance, and echo pulse width
Infrared Laser	905 nm (Eye Safety – Laser Class 1)
Horizontal Field-of-View	145°
Angular resolution – Horizontal Resolution	0.25°
Vertical Field-of-View	3.2°
Angular resolution – Vertical Resolution	0.8°
Vertical Layers – Multi-layer	4 parallel scanning layers: 4 of 0.8° each
Data Refresh Time	40/80 ms
Distance Resolution	4 cm
Accuracy	<0.1 m distance independent
Typical Range	80 m/262 ft @ 6.3% remission 150 m (Passenger Car)
Power Consumption	<7 W
DATA I/O	CAN Interface for vehicle data and Ethernet interface Output: Raw- and object data Input
Package Size	105 × 60 × 100 mm
Embedded Software	Raw data pre-processing All measurements will be classified and tagged as valid/ground/clutter Real-time object tracking Object properties: position, size, speed Ego motion compensation

Figure 3.34 Features of the ibeo ScaLa lidar.

Example: Baraja lidar We come now to our final example. When designing a lidar, one of the main challenges is how to move the laser beam, left, right, up, or down. As we have seen, Velodyne offers a solution that spins several times per second. Luminar uses a pair of oscillating dime-sized mirrors. The main shortcoming with these configurations is that the moving parts create added complexity, and it is not easy for such systems to ensure the mechanical reliability required for automotive applications. We are currently seeing a proliferation of companies offering novel lidar solutions (a trend that can be expected to continue). Each of these companies hope that a major difference will lend them a competitive edge and allow them to conquer this vast market.

The Australian company Baraja has designed a system called "Spectrum-Scan." This particular approach includes the following:

- A prism lens;
- A wavelength-adjustable laser;
- A simpler mechanical way of redirecting the laser beam;
- An optical fiber to carry the light.

It uses a fixed optical prism and continuous scanning of multiple wavelengths in the visible and/or infrared to analyze a scene, with each color in the spectrum corresponding to a specific angle of view. In physics, white light passing through a prism is separated into the colors of the rainbow (see Figure 3.35). The order of these colors is based on the wavelengths of each color in the spectrum: red (around 700 nm) is above orange (around 600), and indigo (420 to 440 nm) is higher than violet (around 400).

This lidar system exploits this phenomenon by firing its single infrared laser beam through a prismatic lens. The prism refracts the separate parts of the light at an angle corresponding to its wavelength, as a prism would do with visible light. Slight variations are made to the wavelengths of the infrared pulses fired by the laser (though all are around the 1550 nm mark). Thus, the shape of the prismatic lens determines the angle at which those beams pass out of the glass, and therefore the direction in space that they travel, and how they illuminate the environment. In addition, in order to focus on one particular element of a scene, the light pulses are simply kept at the appropriate wavelength. This design borrows from a telecommunications technique called "wavelength-division multiplexing," whereby an optical fiber can transport many signals, each on a different wavelength, with the prisms serving to combine and separate those signals. Of course, that applies only to deviation within the vertical plane. The lidar always needs "mechanical" help – a scanner – to shift the laser beam horizontally, from left to right.

Because this lidar sensor (like the semiconductor-based sensors presented above) has a limited field of view, we must have multiple ones at different points on the vehicle to achieve a 360° view of its surroundings. As previously indicated, typical systems achieve this by physically installing numerous lidars, each with their own local lasers. This lidar, though, offers a second unique design point, which sets it some way apart from other possible solutions. The system in question uses only one laser per vehicle, housed in a box around the size of a wireless router, which may be located deep within the vehicle, because the light pulses are sent, via optical fibers, to multiple prismatic *sensorheads* in various locations on the outside of the vehicle (see Figure 3.36).

The system can be configured in a modular and/or evolutive way, using two, three, or four sensorheads in various locations on the vehicle. This solution is less costly than

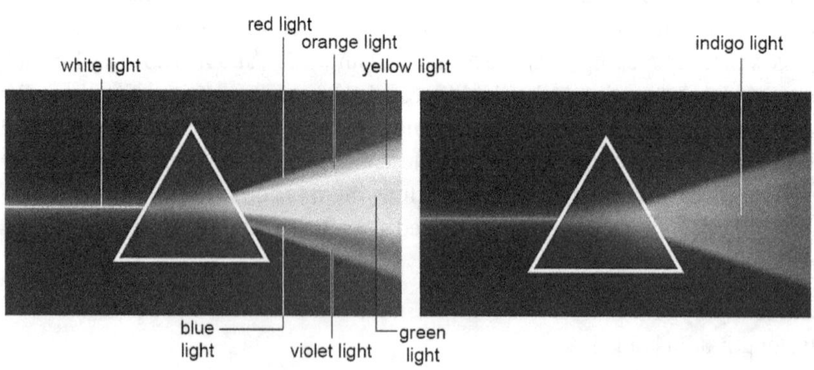

Figure 3.35 Separation of light in a prism.

Figure 3.36 Baraja's concept of lidar using fiber-optical connections.

installing multiple conventional lidar sensors, and much less costly to repair in the event of an accident.

This lidar is capable of intelligently adjusting the scanning modes in real time to adapt dynamically to complex road conditions, thanks to a computation unit that hosts digitization techniques, and the algorithms used to manage and interpret the signals. The system is managed by its central brain, the Lidar Engine, containing all its electronic and photonic intelligence. This Engine controls up to four sensor-heads to provide 360° coverage. The optical fiber connecting them may be up to 10 meters long, so the lidars can be installed both in compact vehicles and trucks. The Engine is designed to be installed in the center of the vehicle, maximizing its robustness and negating the need for additional cost and complexity at each corner of the vehicle.

Finally, this technology can be used to design a high-performing lidar with simple, fixed components, thus eliminating all the moving parts found in traditional lidars, which result in reliability problems when installed in vehicles, due to the vibrations and shocks to which they are inevitably subjected. Its modular and flexible design makes this system easy to integrate into a vehicle. Ultimately, this lidar was designed to be built into vehicles, addressing the constraints of manufacture and reliability associated with large-scale production.

Certainly, development in this field is far from over – more than likely, it has only just begun!

Advantages

In the automotive industry, lidars offer the following benefits:

- Lidars use light, so operate independently of any ambient light. Thus, they can be used at any time of the day.
- They can perceive more or less equally in all conditions, in darkness and in daylight, in cloudy and sunny weather, in shadow or in sunlight.
- They are capable (depending on the models) of "seeing" and reliably detecting objects/targets across distances of up to 300 m.

They can interact with extremely small objects, because the wave emitted is much more directive, is more reliable, and carries more energy. Therefore, non-metallic objects such as humans, wooden posts, etc., are better reflected.

- Lidars have a wide field of view and high angular resolution, which are necessary for precise and reliable detection over medium distances.
- The digitized scan data can be recorded using software, and used to create a model of the vehicle's environment.
- Lidars can help to determine an object's position, shape, speed, direction, yaw, and many other attributes, using the tracking algorithms built into the sensor software.
- They can be used to measure distance and speed.
- They are capable of detecting free spaces within their field of view.
- They offer the perfect marriage between short-range and long-range sensor systems.
- They help to generate accurate maps of the vehicle's environment.
- It is easy to isolate an object behind (or ahead of) the vehicle, the position and speed of other road users, but also to garner information about road infrastructure.
- They offer robust resistance to parasitic influence.
- Their resolution is much higher than that of a radar.

Limitations

Of course, lidars also have shortcomings:

- Lidars are not the ideal solution for real-time monitoring of other vehicles' speed.
- They are not perfect, and struggle in heavy rain, in snow or fog, as do other light-based sensors, including cameras. Indeed, as the wavelengths used by lidars are near to the visible spectrum, water droplets have a direct impact on the waves emitted.
- They require a clear point-to-point view of objects around them in order to accurately identify their distance, size, and shape.
- They can also sometimes see objects that are invisible to the human eye, such as exhaust gas from other vehicles.
- Their cost is still relatively high, which represents a barrier to their use by the general public. At the beginning, a spinning lidar sensor could cost US$70 000! Today, the price of a lidar has just reached levels that are acceptable in the automotive industry.
- Their resolution is fairly poor (images captured with only around 64/128 pixels in the vertical, but many more in the horizontal, at a refresh rate of around 10 Hz).
- They have a limited distance range. Typical lidars can see clearly at around 70 m, but obtain poorer results with larger objects, such as vehicles, at a distance of around 100 m.
- Certain lidars with moving parts can scan the environment. Solid-state lidars avoid the problems posed by moving parts.
- The refresh rates of solid-state lidars tend to be slightly lower. During their normal operation, partly due to the vehicle's own motion and partly due to the motion of the targets being analyzed, the scene undergoes distortions: one part of the image is scanned at a different time, because the vehicle and the detected object have both moved.
- Lidars work best when mounted on the exterior of the vehicle (in the bumpers, radiator grille, front and rear headlights, windshields, wing mirrors, on the roof, on the uprights, etc.), because, in principle, they need every photon, and therefore do not cope well with tinted windshields!

Application aspects

In the land of the autonomous vehicle, the vehicle with the best lidar sensor is King! In general, before level-3, level-4, and level-5 autonomous vehicles are tested, an "ordinary" vehicle is driven along the routes and maps out the roads, driving conditions, including posts, road markings, traffic signs, and more. The generated map is then loaded into the vehicle's software, to help it subsequently identify what is an integral part of the route. Then, when the autonomous vehicle is driving along, with its lidar system:

- It helps to detect the edges of the roadway and identify lane markings by bouncing light pulses off the vehicle's surroundings.
- It supplements the map in the onboard software.
- It supplies additional positional data using the data collected by a GPS and an inertial navigation system.
- It detects both static and moving obstacles to activate and assist functions such as blind-spot alert and emergency braking.
- It models, constructs, and regulates a detailed 360° 3D view of the environment through which the autonomous vehicle is being driven.
- A system using up to six lidar sensors can create a 360° view around the vehicle for environment monitoring and object tracking.
- The lidar system in an autonomous vehicle couples the lidars with optical cameras, an advanced algorithmic vision processing system, and a GPS to pinpoint the vehicle's global position.
- The system compares the perceived map against the pre-existing map to detect non-standard aspects of the road, notably identifying all sorts of obstacles on the road, including pedestrians, cyclists, other vehicles, and dangers on the road, and attempts to avoid them.
- It becomes a potential automated driving engine.
- It offers new benefits and also is able to make additions to a map generated by a radar. Indeed, by scanning the environment and sending laser beams in all directions, bouncing them off the surfaces surrounding the vehicle, measuring the reflected energy, and quickly processing the data with a computer, it is possible to generate a virtual 3D map, allowing the vehicle a "real-time view" of the objects in its environment. On the basis of these data, it is possible to precisely determine the distance and profile of those objects, and analyze and anticipate potential events and dangers that they represent.
- Thus, it is used to detect all static or moving obstacles, such as vehicles, motorbikes, and pedestrians, and improves active safety, by helping to activate functions such as blind-spot alert, triggering evasive measures, or emergency braking.
- Example: see Figure 3.37.

3.2.6 Ultrasound (sonar) sensors

Operational principle

Ultrasound sensors, or sonars, use the same operational principle as radars and lidars – namely the time of flight of a reflected wave to determine the distance from an obstacle. Unlike those technologies, though, they measure distance using the propagation speed of ultrasonic sound waves (between 30 kHz and 5 MHz) over much smaller distances – between one and a few meters. Ultrasound sensors therefore function much like short-range radars.

Figure 3.37 Example of an image captured by a lidar on the roof of a Google Car.

Advantages

The advantages are that:

- Its field of view is narrow;
- With appropriate corrections in temperature and humidity, it is possible to obtain sufficiently accurate results for a very low cost;
- The main advantage of this type of sensor is its cost.

Limitations

The main limitations of sonar are that:

- Its field of view is narrow;
- Its range is relatively short (around 6/10 m);
- It suffers from problems of accuracy in measurement, due to the variation in the speed of sound with changes in temperature and humidity;
- To obtain a sufficiently wide scan angle, it is necessary to use multiple sensors, which are typically built into the bumpers or undercarriages.

Application aspects

With sonars, it is possible to build redundant systems, which enable the vehicle, in real time, to effectively cross-reference data collected by other systems:

- To activate, amongst other things, the brakes and the safety belt pretensioners in anticipation of impact or to avoid a skid or obstacles;
- They are (still) very widely used for parking assistance systems, detecting and measuring the (short) distance from obstacles (many vehicles are also equipped with reversing radar to assist with parking); etc.

3.2.7 Position

Being able to see is all very well, but knowing where we are at the same time is better! Hence, the next step after vision is to estimate the vehicle's position, its veracity, and its accuracy. For this purpose, we need a certain set of tools and ultimately, we will need to fuse their data with those collected by the previous tools.

GPS

The *global positioning system* (GPS) is a satellite-based system that helps vehicles to navigate. Those satellites, orbiting in space, provide information about the time and location anywhere on Earth. With a GPS receiver, the precision is around 10 m, which is not terribly practical in accurately locating an object the size of a vehicle, between three and four meters in length!

The vehicle's GPS inertial navigation system works with sensors to allow the vehicle to find its own location, but the estimations of the GPS may be skewed, or may be off by several meters, due to signal interference and atmospheric interference. To minimize the degree of uncertainty, the GPS data are compared against a map previously built in the same location. When the vehicle moves, its internal map is updated with new positional information displayed by the sensors.

DGPS

A differential global positioning system (DGPS) is an improvement to GPS, which improves the localization accuracy from ± 10 m to around 10 cm. The DGPS correction signal loses around 1 m of accuracy per 150 km. Shadowing by buildings, underpasses, and foliage may cause temporary signal loss.

GNSS

Unlike the sensors presented in the previous sections, a *global navigation satellite system* (GNSS) system does not survey the environment, but measures the vehicle's own behavior. GNSS is a system that can pinpoint the vehicle's 3D position in space, based on the ToF of electromagnetic signals emitted by at least four different satellites in orbit around Earth. If the vehicle's position is known to a sufficient degree of accuracy (to within around 10 m), it is possible to cross-check that information with a map, find the road the user is on, and thus display, amongst other things, the speed limit for that road. Using the time data, the GNSS can also calculate the vehicle's instantaneous speed, and thus warn the user if they are speeding. Today, many, many vehicles are equipped with GNSS. Above all, it is used for guidance. It can also provide global time with a high degree of accuracy. The main drawbacks of this type of sensor are its poor spatial precision and the fact that the information is not always available (for instance, during urban driving, or driving in a tunnel, the signal may drop out).

Inertial navigation systems (INS)

For navigation, the INS analyzes the vehicle's movements and uses signals from sensors comprising:

- Three gyrometers to compute and determine an object's orientation, speed, and true position;
- Three accelerometers that measure the three components of the specific vector, i.e. the displacement in an orthonormal, three-dimensional space;
- Angular velocity sensors using three gyrometers, which measure the speed of rotation and read the direction and curvature of turns.

Long before the advent of GPS and its use in daily life, INSs had, for years, been in use for vehicle navigation. Today, the inertial navigation system plays the same role as a

GPS. However, it is crucial to use the INS when the vehicle's GPS encounters operational problems or network connection problems. The INS fills in data for the GPS, which sometimes malfunctions because of a temporary loss of signal (obstruction by buildings, underpasses, and foliage), to determine the vehicle's position. Thus, with a correctly identified starting position (found by the conventional GPS) and a well-designed INS, it is possible to maintain a highly accurate idea of the position for a short period, even with loss of GPS coverage due to terrain or other environmental factors.

Examples:

- Quickly orientating ourselves on exiting a circular street, coming out on one of the access roads;
- Knowing exactly where we are after skidding, with the wheels locked, on black ice.

3.2.8 Measuring the vehicle dynamics

Other sensors are also present in the vehicle and can be used to determine its dynamics. Examples include the wheel speed sensors, steering angle sensor (SAS), lateral and longitudinal acceleration sensors, angular velocity sensor, odometer, etc. The information gleaned is used, in particular, for safety systems such as the anti-lock braking system (ABS), the electronic stability program (ESP), airbag triggering, etc. These sensors are also used for ADAS applications. The smart camera, for example, uses the vehicle's odometer to determine the distance traveled between two successive images. In addition, for applications in AEB, measured acceleration is exploited to regulate the deceleration applied in emergency braking.

Odometer

An odometer (a kilometric counter that records the distance a vehicle has traveled), in relation with a displacement and/or motion sensor on one of the wheels, can increase the accuracy of the vehicle's location services. It measures the speed or distance covered by counting the number of rotations of a wheel and knowing the circumference of that wheel. It may be fooled by skids, on ice, for example, when the wheels are locked but the vehicle is still traveling forward.

3.2.9 In summary

Figures 3.38 and 3.39 offer an overview of the main performances of the main type of sensors – cameras, radars, and lidars – that we have examined in this lengthy chapter, and applied to autonomous vehicles.

In conclusion, nothing is absolutely perfect (see Figure 3.40), but:

- Cameras are the best types of sensors for detecting textures, the colors of road markings, and other traffic signs (such as speed limit signs, etc.);
- Radars are the best types of sensors for detecting the positions of obstacles and the vehicle's speed, but are not accurate enough to define the obstacle's shape;
- Lidars are the best types of sensors for accurately detecting the shapes of vehicles, pedestrians, kerbs, non-motorized areas, and other structures.

Thus, there are certain applications for which lidar is beneficial, certain ones for which cameras are best, and others for which radar is the most appropriate solution

Rating: H = High, M = Medium, L = Low	Camera	Radar	LiDar	Autonomous Requirement
Object Detection	M	H	H	H
Classification	H	M	L	H
Close-Proximity Detection	M	H	L	H
Speed Detection	L	H	M	H
Lane Detection	H	L	L	H
Traffic Sign Recognition	H	L	L	H
Range	H (200 m)	H (250 m)	M (120 m)	Full range
Work in Rain, Fog, Snow	L	H	M	H
Work in Low Light	L	H	H	H
Work in Bright Light	M	H	H	H

Figure 3.38 Comparison of performances (*source*: Denso International).

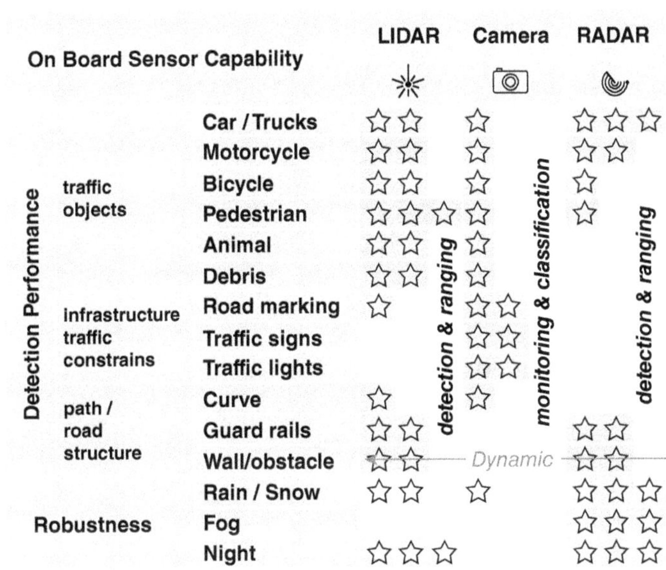

Figure 3.39 Another comparison of performances (*source*: Denso International).

(Figure 3.41). If we were to integrate only two of these systems into a solution, we would produce a vehicle that, though it might be capable of navigating and solving problems to avoid fundamental obstacles, would be intrinsically less safe. Autonomous vehicles are generally quick enough to predict the movements of obstacles but, for example, the movements of a deer, on a snowy night, with a little fog, etc., are totally unpredictable, by nature! Having headlights allows the driver and passengers the opportunity to act when the vehicle is not able to accurately predict what is going to happen.

Figure 3.42 presents an overview of all the technical solutions for sensors that must be used when designing the ADAS for an autonomous vehicle.

- Figure 3.43 gives a brief summary of the quantified datarates we have evaluated for the different elements (sensors, etc.) in this Section 3.2. We can now use these figures as the basis for choosing our communication and data transport protocols, and begin constructing the network architecture for an autonomous vehicle.

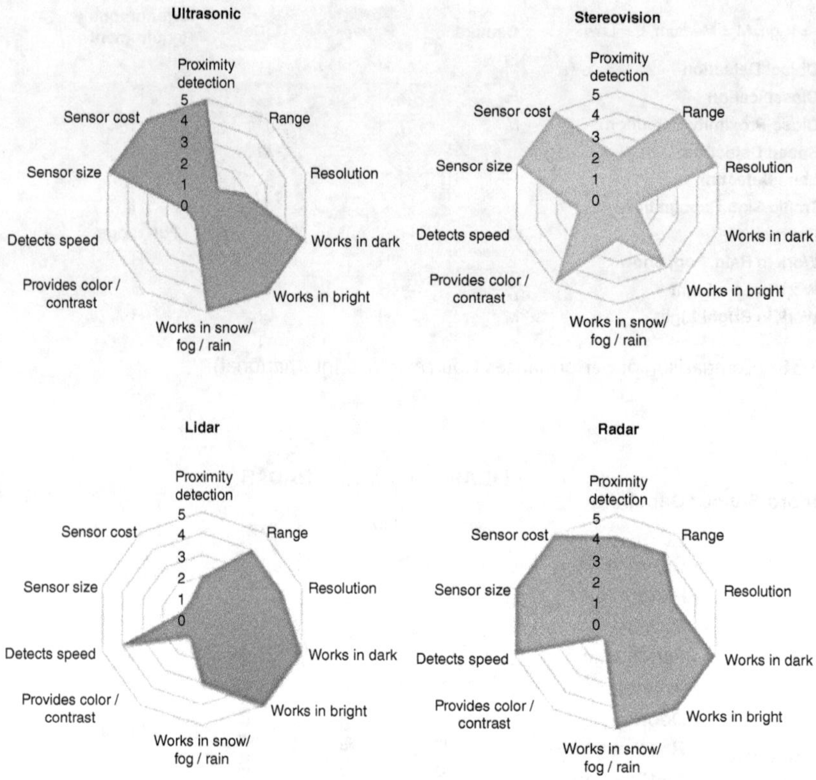

Figure 3.40 Characteristics of different sensors.

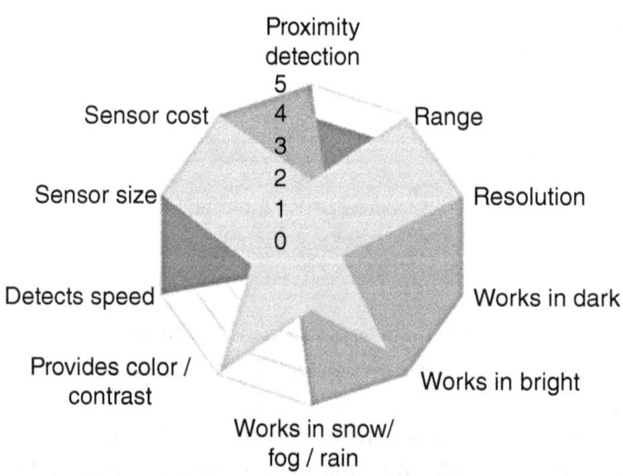

Figure 3.41 Comparison of performances of the different systems (*source*: INSA Rouen). Light gray: stereovision camera; medium gray: sonar; dark gray: lidar).

We have now come to the end of the list of the main tools available to build driver assistance and safety systems.

The final point, to determine where such multiple systems should be placed in vehicles, is the need for complementarity and redundancy. Figure 3.44 offers an example of a solution.

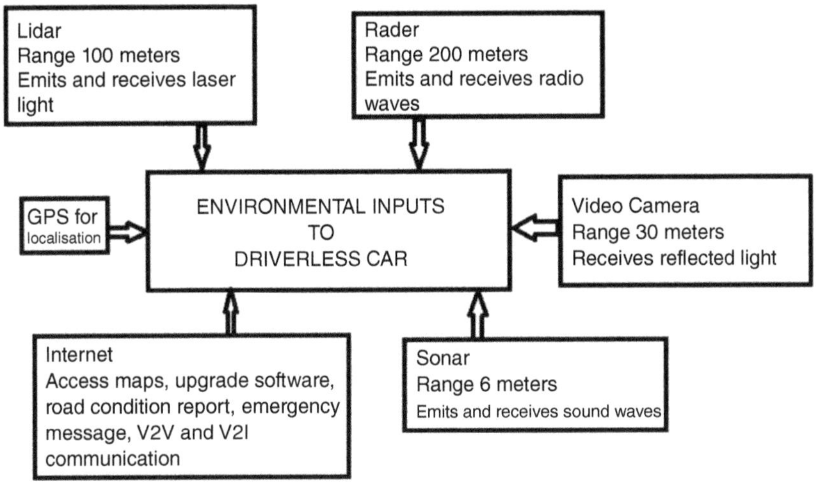

Figure 3.42 All solutions that can be chosen for an autonomous vehicle.

Types of sensors	Camera	Radar	Lidar	Sonar	GPS	Audio video
Raw digital datarate needed	~20 to 40 Mbit/s	~10 to 100 kbit/s	~10 to 70 Mbit/s	~10 to 100 kbit/s	~50 kbit/s	~200 Mbit/s

Figure 3.43 Summary of datarates for different sensors.

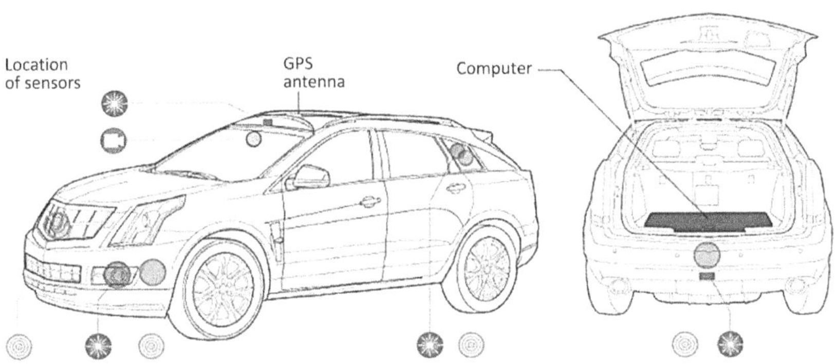

Figure 3.44 Examples of physical placements of the different sensors in a vehicle.

Before combining many of these sensors to perform the functions of an ADAS for level-3 to level-5 autonomous vehicles, let us look at some concrete examples of applications.

3.2.10 Examples of applications to autonomous vehicles

Numerous technological challenges need to be taken into account when designing autonomous vehicles. For example:

- With the passing of time, smaller and smaller sensors are needed, to adapt to any type of vehicle without the vehicle looking like a Christmas tree, as was the case with the earliest prototypes (see Figure 3.45)!

Figure 3.45 Earliest implementations of sensors in experimental vehicles. Steve Jurvetson / Flickr / CC BY 2.0.

Figure 3.46 Examples of esthetically pleasing ways of integrating sensors.

- The sensors must be able to be integrated into the bodywork in an esthetically pleasing manner. They must also be so well camouflaged that, to the untrained eye, they are scarcely visible (see Figure 3.46).
- The sensors' performances in terms of field, angle, and distance of detection must be such that the driver can be certain of detecting a pedestrian ahead at a specific distance. That might be 50 m, or even 100 m. Plainly, that distance is crucially important because, the further away an obstacle can be detected, the faster we can safely travel (on the highway, at least, up to 130 km/h).
- The vehicle needs to be made smarter still in its understanding of its environment, interactions, and decisions.
- Numerous major players in mobility – in particular, GAFA – have entered the marketplace, shaking up the current pace of development. In fact, their arrival has brought about a marked change of pace in the automotive industry. The whole industry (automakers and OEMs included) have adapted to this frenetic pace. There is the idea of a race to bring about the "automobile revolution, or transport revolution," or of a "gold rush toward the goal of autonomous vehicles."
- There is the notion of desire on the part of consumers, whether wrongly or rightly, and acceptance of this type of new technology as part of their daily lives. For example: Google began talking, at a very early stage, about making autonomous vehicles. For this reason, the general public has the impression that it is a realistically achievable goal and certain people's response is: "if it is possible, I want it!"

Examples

Let us now look at some representative examples that illustrate the foregoing discussion.

Example 1 – Valéo, with the Volkswagen Golf

- Long ago, in 2013, one of France's main players in autonomous vehicles, the OEM Valéo, launched an initiative called "*intuitive driving*" to design, and provide automakers with, solutions for autonomous driving at levels 3, 4, or even 5, in prototype vehicles. Much like the Audi A8 – Audi was the first carmaker to offer a model with level-3 autonomy – in 2016, Valéo's prototype, based on a Volkswagen Golf, was able to drive autonomously without human intervention (with special dispensation) in a controlled environment (notably, with no pedestrians or traffic lights – conditions such as those found on a highway or a major artery of a city). Outside of these areas, the driver assumed control once again. For the purpose of this test, the demonstrator was equipped with:

- Five cameras:
  - One in the front;
  - One behind the windshield;
  - Two in the rear;
  - One on the steering wheel to monitor the driver.

- Six radars hidden behind the bumpers:
  - Two in the front;
  - Four in the rear.

- Six lidars arranged around the vehicle:
  - Two on the sides (one on each side);
  - Four at the corners (one at each corner).

Numerous computers were hidden in the trunk and behind the rear seats (see Figure 3.47).

Figure 3.47 The hidden part of the iceberg of electronics managing the sensors and artificial intelligence in a nearly autonomous vehicle.

Today, these technologies have become standard (2021), and are used in level-3 private-use vehicles.

Example 2 – Valéo with Range Rover

Valéo's next step was the Cruise4U: the autonomous vehicle is capable of navigating from point A to point B without human intervention, in "all driving conditions" (it is at least a level-4 vehicle). In order to do so, new technologies must be integrated and operated using algorithms and artificial intelligence. This phase culminated in a demonstration with a Range Rover vehicle (in September 2018) (see Figure 3.48). The vehicle was equipped with:

- Five cameras;
- Eight lidar scanners;
- Twelve sonar sensors.

This solution allows level-4 autonomous driving, taking responsibility away from the driver in certain situations but, according to an interview with the CEO of Valéo in 2018: "It will be at least 10 years before we see a true level-5 vehicle on the streets on an industrial scale – so in 2028." Maybe it will. Who knows? We shall have to wait and see!

Example 3 – Tesla Model 3, Model S, and Model X

Of course, in this short series of examples, we cannot fail to mention Tesla, with the Model 3, Model S, and Model X (see Figure 3.49) in which "autopilot" (still a commercial term) was first put into action. As at January 2019, these vehicles had:

- Eight cameras:
  - One long-range camera, with a range of 250 m;
  - One main front fisheye camera, with a 120° degree field of view and a range of 150 m;
  - One wide-angle camera, with a 60 m range;
  - Two front-facing lateral cameras, with a field of 90° and a range of 80 m;

Figure 3.48 2018, a demonstrator vehicle kitted out by Valéo.

Figure 3.49 TESLA 3S with "autopilot".

- ◦ Two rear-facing lateral cameras built into the wing mirrors, which have a 75° field and a 100 m range;
- ◦ One camera in the rear bumper, with a 120° field of view and a range of 50 m.
- One long-range radar with a 40° field of view and a 180 m range;
- Twelve sonar sensors.

In principle, as in the previous example, subject to the regulatory restrictions mentioned in Chapter 2, we need only tell the vehicle where we want to go. The vehicle will then drive to the destination by the optimal route, navigating on the roads (even without road markings), weaving in and out at complex intersections with traffic lights, stop signals, or roundabouts, and knowing perfectly well how to cope on a highway with large numbers of fast-traveling vehicles. Upon arrival, after the driver gets out of the vehicle, it will search for a parking space on its own (see Section 2.3 on the Highway Code). Then, you need only tap a button on your mobile phone, and the vehicle will come back for you. In short, this is the dream!

It should be noted that you are allowed to use the vehicle for autonomous driving for carsharing, or for rides on demand, with your friends and family.

Subject to regulatory approval, other functions are sure to emerge. Of course, the driver remains responsible for their own vehicle, and must remain alert even when the autopilot system is active – they must be ready to intervene at any moment. Thus, the technology is still at level 4, not yet level 5. Tesla has stated that, on a hardware level, it is ready to advance to level 5, but still has work to do on the software; also, Tesla is waiting for authorization from the authorities. It should be noted that the level of autonomous driving is highly dependent on the validation of the software and on the regulations in force. Those regulations may vary depending on where the vehicle is to be used. At present, it is impossible to say exactly when each element in the function illustrated above will be available, depending on the local regulations. For the moment, everybody is in the starting blocks, awaiting the starter pistol. Thus, it is now time to examine the functions that are offered by each potential player.

Example 4 – industrial solutions from automakers

In closing, let us examine one of the ways in which historical industrial players/automakers (The PSA Group, Renault, Volkswagen, BMW, etc.) dealt, on a day-to-day basis, in 2018–2019, with the demands for different kinds of vehicle autonomy (the levels), advancing gradually from standard ADAS solutions to increasingly sophisticated ADASs (see Section 3.3), with the intention of achieving industrial levels 3 and 4 by 2020 or 2022. These graded solutions are generally available across the range of a maker's vehicles, with suites of "options" (or "packs") that can be added, starting with the basic vehicle, then moving up to the middle of the range, and then the top-of-the-range vehicle.

Again, let us give a few examples. In addition to ordinary standard equipment (ASR, ABS, REF, EBS, etc.), there are generally a range of options (also see Section 3.3 for numerous details):

- Driving assistance and assisted maneuvers:
  - Adaptive speed regulator (radar camera);
  - Start–stop function;
  - Lane assist (camera);
  - Video parking assistance (lateral cameras to the front and rear).
- Advanced safety features:
  - Night vision (forward infrared camera);
  - Collision risk warning (front radar camera);
  - Automatic safety braking (ASB) (front cameras and radar);
  - Line and verge alert system (front and side cameras);
  - Blind spot alert (side camera and/or sonar);
  - Automated traffic light switching;
  - Driver attention alert (camera);
  - Road sign recognition (or at least, recognition of some road signs) (front camera).

With all these options already available in large-scale production, numerous designs are already in place. All that remains is to take care of how to fuse all these data to achieve levels 4 and 5 of autonomy.

3.3 ADAS and C°

3.3.1 DA – driver assistance

Let us start with DA (*driver assistance*) systems. For many years, simple supports have been in place. Examples include automatic indicator cancellation, automatic windscreen wiper activation, and more. All of these are dedicated to known, simple tasks that have no direct relation to other functions.

3.3.2 ADAS – *advanced driver-assistance systems*

Now, let us move on to look at Advanced Driver Assistance Systems, which represent a step forward in terms of desires and functionality. Examples include lane departure warning systems, which sound an alarm to alert the driver, but can also take control of the vehicle's steering to correct the line, or even accelerate or brake.

3.3.3 HADAS – *highly advanced driver-assistance systems*

With HADAS, we are dealing with more exacting wishes, and higher levels of achievement in terms of real-time data fusion and mixing of data types. For example: lane departure alert, with connections with other systems in the vehicle, to automatically correct its course with a camera's view of the surrounding environment.

We shall now take a more detailed look at the state of the art in the area of ADAS.

3.3.4 ADAS – further detail

Given that the majority of traffic accidents are caused by human error, automated ADASs are designed to help the driver and make it easier to drive the vehicle, throughout its journey, and increase safety. More generally, they have been developed to automate, adapt, and improve the vehicle's safety systems so that, with a man–machine interface, human error can be reduced as far as possible, and thus avoid or reduce the number of collisions or fatal accidents. These devices offer technologies that alert the driver to potential problems, or prevent collisions by implementing protective measures or taking control of the vehicle.

To cite a few examples: adaptive functions can automate the lights, control and prearm the adaptive speed regulator, brake automatically, incorporate the GPS/traffic alert functions, connect to smartphones, trigger alerts in other vehicles, automatically center a vehicle in its lane, etc.

With ADAS, a vehicle may begin to approach levels 3 and 4 of autonomy, and allow the driver some free time and make their life easier. In spite of a very long list of driving aids already commercially available, they will be rolled out a long time before they are operational in all situations (for a level-5 system), because the technology needs to perform even better than a human driver.

It should be noted that an indication of the time needed for the development of autonomous vehicles is shown on the emerging technology maturity curve (Gartner's Hype Cycle). In 2012, the Hype Cycle indicated, as seriously as possible, that autonomous vehicles should be rolled out within 5–10 years. However, the 2018 cycle indicated that it would be another 10 years yet. There is still a great deal of development to be done for better autonomy, and to completely deal with totally unexpected situations in traffic (fishtailing, diversions, objects or roadworks, swirls of dead leaves, traffic conditions such as those encountered on Place de l'Étoile in Paris, or a crossroads in Mumbai with cattle, mixing of autonomous and non-autonomous vehicles for many years, failure to respect the Highway Code on the part of others, while the autonomous vehicle is perfectly compliant, respecting the right of way regulations, etc.) (see Section 2.3 – the Highway Code).

As vehicles are handed increasing levels of responsibility for driving themselves, from levels 2 to 3 and then 4, developers face the eternal problem of the driver resuming control of the vehicle, and the transfer of liability from the driver to the automaker. As noted at great length (Section 2.2), the regulation also needs to change: certainly, it is already recognized that soon it will be possible to take your hands off the steering wheel of a vehicle equipped with lane-keeping assist (but that is only at level 3). The rising number of modern vehicles has brought about progress in driver assistance systems such as anti-aquaplaning systems, anti-wheel lock braking systems, lane departure alert, adaptive speed regulators, and electronic stability control. It must also be noted that some of these systems are directly affected by regulations concerning mechanical alignment. This has led numerous automakers to require electronic reinitializations of these systems, to ensure that the intended mechanical wheel alignment is able to meet the safety requirements.

In addition, there are many different types of ADAS: certain functions are built in directly when the vehicles are manufactured. Others are available as an aftermarket package of additional components. There are also replacement solutions available.

ADASs draw on data from a wide range of sources. For example: automobile imaging, lidars, radars, cameras, image processing, computer vision, and in-vehicle networking, but also additional inputs from sources outside of the vehicle platform, such as other vehicles, known as "vehicle-to-vehicle" (V2V) or road infrastructure "vehicle-to-infrastructure" (V2I).

To improve vehicle safety and reduce road accidents, from an automaker's point of view, the tendency toward autonomous driving, coupled with the security of the vehicle systems, is defined by the degree of adoption of ADASs. ADASs represent one of the most dynamic areas in automotive electronics, with the constant increase in the degree of uptake of industry-wide quality standards, in automotive safety systems (ISO 26262), in the development of technology-specific standards, such as IEEE 2020 on image sensor quality, and communication protocols, such as the vehicle information API. The ADAS generation will make increasing use of wireless connectivity to provide the best value, with V2V and V2I communication being key parts of this.

Smart sensors, detection, fusion, ADAS, and HADAS

ADASs are based on complementarity with other sensors and fusing their data, which we shall examine in greater detail in Section 4.3. For example: a sensor converts the luminosity and color information into "image" electrical signals. These signals are then processed by a processor built into the camera casing, without the need for a separate controller. The system processes the image, recognizing, classifying, and implanting a broad range of situational factors around the vehicle with high degrees of precision and reliability, notably for objects, such as pedestrians, vehicles, road markings, light sources, and road signs. In addition, the algorithms are designed and optimized to provide the best possible yields, while minimizing the memory occupation and the size of the hardware. This is what we shall now discuss.

Object detection The object detection system is based on predefined classes of objects (see Section 4.3), designed to recognize and distinguish pedestrians, cyclists, motorbikes, cars, trucks, etc. Attributes are assigned to the detected objects, such as velocity, distance, angle, lateral position, etc., and, where applicable, the probable time to collision. The detection range depends on the object's size – it can be 120 m or further in terms of detecting vehicles, but only around 60 m for detecting pedestrians. The access path for functional pedestrian detection was developed in accordance with ISO 26262, in relation to risk class ASIL A (automotive safety integrity level A).

Data fusion Data fusion combines the benefits of different sensors and the most effective measuring techniques possible. It produces new data that individual sensors, working in isolation, cannot generate. By fusing the data from multiple sensors, the range of measurement, reliability, and accuracy are increased. A multi-application camera can be used in conjunction with other surround sensors, such as radars and sonars. With automatic emergency braking (which independently applies the brakes if the driver is unable to react in time to an imminent collision), the multi-use camera fuses its output with a radar sensor, or else uses a stereo video camera. When fusion takes place, complete automated deceleration is triggered only if both sensor systems detect the critical object.

Data fusion can help to considerably improve the performances of comfort-related functions. In light of the accuracy of the lateral measurements, the ACC function is capable, amongst other things, of early detection of whether vehicles are pulling towards the inside or outside of the roadway, and responding dynamically. The system also ensures that oncoming vehicles are assigned to their correct lanes, which further improves the function of the ACC – particularly when going into a curve or swerve.

Applications of ADASs

We shall now list a few (though by no means all) of the applications for an ADAS:

- Dashboard display of a 360° surround-vision view, to give the driver an overall view of the outside of the vehicle, to help prevent accidents;
- Rear view of events outside the vehicle, to avoid injuring pedestrians or damaging the vehicle when reverse parking, on parking in a narrow space (see Figure 3.50);
- Infrared cameras providing a dashboard display of night vision of the road ahead;
- Construction of 3D images of the route, so that drivers can more easily perceive hazards;
- Systems enhancing the safety of the vehicle and its passengers;
- Alerts when objects enter into the vehicle's blind spot;
- Multiple cameras used to quickly detect objects, obstacles, or people require the transfer of uncompressed data (thus, high datarates are required, whether the data are analog or digital), so as to avoid artifacts and errors due to the compression system:

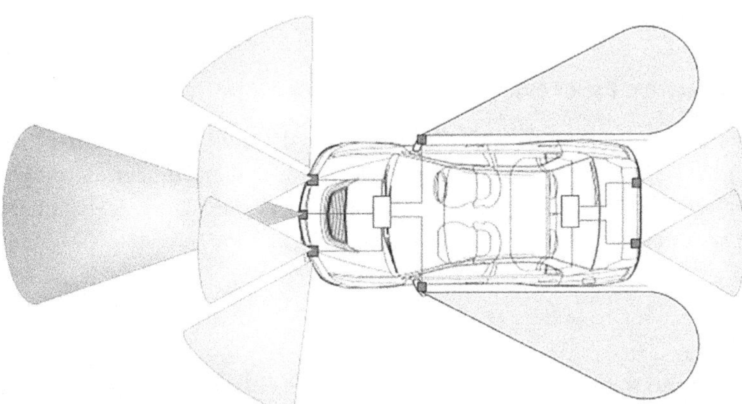

Figure 3.50 ADAS with cameras and radars for driving assistance systems.

  - For example: for night-time detection of pedestrians, dressed in dark clothing, etc., the luminance signal from monochromatic analog videos must be processed, which requires bandwidths (in ordinary resolution) of 6–8 MHz. Alternatively, the system must transport a sampled digital signal at a rate of at least 14 Mbit/s, or even up to 100 Mbit/s in video streaming from a camera to an ECU;
- Sensors that are able to detect the presence of fingers on a pane of glass, and deactivate the electric window winder (to prevent trapped fingers);
- Posture detectors to ensure the correct airbags are triggered in the event of a crash;
- Automatic emergency braking;
- Reporting of the vehicle's position using the GPS, so as to quickly obtain assistance in an emergency;
- Hands-free and eyes-free applications for safer, more pleasant driving (for example: text messages and e-mails that come to the driver's mobile phone can be read aloud by the onboard computer);
- Information systems;
- Monitoring of traffic situations and traffic jams;
- Real-time information on availability of parking spots;
- Smart systems under the hood and in the trunk to improve fuel consumption, diagnose potential issues, etc.

Validation of ADASs

Safety is the most important issue. Conventional validation methods cannot be used, because, according to certain automakers, they would require 250 years' worth of data from a fleet of 300 vehicles! Thus, the solution (which does have its limitations) lies in the development of simulation tools using databases, use cases that are as detailed as possible (e.g. white lines have been removed, works are underway, people on a 4-lane road, etc. – see Section 6.1). Thus, that database must be shared with all actors: automakers, OEMs, and other bodies; however, it is nonetheless crucial to perform tests in order to validate the systems (see Section 2.15). The challenge facing autonomous vehicles requires new skills, which are still in short supply, a shift in the electronic architecture, and various development teams to collaborate effectively.

With this brief introduction done, let us now review the long list of ADASs that are well known and conventional.

3.3.5 Non-exhaustive list of examples of some representative ADASs

This section presents a few examples of ADASs (classified by general area rather than alphabetically) that are commonly found in the market and have been developed through perfect collaboration between automakers and OEMs. Thus, we shall see many familiar systems, and a few novel ones as well.

Vision and lighting

Blind spot monitor

The blind spot monitoring system helps the driver, when changing lanes, to monitor and detect vehicles that are in the blind spots. It is generally based on (see Figure 3.51):

- Radar scans of the left and right rear quadrants of a vehicle, which sound a warning or even intervene (the radars are generally located in the rear bumper or in the wing mirrors);

Figure 3.51 Blind spot monitor.

- Cameras located on the sides of the vehicle or in the wing mirrors; or
- Sonar sensors in the wings and undercarriage, to the front and rear.

If the system detects a static or moving vehicle in the driver's blind spot, a warning light comes on. If the driver turns on the indicator, a beep is heard. If the driver carries on regardless, and begins changing lanes, the vehicle gently applies the brakes on the opposite side to re-center it in its own lane.

Automotive night vision – light source detection Smart light source detection opens up a range of new lighting functions that make driving safer and more comfortable. In addition to detecting light sources, the purpose of this type of ADAS is to recognize and classify the sources of individual lights, pairs of lights, and dense clusters of lights, at dawn or dusk, and in darkness. The algorithm measures the horizontal and vertical angular positions and distance of the detected light sources. It differentiates between rear lights and front headlights, so as to be able to distinguish oncoming vehicles from vehicles ahead. The headlights of oncoming vehicles are detected and classified from a distance of around 800 m, and rear lights can be detected from around 400 m. The algorithm is also capable of detecting and classifying elements of road infrastructure such as streetlights, etc., and, by using these data alongside information concerning the ambient light conditions, evaluates whether or not the vehicle is traveling in an urban environment, and decides whether the high beam lights can or should be switched on.

Able to provide a broad range of other smart lighting functions, this ADAS addresses the numerous constraints imposed on modern headlamp technologies, including systems such as headlight control, adaptive light control and continuous lighting.

Intelligent headlight control – adaptive light control – swiveling curve lights Anti-dazzle headlamp technology, using adaptive LED projectors, helps avoid dazzling other road users, whilst maintaining as much light as possible on areas that the driver must be able to see. Intelligent headlight control allows the driver to use the high beams as much as possible to improve visibility when driving at night, without constantly having to switch the high beams on and off manually. This ADAS activates the high beams when there are no other vehicles detected in the vicinity; if a vehicle is detected, it switches the high beam off. In addition, for generating light in LED headlamps, it is possible to control the distribution of the light beam in different zones and segments. This means that during night driving, the vehicle's high beams can be on constantly, improving visibility without blinding or dazzling oncoming drivers (*glare-free high beam and pixel light*).

Such adaptive headlight control not only regulates the segmentation of the light, but also adapts the width of the beam depending on traffic conditions. Thus, curves and turns can be better illuminated as the vehicle approaches, and a wider light cone can be used to better illuminate the edges of the roadway in urban areas, helping the driver to see potentially vulnerable pedestrians.

Tires

In addition to the now classic *tire-pressure monitoring system* (TPMS), puncture detection, run-on-flat systems, etc., let us now look at an ADAS that warns of the risk of aquaplaning.

Aquaplaning

Besides rain sensors, which have been a fixture in vehicles for decades, on extremely wet roads, the loss of grip considerably reduces vehicles' capacity for control, and presents a major accident risk. Even with the best tires, it is important to detect, very early on, situations where aquaplaning is imminent, so that drivers are aware and can reduce speed, as aquaplaning can pose a risk of an accident. For example, the tire specialist Continental provides an ADAS designed for this purpose (see Figure 3.52).

Aquaplaning occurs when the tire's tread cannot drive the water off the road surface quickly enough. This happens when:

- The tread depth is insufficient (a tread depth of at least 3 mm is required with summer tires, because anything less presents a greater risk of aquaplaning);
- There is a very thick layer of water on the road surface;
- The speed is excessive on wet routes and in rain.

Wide-angle cameras are installed in the wing mirrors, the grille, and at the rear of vehicles to detect any excessive movement of water and trigger an aquaplaning alert. When there is a great deal of water on the road, the camera images may show a specific type of splash and spray pattern, which can be detected as an early warning of aquaplaning. The vehicles also have a smart detection and alert software system for when aquaplaning is imminent, to alert the driver in time and prevent accidents. The data from the camera and the information coming directly from the tires are fused to identify the risk of aquaplaning. The signals from a sensor directly built into the tires (the specific signal from the eTIS accelerometer – electronic-tire information system – see Figure 3.53) can also identify the remaining tread depth. These data can be used to determine a safe speed for the specific road conditions, and display it to the driver. Aside from the normal functions and benefits of a tire pressure monitoring system, the eTIS offers functional scalability.

Figure 3.52 Aquaplaning detection ADAS (*source*: Continental).

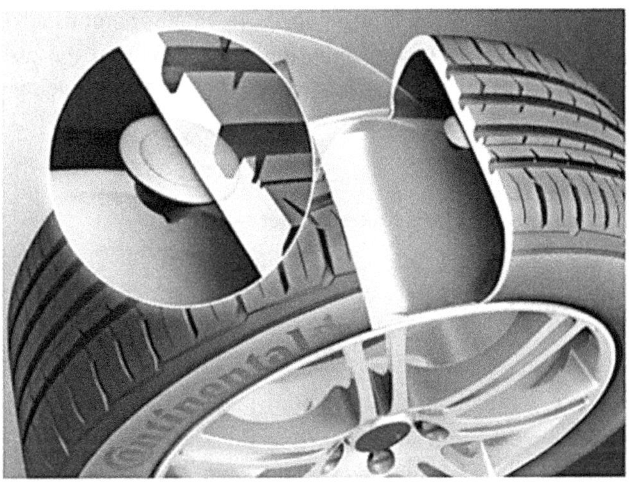

Figure 3.53 Electronic-tire information system (*source*: Continental).

With this arsenal of technologies:

- Tire sensors detect the risk of aquaplaning;
- The aquaplaning alert system is based on data fusion, combining data from the camera and from the tire sensors;
- Automakers can choose their own functional approach depending on their specific requirements:
  - Via a CAN bus, the tire-sensor receiver unit communicates the tire-specific data to the ECU, to be fused with data from the cameras. An example of such an ECU might be a stability control unit. It is possible to assess all the data from the sensors to feed into the aquaplaning alert system. If the system detects that there is a danger due to the current speed, the driver will be advised of a safe speed.

  - Vehicles that are still a significant distance away from the potential aquaplaning danger spot can be alerted immediately by means of V2V communications, via the electronic navigation digital map. V2X traffic control systems can also receive information about these danger zones. The aquaplaning alert system is particularly important for autonomous vehicles, which must be able to avoid the danger without the intervention of a human driver.

Speed

Adaptive cruise control Adaptive cruise control (ACC) is an ADAS that is an intelligent form of cruise control. The system speeds up and slows down your vehicle to maintain a constant, safe distance from the one in front. Generally, a radar behind the grille or under the front bumpers measures that distance. Certain vehicles use a lidar, while others use a system based on stereoscopic cameras. ACC is ideal when the accordion effect is seen in traffic, and during rush hours when traffic oscillates between 80 km/h and a dead stop.

Whatever the type of technology used, ACC works day and night, but its performances can often be adversely affected by rain, fog, or snow. In an autonomous vehicle, the ACC connected to a GPS must track not only the vehicle in front but also vehicles in the adjacent lanes, in case it becomes necessary to change lanes or a vehicle cuts into your lane.

Other ADASs linked to speed control

We shall now briefly recap some other ADASs linked to speed control:

- Hill descent control;
- Intelligent speed adaptation (ISA);
- Crosswind stabilization.

Safety

Forward collision warning As part of a strategy focusing on accident prevention technologies, one of the recommended functions is forward collision warning. If the system detects an imminent collision with an obstacle in front, it will alert the driver by means of a warning light, an audible alarm, and/or a haptic alert such as a vibration. The system does not act independently to prevent the crash, but warns the driver to brake.

Collision avoidance system (pre-crash system) A collision avoidance system is designed to reduce the seriousness of a collision. It uses the radar (in all weathers), and sometimes a lidar and a camera (but these two types of sensors are less effective in bad weather) to detect an imminent collision. When such a situation is detected, these systems generate an alert for the driver or, if need be, they can take action autonomously (automated emergency braking and/or steering correction). Collision prevention by braking is suitable for vehicles traveling at slow speeds (e.g. slower than 50 km/h), whereas at higher speeds, it is easier to avoid the collision by correcting the steering.

Pedestrian protection system – pedestrian warning The system continually analyzes the space in front of the vehicle to detect imminent collisions with pedestrians who are either in the vehicle's path or approaching it in a way that is likely to present a danger. When traveling at up to 60 km/h, if the system recognizes any imminent danger to pedestrians, it can alert the driver and, working with a radar sensor, can also activate the emergency brake if necessary.

eCall

In 2018, the European Commission, in pursuit of its "Vision Zero," which aims to achieve transport with no road accidents, made it compulsory for vehicles to have an eCall system. Vehicles are required to have an onboard emergency call system which, in the event of an accident, can/must automatically send an alert and request assistance, notifying the nearest emergency services of the vehicle's exact location. The eCall system is activated by the triggering of the airbag control unit. The vehicle's inbuilt telematics unit comprises essentially a modem, a satellite positioning antenna, the electronics needed to connect with the vehicle, and an emergency battery. It is also possible to manually trigger an emergency call – for example, if a driver witnesses a serious accident. The system uses the standard European emergency number (112), reducing the response time by up to 40% in urban areas, and even up to 50% in rural areas. A shorter response time means that the victims receive help more quickly, which helps prevent further harm.

Per the European norm, a dataset comprising certain information as a minimum must be sent to the emergency services. This minimum dataset, measuring 140 bytes, contains information selected for relevance, including the time of the call, the vehicle type and its position, the number of passengers, and the direction of travel.

Other ADASs connected to safety management

In closing, let us point to a few other ADASs connected to safety management, which have been in use for a number of years:

- Anti-lock braking system (ABS);
- Driver drowsiness detection system;
- Traffic information;
- Emergency brake assist (EBA);
- Emergency driver assistant;
- Intersection assistant;
- Auditory alerts to the approach of electric vehicles; etc.

Trajectory tracking

Electronic stability control

Electronic stability control (ESC), also known as an electronic stability program (ESP) or dynamic stability control (DSC), improves the vehicle's stability by detecting and reducing loss of traction (skidding). When the ESC system detects the loss of control of the steering, it automatically applies the brakes to help "steer" the vehicle in the direction in which the driver is trying to go. The brakes are applied individually to each wheel, so that the front outside wheel can correct an oversteer, or the inside wheel can correct an understeer. Certain ESC systems also reduce engine power until control of the vehicle has been regained. ESC does not improve vehicle performance in the event of skids, but does help to minimize the loss of control.

Lanes

This section covers numerous ADASs linked to the lane in which the vehicle is traveling:

- Lane detection;
- Automatic lane centering;
- Lane-change assistance;
- Lane departure warning;
- Lane keeping and lane guidance support.

Lane detection

The lane detection algorithm records and classifies all lane markers up to a distance of around 60 m (or up to 100 m in excellent visibility), whether they are solid or broken lines, white, yellow, red, or blue. Not only does the system detect the lateral and longitudinal geometry of the lines, but also the gradient of the road (slope), to define whether the lanes lead uphill or down. If there are no clear road markings, the system relies on secondary information, such as the lines of grass along the edge of the road, to determine whether the lane is continuous. The algorithm is also capable of determining the vehicle's lateral position and angle in its own lane, to a high degree of accuracy, which is crucial for functions such as lane departure warning or lane keeping/lane guidance. Even if the road markings have been temporarily removed or are not visible on a particular section of road, the steering assistance functions use the lane detection algorithm and remain perfectly functional, ready to step in and assist at any moment.

Lane departure warning (LDW)

The lane departure warning system is a mechanism designed to alert the driver when the vehicle begins to stray out of its lane on urban roads and highways (unless a turn signal in that direction has been switched on). These systems are designed to minimize the main causes of collisions (driver error, distractions, and drowsiness). They alert the driver, using lane sensors, when the vehicle strays from its lane or changes direction without signaling, by vibrating the steering wheel or displaying a visual alert on the dashboard.

Though lidar performs better in precisely mapping the environment, its ability to accurately monitor the speed of surrounding vehicles, in real time, is limited. This function is performed, primarily, by four radars on/in the bumpers (two in the front and two in the rear). The radars send a signal to the onboard computer, allowing the vehicle to brake or change direction and avoid impact, where there is a risk.

Lane warning – Lane keeping system (LKS)

Often, the lane departure warning system described above is accompanied by a lane keeping system, which applies mechanical torque to the vehicle's steering to keep it safely in its own lane. This system compares the road markings and the vehicle's position in its lane, using:

- Cameras (typically behind the windshield in the rearview mirror);
- Laser sensors on the front of the vehicle;
- Infrared sensors either behind the windshield or underneath the vehicle.

At speeds of more than around 60 km/h, if the system detects that the vehicle is moving too close to the lane divider, or is at risk of accidentally straying from its lane, this ADAS raises an alert signal – visual, audible, and/or haptic, such as steering wheel vibration. These alert the driver to the fact that the vehicle is drifting. After a brief delay, the system gently counteracts the steering to keep the vehicle in its lane and avoid danger. The driver can individually adjust the threshold at which steering correction is applied, and how hard it is applied, using options ranging from very early, gentle intervention to a later, but more drastic countering. The system can act directly on the electric steering, or act indirectly by applying the brakes to one side of the vehicle. The driver can, at any time, take over this function, retaining control of the vehicle. Of course, when the driver turns on the indicator, deliberately signaling the intention to turn or change lane, the system does not throw an alarm, and takes no corrective action.

Parking

Parking assistance and valet parking

In the wake of sonar parking sensors that give audible or visual alerts of the remaining distances during parking maneuvers, modern parking assistance systems are able to automatically park vehicles in tight spaces, in perpendicular spaces. and at an angle. Automatic park assist can also be activated even when the driver is outside the vehicle. This technology uses sonar to find a free parking space even at speeds of up to 30 km/h. When the vehicle detects an appropriate space, it alerts the driver, who can either stay in the vehicle or get out and use the remote control to accomplish the task of parking. The vehicle will then park itself in the space (see Figure 3.54).

Automated parking is an autonomous maneuvering system that moves a vehicle out of a traffic lane into a parking lane to parallel park, perpendicular park or park at an angle. The automated parking system is intended to improve driving comfort and safety in tight environments where a great deal of attention and experience are needed

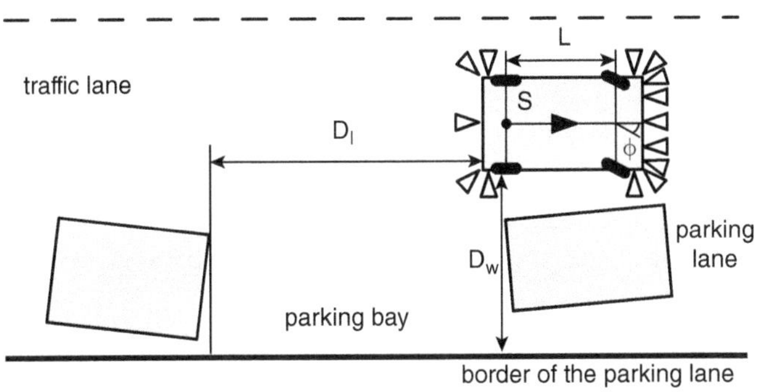

Figure 3.54 Parking assistance ADAS.

to steer the vehicle. The parking maneuver is performed by coordinated control of the steering angle and speed, taking account of the real situation in the environment to move in the available space without collision. It is not necessary to own fully autonomous vehicles to get parking assistance. Carmakers are already producing vehicles with parking assistance as standard. However, once you are in a desired parking lot, you allow the vehicle to find its own parking space (see the comment on the Highway Code in Section 2.3), which then becomes automatic parking, and in the case of higher-level autonomous vehicles, valet parking.

Road signs

Road sign recognition A road sign recognition ADAS detects and classifies the circular, triangular, and rectangular signs at the start and end of sections where a speed limit is in place, or there is no crossing allowed. It also recognizes signs such as: "No entry," "Stop," "Yield," "Road works," etc. The road sign recognition system also categorizes relevant additional signs, such as signs only applying to particular types of vehicles, or directional arrows. The device can reliably detect road signs and the variable messages if they are placed on systems or porticos, offering a high level of international coverage for a broad range of different types of signs.

Road sign assistant The road sign assistant assesses the recognition data and displays the relevant information on the dashboard. The road sign display can be used for a range of alert functions. For example, it can alert the driver before the speed limit is broken, warn them against entering a street where certain maneuvers are forbidden, or alert them of having missed a "Stop" sign or "Entry" sign, and detecting the speed limits that can then be used by the ACC system, to automatically adjust to the speed limit in force on the road. To improve reliability and supplement the data used with information that generally cannot be detected by a camera, the system can process data from the navigation system – for example, to distinguish urban and rural roads, or textual interpretation of additional signs, such as the periods of validity of the speed limits. These ADASs are often coupled with turning assistants, vehicular communication systems, wrong-way driving warning, etc.

Technological example of a real-world ADAS

Without wishing to provide undue publicity, let us consider, among other examples (in 2018), a conventional top-of-the-range commercial vehicle: PSA 508 – First Edition, which included a wide range of ADASs. It carried:

- Four cameras:
  - One front camera behind the windshield (to track trajectory, lane, etc.);
  - One front infrared camera to detect pedestrians, animals, etc., during night-time driving;
  - One front camera for parking assistance;
  - One rear camera for parking assistance.
- One front lidar for the anti-collision system, to measure distance between vehicles for the cruise control system;
- Twelve sonars:
  - Four in the front, for parking assistance and close-range collision warning;

- ◦ Four in the rear, for parking assistance and close-range collision warning;
- ◦ Four on the sides (one per side), for blind-spot assistance while driving.

Next steps: the ultimate ADAS

Consider, for a moment, what the advent of a truly connected, genuinely intelligent vehicle could mean. To begin with, it could mean greater safety and fewer traffic jams. It could also offer better energy efficiency and, as a corollary, less pollution.

In parallel, with the rapid convergence of information technology, the automobile industry is undergoing a major transition, affecting everything from vehicle design right up to the assembly line. This union between automotive professionals and computing experts offers the opportunity to adopt a fundamentally different approach to vehicle function, design, and construction (even the materials used could change, so we could see plastic vehicles instead of metal ones on our roads, or even biodegradable ones). Today, vehicles are transport machines containing computers. Tomorrow, they will be smart, data-driven systems that will actively help make the most of every second on the road. Does this sound too like science fiction? Consider the many ADASs that are already in use today. Twenty years ago, adaptive cruise control, automated parking and braking, blind spot detection, collision prevention, and self-driving vehicles were nothing more than engineers' daydreams – and now they are a reality. With that in mind, let us come back to the question at hand: what will the ultimate ADAS look like? Among automakers, there are two opposing concepts of what the ultimate ADAS will be.

The **first camp** focuses on *vehicle message passing*, whereby as it passes by, the vehicle communicates wirelessly with:

- Other vehicles (V2V);
- Infrastructure, such as signs and traffic lights (V2I);
- Pedestrians and cyclists via their smartphones (V2P);
- The cloud (V2C).

As an umbrella term, we refer to "V2X" (Vehicle to X).

The main question in relation to V2X is: "What do we do with all the incoming data?" Many of the proponents of this type of solution believe the best approach is to compile the data and display the results to drivers, so that they can interpret and take the appropriate action. But – and this is a big but – there is a real danger of information overload! In addition, we know from our discussion of "cybersecurity breaches" (see Section 2.9) that it is risky to deal with all incoming threats in the same way. There must be different levels of priority attached to different datasets, so that users are only alerted where there is a pressing problem. In addition, there is still a long way to go before we have standardization of V2X communication protocols (see Section 4.2.2). To achieve a major systemic gain with such technology, all vehicles need to speak a common language. Today, though, the disparate data formats used dilute the safety benefits, and sow confusion among consumers. Finally, V2X technology is, itself, dependent on true connectivity with robust networks. Unfortunately, not all wireless communication protocols are equally efficient at passing secure, deterministic messages. It seems a major challenge to bring about ubiquity and an effective economy of scale.

The **second camp** is made up of those aiming to increase road safety, enjoyment and productivity. The tendency here is to take critical tasks away from the driver and entrust them to an onboard computer, creating what is essentially an autonomous vehicle. This means that the autonomous vehicle reads and reacts to its environment, so is less heavily dependent on connectivity and incoming messages. The security concern is the same with autonomous vehicles as with V2X: we do not want the connections or the onboard systems to be vulnerable to hacking! In this second case, the proliferation of electronic control modules on board an intelligent vehicle offers a tempting range of potential ways in for attackers, particularly as many of the systems come from third-party suppliers, and may go for years without being replaced. Another challenge lies in convincing the public to buy autonomous vehicles, because currently we have a driving culture that favors the driver being in ultimate control. Ask the "average" driver whether they want a computer to take control of their vehicle in an emergency and many will say an emphatic "no."

Thus, the question remains unresolved. One potential solution, though, would be to combine the two approaches, creating an intelligent vehicle equipped with ADASs that can interpret and react to the environment, whilst also increasing connectivity to the cameras, lidars, and radars (see Section 3.2). Why could we not have a vehicle that supplements and improves human cognitive capacities, rather than attempting to replace them?

In order to build such an intelligent vehicle, we need to activate fusion of the signals from the sensors and the vehicle's movements, passing all this information to a central computer that can imitate a human brain. The task would require an enormous amount of computational power within a small thermal casing. It would also require scalability from vehicle to vehicle and from one generation to the next. In addition, we need to achieve a better ratio of performance per watt, per unit cost, and improve functional security. In view of all the above, what is needed is a major reinvention of vehicle architecture and unprecedented collaborations between industries, governmental organizations, consumers, and special interest groups. For over a century, research and development in automobiles have focused on the engineering of the mechanical systems (automatic gearboxes, air conditioning, more efficient power train, etc.). However, in the coming decade, the emphasis will shift, and the automotive and information technology spheres will collide. Most research into vehicles today focuses more on software than on hardware. The vehicle should now be viewed as a holistic platform, and we should explore the most effective ways to capture data and then safely send those data to the central brain in real time, so as to create a whole vehicle that becomes one huge ADAS, in which drivers are not obsolete, but are still decision-makers, as is the case with airplanes (there are onboard computers, and autopilot, but there are still pilots in the cockpit).

Notes

1 See Dominique Paret, *RFID at Ultra and Super High Frequencies: Theory and application*, Wiley, 2009.

2 For information, in 2021, the lidar market for automobiles is dominated by a few companies, including Delphi Automotive, PLC (UK), Continental AG (Germany), ZF Friedrichshafen AG (Germany), Infinéon Technologies AG (Germany), Velodyne LIDAR, Inc. (USA), Texas Instruments, Inc. (USA), LeddarTech, Inc. (Canada), First Sensor AG (Germany), Quanergy Systems, Inc. (USA), and Innoviz Technologies, Ltd. (Israel), etc. There are also numerous startups operating in this area.

4

Networks and Architecture

4.1 The various networking options

We are finally about to dive into the core of this book's USA subject and look at the insides of a vehicle that is intended to be "autonomous" (at least level 5 – driverless – in mass production and driving on all the world's road networks, though it is likely to be 2030–2035 before this becomes a reality) and at its functional architectures.

Over the past few decades, functional architectures have changed markedly. Today (in the 2020s) and in coming years (in the 2030s, for example), they are divided into different *in-vehicle networks* (IVNs). The main tasks entrusted to these subsystems are:

- Power train management: combustion engine, hybrid, electric, transmission, and its environment (CAN, CAN FD, CAN XL, SENT, etc.);
- Chassis management (suspension, road handling, relations with tires and tire pressure levels – CANx and FlexRay);
- Vehicle body management (LIN and CAN Low Speed and, later, CAN FD and CAN XL);
- Comfort management (heating, air conditioning – CAN, CAN FD, and CAN XL);
- *Infotainment* management – audio, video, etc. (D2B, CANx, and later MOST, LVDS, and now Ethernet);
- Communications management (V2V, V2I, I2V, V2X, and soon to be 5G);
- Safety management (braking, lighting, etc., CAN, CAN FD, CAN XL, and Ethernet, safety belts, and safe-by-wire in IP5);
- ADAS management (CAN, CAN FD, CAN XL, and primarily Ethernet);
- Supervised and organized management of the whole, marvellous system through Ethernet, etc.

All these specific networks, which have their own features (bitrates, lengths, media types, topologies, means of data collision management, etc.), overlap, complement one another, interlink, constantly exchange data, and, in places, shadow one another to provide redundancy for functions and safety, etc. In short, in addition to their proper operation and to the problems of configuring the necessary gateways for

Autonomous and Connected Vehicles: Network Architectures from Legacy Networks to Automotive Ethernet, First Edition. Dominique Paret and Hassina Rebaine.

network-to-network interconnection, an autonomous vehicle system includes vast lengths of wires and cables of differing types and vast quantities of copper, which represents an enormous amount of energy consumption, and therefore pollution. All of this means that the science of vehicle architecture is in motion and will continue to evolve from day to day, on the basis of automakers' sensibilities, technological progress, and developments in connections.

Above are the names of numerous communications protocols and networks. Figure 4.1 offers a snapshot of one possible example of the network architecture in a conventional vehicle.

Let us now examine the above list in detail, point by point.

4.1.1 Power-train management

Engines and propulsion groups – be they combustion engines, hybrid or electric motors, etc. – have historically been managed by CAN High Speed (CAN HS) networks or, now, CAN with Flexible Datarate (CAN FD) and CAN XL. The datarates of these networks (between 1 and several Mbit/s) are generally sufficient for these applications. The same is true of the management of gearboxes and transmissions, etc. Huge numbers of sensors employed are in the environment of the engine, sensing temperature, pressure, absolute magnetic position, linear or angular sensors (notably the engine actuators, the pedals or gear levers), and rotary sensors (number of rotations per minute). They often use CAN High Speed, FD, or XL, and, in places, for simpler functions and for reasons of cost, there is a certain tendency to move from LIN protocol to SENT (*single edge nibble transmission* [SAE J2716]) – a new development that is described in Section 4.5.2 – an improvement on LIN and its direct competitor. Ethernet serves as the global supervisor and vehicle management system.

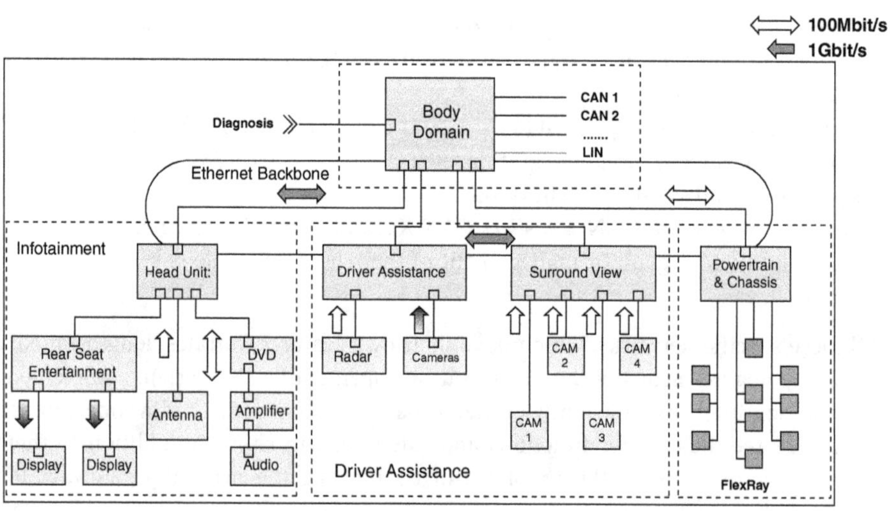

Figure 4.1 Snapshot of an example of communication network architecture.

4.1.2 Chassis management

Dynamic management of the chassis – of the suspension, road handling, relations between the tire types, their pressure levels and the suspension system, etc. – has typically been performed as follows:

- In Germany, CAN High Speed or FD or XL and FlexRay (10 Mbit/s), which BMW helped to create and is the main promotor of (German automakers develop "models");
- In France, CAN High Speed and CAN FD, whose datarate is between 1 and some 2.5 Mbits (French automakers mainly develop "platforms"). In fact, at a certain time (around 2012), French automakers lacked the courage or the ambition to make the small transition to FlexRay, prior to the advent of Ethernet.

4.1.3 Vehicle body management

In a vehicle, the surface area or volume covered by the term "body" often varies. It tends to be made up of different zones:

Cockpit

The cockpit normally accommodates the driver, and contains all the controllers for driving the vehicle (steering wheel, indicator controls, accelerator pedal, brakes, etc.). It also contains all the on-board instruments that provide the driver with necessary information (speedometer, tachometer, control screens, etc.). To retain consistency with earlier chapters in this book, when we speak of truly autonomous, level-5 vehicles, the cockpit should be done away with in driverless vehicles, being replaced by something else. What might that "something else" be? It is a mystery! Thus, in principle, we would say goodbye to the CAN High Speed, FD, XL, and Low Speed networks connecting to the dashboard. However, the question remains open in relation to level-4 vehicles, in which the driver must be able to take the wheel or resume control swiftly in the event of an incident. The question then becomes: "What actually needs to be kept in the cockpit?"

"Passenger" compartment

The classic controls that are necessary for the usual "working parts" of a vehicle (doors, windows, trunk, roof, etc.) and to start and stop the vehicle (immobilizers, using LF, UHF, HF (NFC), UWB, or other solutions via a mobile phone) will, of course, be present. In addition, the passenger zone will be enriched with a host of options, including controls for articulated seats that can fold up on themselves entirely. (An example can be found in Renault's futuristic so-called *concept car* SYMBIOZ – see Figure 4.2 – which "delivers on the idea of a traveling home. Its multi-use cabin offers a completely new experience, with a vehicle that is built around a human-centered ecosystem.") All that is missing is a dance floor and a bathroom!

Coming back down to Earth, in the passenger zone where the datarates of the controls are medium and latency is not a highly critical issue, the main protocols used will remain, with CAN, CAN FD, CAN XL, LIN, and SENT, except for the purposes of infotainment, which will come through Ethernet.

Figure 4.2 Symbioz: Renault's concept car.

"Comfort" zone

The controls for the comfort of the cabin (heating, air conditioning, ventilation, etc.) do not require high datarates, and generally use CAN Low Speed at 125 kbit/s or standard CAN, FD, and XL at around 500 kbit/s in some cases.

"Infotainment" zone

To understand the current intrigue and the future of how the *infotainment* zone is managed (the term is a contraction of *information* and *entertainment*), we need to look back quite some distance in time, to around 1970. *Once upon a time*:

- Around 1970, standard vehicles were not equipped with radios. Drivers had to buy a radio and affix it with a metal stirrup screen and two screws under the dashboard. It was a luxury item, and therefore very much in demand. As a result, a great many cars were broken into and the radios stolen;
- The basic car radio (a radio receiver giving "info") quickly turned into a minicassette player (this was the beginning of the "tainment" part of infotainment). Of course, this made the systems even more tempting for thieves and the number of thefts soared;
- Automakers soon got the message and began building car radios into the body of the vehicle – for example, as part of the dashboard. In spite of a number of clever tricks (such as removable controls), the thefts did not abate;
- Then, a genius idea emerged: to break with the conventional architecture of a car radio. Instead, a screen was placed on the dashboard, displaying the name of the station being received or the title of the cassette being played (seven-segment display), and the radio receiver itself was integrated into the innards of the vehicle, which put an end to thefts;
- The screen was made bigger, and was able to display more text (greater ranges of characters);

- The display unit then became a mini-screen, much like a television, which began to serve for multiple functions – for things other than the radio and cassettes. This is where we see infotainment truly come into its own;
- Around this time, in-vehicle audio began to be a major concern. The numbers of stereo speakers increased from four to six to eight, mounted in the dashboard, the doors, etc.;
- Now, the problem became how to distribute audio from the different sound sources – which, it must be remembered, came in different formats (analog sound, CDs, and other media, followed by MP threes and USB sticks) to the different passengers who might be on board the vehicle. The proprietary ring network MOST, which was capable of a datarate of 50 Mbit/s, then came onto the market for top-of-the-range German vehicles;
- Subsequently, vehicles began to display all sorts of videos on the dashboard screen and on screens built into the front-seat headrests, and the MOST protocol was upgraded to 100 and then 150 Mbit/s, using fibre optics and later copper cables;
- Finally, we come to the 2020s, but it is necessary to manage all possible or conceivable audio and video sources for infotainment, and all extant formats and digital datarates (e.g. MPEG 4, H264, video from the cameras that are part of the ADAS, different transmitted formats, and different aspect ratios – 4/3, 16/9, 720 × 540, 1280 × 1024, etc.) and synchronize those streams. This is why Ethernet, with all its variants, began to be used in the infotainment zone.

4.1.4 Communications management

We shall examine these topics in depth in a dedicated section (Section 4.2.2 – V2V, V2I, I2V, V2X, etc.).

4.1.5 Safety management and ADAS management

In autonomous vehicles (and, of course, in non-autonomous vehicles), passenger safety is ensured in a range of ways:

- Conventionally, by the driver's direct action (levels 3 and 4);
- Management of the very specific operation of the seatbelt pretensioners and airbag deployment during a collision. Nowadays, these tasks are often handled by a specific protocol: PSI5 (peripheral sensor interface);
- By means of assistance systems varying in complexity, which offer support to the driver and to the vehicle's ADASs.

The coming sections go into greater depth on this point, discussing, for example, braking, lighting, CAN, FlexRay, and Ethernet.

4.1.6 Supervised and organized management of the whole system

All these different management systems are beginning to be supervised by means of a switched Ethernet network at a (very) high datarate. The ins and outs of how this works are explained in the lengthy Chapter 5.

4.2 The connected vehicle

4.2.1 Big data – golden data and automobiles

In the age of autonomous and connected vehicles, as services evolve and cybersecurity becomes a concern, *big data* is one of the major issues for businesses, at all stages in their process. It is for this reason that a great deal of importance is attached to case studies and feedback.

In the world of automobiles, big data represents a major shakeup for all actors, automakers, OEMs, startups, engineering firms, IT firms, research labs, and also schools and universities. It has become a tool that is capable of modifying the services and offers of automotive products from one day to the next, combining artificial intelligence and computing techniques with data analysis, to handle vast quantities of heterogeneous data. Autonomous vehicles will gradually become a digital service like any other, based on the creation and constant use of data. In addition, big-data techniques require new skill sets and organizations to develop vehicles and associated services in a different way. In the age of big data, it is essential to adapt very quickly and address the following questions:

- What must be done to manage a big-data application?
- What methods and skills need to be mastered?
- How can these new skills be brought to bear as part of the automotive sector?
- How can companies establish win–win relationships with the newly emerging actors in the sector: startups, insurance companies, the Big Four (GAFA), and other actors in the digital revolution?

Big-data applications in the automobile sector

The reach of big data is extremely broad and is being increasingly widely used, causing breakthroughs in commerce and automotive engineering – e.g. in terms of:

- Predicting customer behavior to create tailor-made offers, and thus steer their choices of options of in-vehicle systems;
- Analyzing customer testimonials to detect their level of satisfaction;
- Projecting sales;
- Optimizing maintenance and predicting breakdowns, etc.

At this juncture, it is opportune to point to a few applications that are more specific to the world of autonomous and connected vehicles:

- Updating maps and road states;
- Dynamically creating and constructing databases for digital simulations for the testing phases (see Chapter 6.1);
- Support for the associated simulation techniques;
- Statistical analysis of acquired driving experience and projection of prospective likely behaviors;
- Statistical studies of quality and reliability;
- Summarizing big data produced by measurements;
- Statistical studies of feedback, by *text mining* as applied to repair workshops;
- And many other applications.

Note that text mining is a branch of data mining, which is part of artificial intelligence. It refers to computerized processing to extract knowledge on the basis of a criterion of novelty or similarity.

4.2.2 Communication

Types of communications in an automobile

In automotive applications, there are numerous types of communications. They can broadly be classified into the following categories:

- **in-in**: communications that only take place within the vehicle (in the broadest sense) and are not intended for transmission to the outside world;
- **in-out**: communications that are designed to be sent into the outside world (for example: eCall);
- **in-out-in**: bidirectional exchanges of data between the vehicle and the outside world.

These communications can also be classified into other well-known categories:

- *Vehicle-to-vehicle* (V2V): V2V communications are used in safety systems with *non-line-of-sight* (NLOS) transmission, and in latency-sensitive anti-collision systems;
- *Vehicle-to-infrastructure* (V2I): V2I communications apply to a variety of highly interactive security/safety systems and applications, including interactions with road signs to gain information about the road, with traffic lights for information on their timings, monitoring GHG emissions, communication with traffic lights to ensure that vehicles are crossing intersections safely, and so on;
- *Vehicle-to-pedestrian* (V2P);
- *Vehicle-to-device* (V2D): to any device;
- *Vehicle-to-grid* (V2G);
- *Vehicle-to-cloud* (V2C);
- *Vehicle-to-everything* (V2X): an automotive communication system where vehicles transmit and receive information about their surroundings, for effective operation.

Figure 4.3 offers a simplistic view of most of these concepts.

These communications can also be classified on the basis of their connectivity, the technologies used or the market they serve:

- In terms of connectivity type:
  - Dedicated short-range communication (DSRC);
  - Cellular connectivity.
- In terms of technology type:
  - Automated driver assistance;
  - Intelligent traffic systems;
  - Emergency vehicle notification;
  - Passenger information system;
  - Fleet and asset management;
  - Parking management system;
  - Line of sight;
  - Non-line of sight.

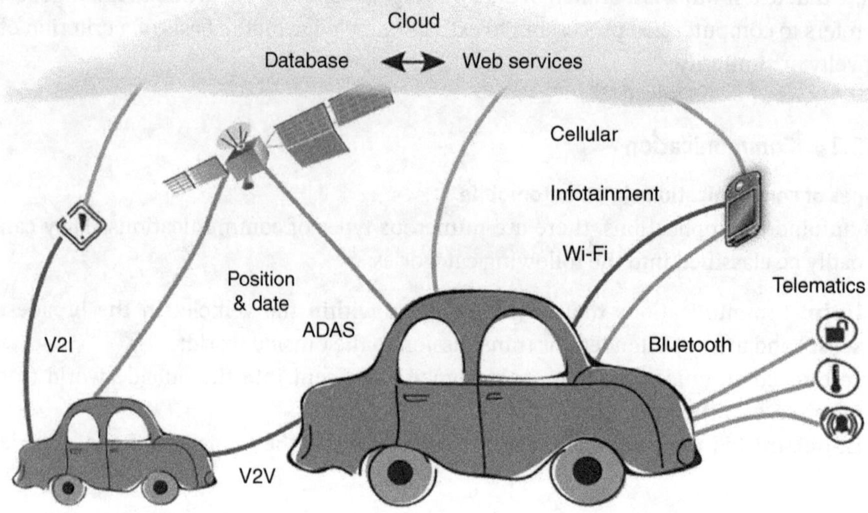

Figure 4.3 Example of V2X communications around a connected vehicle.

Communication standards

There are multiple bodies managing standards in this domain. The most prominent are the ISO, the IEEE, and the ETSI. The ETSI is the main standardization organization in this field, regulating all 5G technology, ITS (Intelligent Transport Systems), and Communications Architecture in the 5.9 GHz frequency band (ETSI EN 302 665 V1.1.1 [2010-09]).

V2X (vehicle-to-everything)

It must be remembered that according to figures from the World Health Organization (WHO), over 1.2 million people die in traffic accidents every year. As there are increasing numbers of sensors in vehicles, generic V2X communication (whatever the technology) offers considerable progress and an improvement to road safety. It allows vehicles to communicate with other vehicles, infrastructures, etc. With this technology, vehicles can communicate with one another (with mopeds, emergency vehicles and infrastructures, pedestrians, traffic lights, and digital road signs), even if they do not have a direct line of sight, and "see" their surroundings at up to a kilometer away, round corners, around other vehicles, in dense urban environments (see Figures 4.4 and 4.5).

Therefore, there are three main themes that are of interest in the automotive market:

- Autonomous and semi-autonomous vehicles need an effective communication system so they can work properly;
- The increasing problem of road congestion may lead to accidents;
- Automakers need to turn their attention to new functions.

Thus, governments from around the world, the automotive industry, and the telecommunications industry support two different V2X radio technologies, which are briefly presented below.

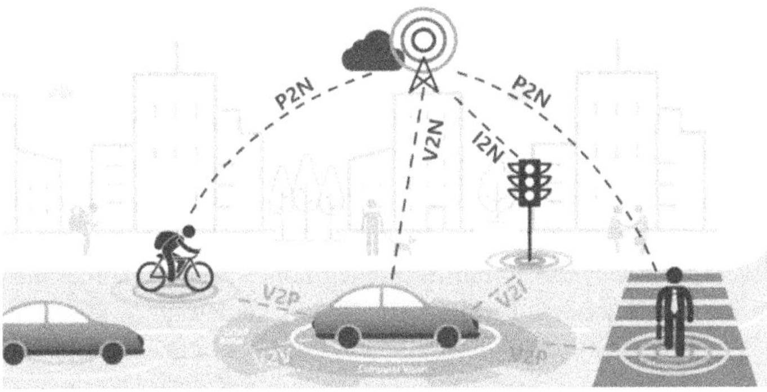

Figure 4.4 Example of V2X and P2X communication (*source*: 5GAA).

Figure 4.5 Specific example of V2V communications.

"DSRC/ITS-G5"

DSRC/ITS-G5 is a technology developed between 2005 and 2015, based on IEEE 802.11p. In the absence of any cellular network, DSRC/ITS-G5 allows for operation on a dedicated channel that offers automatic secure transmission of critical safety messages between objects, and other data, in real time. DSRC technology has proved its worth in autonomous and connected vehicles, as it can "see around corners" accurately. Today, it is in production and many semiconductor companies already have automotive products that are certified compliant with DSRC. In addition, there is a great deal of hardware and software from a range of suppliers, offering a rich ecosystem.

"3GPP C-V2X" and "5G NR C-V2X"

3GPP C-V2X – Cellular-based V2X 3GPP C-V2X (third-generation partnership program cellular-based V2X) is the first step in the evolution and adoption of 5G in V2V communications between vehicles of different makers and different types, V2I between vehicles and road infrastructure such as road signs, work zones, and also other road users (V2P), such as cyclists and pedestrians, with or without a cellular network (and without a mobile subscription). This standard is based on the 3GPP Release 14 and 15 specifications.

In principle, this technology requires the presence and assistance of a communication network. It draws upon numerous advances made in communications and is an important step in strengthening road safety, keeping traffic flowing smoothly, and deploying autonomous and connected vehicles in self-driving mode. For example: in the case of road safety, the vehicles can communicate with one another to report any potential roadside hazards, such as slow-moving or stationary vehicles.

To encourage C-V2X communications, two parallel modes of operation are intended (see Figure 4.6):

- "Direct" mode: the aim here is, without using a cellular network (and therefore without a mobile subscription), to allow vehicles to communicate with other vehicles, with devices belonging to pedestrians, or with road infrastructure equipment (signs, work zones, etc.).
- "Network" mode (cellular): this type of C-V2X works with connections to mobile infrastructures such as LTE and 5G.

The main functions of 3GPP C-V2X The main functions of V2X technology are to:

- Help maintain public safety;
- Help prevent accidents by sharing acquired information with other drivers via sensors installed on the roads and in the vehicles;
- Make the roads safer;
- Improve the driving experience;
- Improve fuel efficiency (reduce traffic jams, thus reducing GHG emissions, leading to a rise in demand for electric vehicles);
- Prevent vehicle theft, if the vehicle has good V2X technology.

In addition, the rise of V2X technology opens the door to new applications of 5G cellular technology, which will evolve with the migration of mobile networks. In 2020, this technology began to be rolled out commercially (in production vehicles), which is deemed

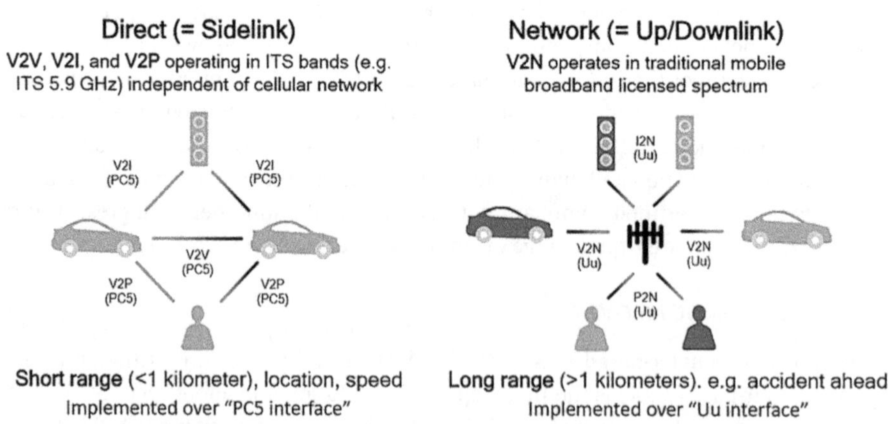

Figure 4.6 The two modes of operation of C-V2X communications.

essential to ensure safer, more autonomous driving. Furthermore, this technology has been approved by mobile network operators.

With the rise in demand, OEMs and automakers are now directly equipping vehicles with V2X technology. In 2015, an initial test phase demonstrated the potential of this technology, with two use cases:

- *See through*: to establish communication between two vehicles traveling along the road;
- *Emergency vehicle*: to report the approach of an emergency vehicle in real time.

However, lack of infrastructure and the high costs of V2X vehicles are barriers to the growth of V2X in the automotive market today.

3GPP "5G NR C-V2X" – 5G networks and C-V2X technologies 3GPP Release 16, or "5G NR C-V2X", was designed on the basis of 3GPP Releases 14 and 15 (also called LTE-V2X), describing the basic security technology C-V2X.

Operating in the 5.9 GHz band, this technology is designed to be used in an intelligent transport system (ITS), with V2X communications among the vehicles and their surroundings, through commercial LTE and 5G networks. The technology also includes 5G New Radio (5G NR) functions. 5G NR offers technical advantages over current networks: high throughput, long range, data transport, and connecting a larger number of objects and devices. Latency is much lower; it provides greater reliability and scalability, as needed for autonomous driving (very high-datarate sensors, sharing of intentions and driving decisions, content sharing, updating of HD maps, 3D, etc.); and, overall, the performances are superior (see Figure 4.7).

These conditions are indispensable to the intended applications of mobility and road safety, such as:

- Connected autonomous driving;
- Ubiquitous service access for drivers and passengers;

NR-V2X requirements for autonomous driving (SA1 TS22.186)

Use Cases	E2E latency (ms)	Reliability (%)	Data rate (Mbps)
Vehicle Platooning	10	99.99	65
Advanced Driving	3	99.999	53
Extended Sensors	3	99.999	1000
Remote Driving	5	99.999	UL:25, DL:1
	Lateral (m)	**Longitudinal (m)**	
Positioning Accuracy	0.1	0.5	

Note: 5GAA may adjust the above requirements according to inputs from car OEMs.

Figure 4.7 Technical requirements relating to 5G-NR-V2X.

- Integration of vehicles into smart cities and intelligent transport systems;
- Creation of high-resolution video streaming between two vehicles for real-time information-sharing and alerts;
- Improved "non-line-of-sight" safety (over a distance of around 1.5 km);
- Better knowledge of the surrounding situation, by also directly communicating with the vehicles around, with (see examples in Figure 4.8):
  - Local updates in real time;
  - Real-time data sharing with infrastructures and other vehicles (e.g. HD 3D maps);
  - Faster sharing of planned journeys and/or trajectories in order to carry out maneuvers safely;
  - Sharing of perceptions of data from high-datarate sensors and the real model;
  - Coordinated driving with exchange of data on intention and sensor data for autonomous driving that is more predictable and coordinated;
  - Sharing of sensor data at high datarates;
  - High datarate and reliability to facilitate data exchange, whether raw or processed data;
  - High datarate to construct dynamic local maps, based on data from the cameras and other sensors; and distribute them at road intersections;
  - Sharing of trajectory intentions;
  - High datarate sharing and URLLC to activate planned trajectory sharing;
  - Advanced use cases for autonomous driving;

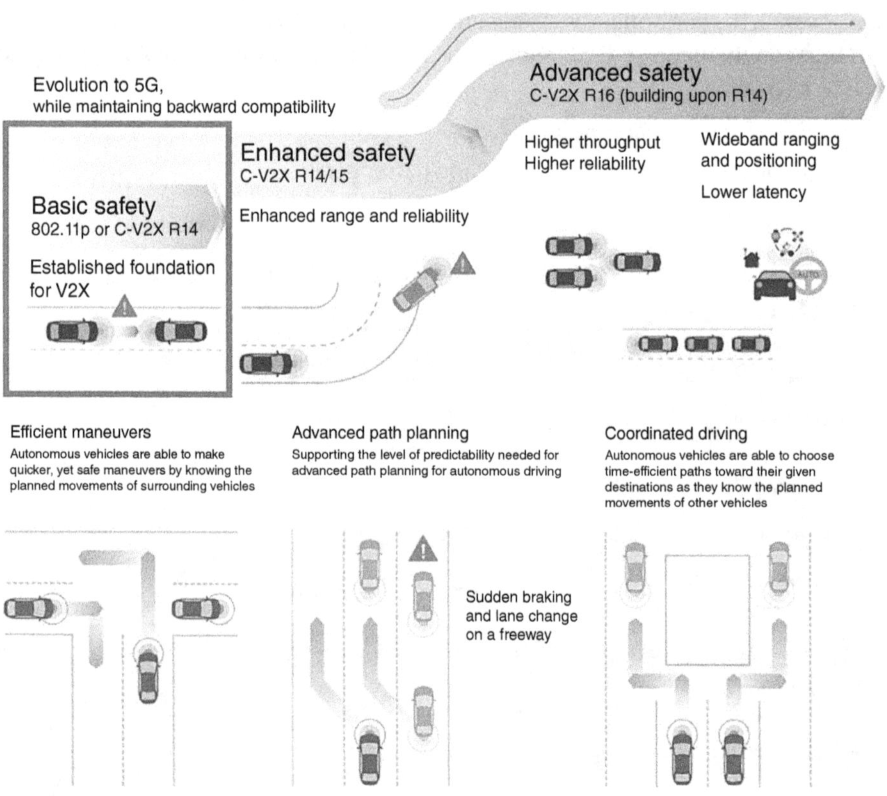

Figure 4.8 Examples of possible applications using 5G NR-based V2X.

  - URLLC and high datarate coordinated driving to exchange information about planning of routes/trajectories in good time;
  - Autonomous vehicles without a C-V2X safety system may require much more time to maneuver;
  - Autonomous vehicles with 5G NR-based C-V2X are able to select quicker, yet safer, paths;
- Fuel savings and reduced travel time.

Other organizations involved

5GCAR The R&D project 5GCAR relates to V2X communications built on 5G. Launched in late 2013 by Ericsson, this project involves numerous partners. The goal is to develop and test a global 5G architecture that can provide end-to-end V2X connectivity and ensure reliable, low-latency V2X services. It is an influential force in building the future standards that will apply to 5G mobile networks. 5GCAR was selected for phase 2 of the 5GPPP initiative (5G public–private partnership).

IEEE and Wi-Fi technologies The IEEE has also formed two study groups to promote and roll out the latest generations of its 802.11 standard (including Wi-Fi) in two specific areas of application: automobiles environments and information broadcasting services. These two groups are making contributions in defining the objectives for future amendments to IEEE 802.11:

- IEEE 802.11 "Next Generation V2X" (NGV) study group.

Evolutions of autonomous vehicle technologies, V2V and V2I connectivity, infotainment services in the automobile environment require higher datarates and improved reliability and efficiency in comparison to what is currently offered by the 802.11 standard. Consequently, this group is investigating the various ways of applying the latest Wi-Fi technologies in new wireless access applications for automobile environments, where the requirements in terms of datarate, reliability, efficiency, and/or range are higher. Though there are many companies working on cellular V2X processes, only IEEE 802.11p – which uses the 5.9 GHz band and is built on an old version of Wi-Fi – was considered by specialists to be a proven, commercially viable V2X technology able to improve vehicle safety, help traffic to flow more smoothly, and reduce pollutant emissions. Today, the Wi-Fi Alliance is presenting the Wi-Fi 6 standard (IEEE 802.11ax) as the most appropriate solution to address the requirements in terms of robustness and latency for automobile users in dense, congested environments. It offers:

  - Wi-Fi connectivity with a datarate of approximately 1.8 Gbit/s in dense environments, by means of DBS (*dual-band simultaneous*) operation, and 1024 QAM modulation in the 2.4 and 5 GHz bands;
  - Diffusion of images from the rear-view cameras.

On the basis of this technology, certain integrated circuits on the market are also capable of:

  - Streaming video content in ultra-high definition to multiple displays;
  - Duplicating smartphone user interfaces on the infotainment system screen, which offers a dual Wi-Fi 6 access point;

- Handling the Bluetooth 5.1 standard;
- Handling audio technology for high-fidelity vocal reproduction and audio diffusion;
- Acting as a mimo (multiple-input, multiple-output) client to extend the range of high-datarate communications with external access points in automotive services such as vehicle diagnostics, software updates, or automated checks when the vehicle is brought into a dealership for a service;
- Implementing the security mechanisms WPA3 and WPA3-Easy Connect;
- Allowing different radio technologies (4G/5G, C-V2X, Wi-Fi 6, Bluetooth 5.1, GNSS) to coexist within the same vehicle;
- Supporting a specific application processor with security accelerators for C-V2X and a technology to fuse data from multiple automobile sensors (geolocation by GNSS, cameras, inertial navigation unit, odometers) to determine the vehicle's position extremely accurately and construct maps on the basis of driving information that is precise to within a meter.

• IEEE Broadcast Services (BCS) study group

The BCS study group, for its part, explores new use cases for broadcasting localized information (in parallel to data) over wireless local networks IEEE 802.11, without the receivers having to actively connect to an access point and without drastic security constraints. The target applications include the broadcast of "local" information about event venues or tourist attractions, and in shopping malls, stations, airports, etc. The technology is also intended to be used for broadcasting more fluid information, such as updates on traffic conditions, available parking spaces, emergency information, or commands from connected objects.

NGMN Alliance The NGMN Alliance (Next Generation Mobile Networks), set up in 2006, includes a number of major mobile operators (AT&T, BT, China Mobile, Deutsche Telekom, KPN, KT, NTT DoCoMo, Orange, Singtel, SK Telecom, Sprint, Tele2, Telecom Italia, Telefonica, Telus, T-Mobile, Verizon, and Vodafone), and today its focus is on 5G. The working group NGMN V2X is defining strategic directions in terms of end-to-end requirements for use in automobiles.

NGMN provided wholehearted support to the C-V2X specification, publishing a white paper on V2X procedures demonstrating that C-V2X technology is superior to IEEE 802.11p standards – in terms of both technical and economic criteria, and its ability to serve applications where secure operation is critical. That white paper is the product of work to evaluate V2X technologies and harmonize the positions of mobile operators in relation to the V2X specifications, based on LTE, then on 5G, and on the older standard DSRC/802.11p.

5GAA The 5G Automotive Association (5GAA) was set up in late 2016 by Audi, BMW, Daimler, Ericsson, Huawei, Intel, Nokia, and Qualcomm to "develop, test and promote 5G communication solutions for automobiles, support standardization, encourage the worldwide expansion of the C-V2X ecosystem, and speed up the commercial availability and worldwide rollout of these solutions." The 5GAA are working closely with the NGMN in the field of V2X solutions built around future 5G technology. This interprofessional organization encourages C-V2X communications, whether direct or passing through networks. Today, it has over 100 members throughout the world, including automakers and their suppliers, mobile telecom operators, semiconductor manufacturers, test material suppliers, telecom network operators, and suppliers of STI software and equipment.

The close cooperation with the 5GAA alliance makes for a particularly effective driving force in the development of the 5G technological platform needed for V2X applications, future 5G-compatible mobile networks, and emerging technologies, such as C-V2X (Cellular-V2X) infrastructures.

It should be noted that, in parallel, the 5GAA has signed an agreement with the EATA (European Automotive Telecom Alliance), which includes six associations and 38 companies (telecom operators, OEMs, and automakers), to cooperate on shared solutions for autonomous and connected driving, and on matters relating to standardization, spectrum resources, and associated use cases.

4.3 Autonomous vehicles: data fusion, AI, and similar technologies

The previous sections have given lengthy inventories relating to the numerous sensors that are needed and available, to draw up the long list of dedicated unitary functions in ADAS, all sorts of accessible systems for internal and external connectivity for a V2X vehicle, big data that can be gathered or recovered (e.g. the state of the road surface), and so on. We now find ourselves at another crossroad.

In order to achieve the desired level of vehicle autonomy, what is required is to dynamically fuse all these data, manage them using appropriate artificial intelligence, define and make appropriate decisions, and apply them to the different systems to deliver either a more comfortable or a safer driving experience. These choices and solutions will, sooner or later, involve burgeoning rates of data processing (Gbit/s, Tbit/s, TOPs), which, of course, the vehicle's internal communication networks will have to be able to handle.

Let us begin this section by looking at a block diagram (Figure 4.9, from C3Car and NXP), which shows a realistic way of fusing data from numerous sensors, cameras, radars, lidars, and ultrasound sensors.

This figure divides the overall architecture into three domains:

- Sensors of all types, and local processing of the raw data they produce;
- Intelligence – a zone containing all the particular processors, specific to the processing of signals from each of these types of sensors;
- Then, in the center of the figure (see Figure 4.10), the processor that fuses all these data and prepares them for transmission to the vehicle's action systems;
- Finally, actions – a zone containing the mechanical systems (the engine, brakes, steering, etc.) and notifications given to the driver (dashboard, HMI, etc.).

4.3.1 Fusion of sensor data

To begin with, are the data from the sensors raw or have they already been refined? That is an excellent question. This gives two totally different angles from which to approach the conception and solution of the same problem, depending on the strategies and industrial policies adopted.

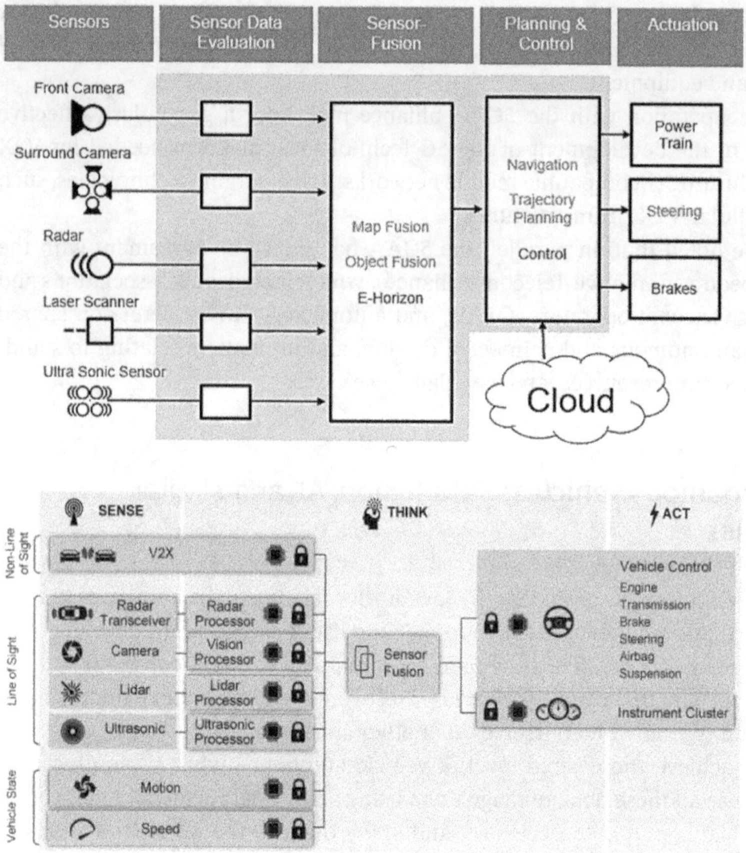

Figure 4.9 Overview of how to fuse data from numerous sources.

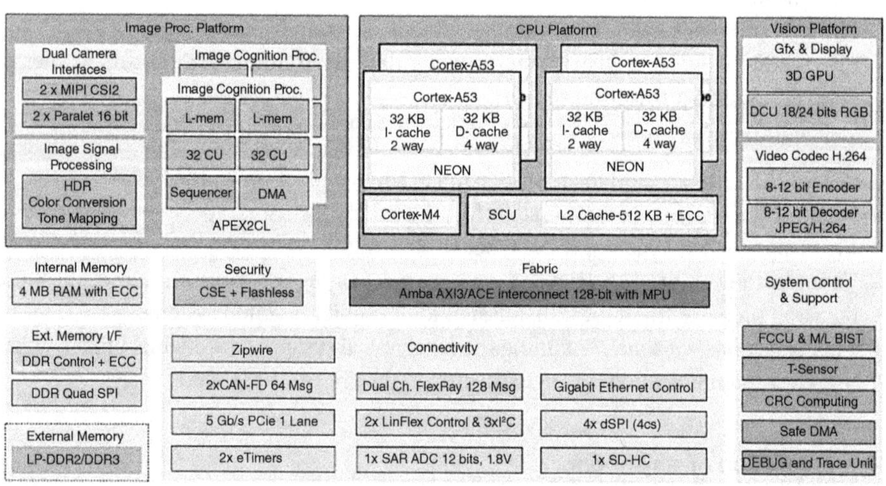

Figure 4.10 Example: block diagram of the S32V234 from NXP.

Raw data

With raw data, the vehicle networks must include resources operating at a high datarate to ensure synchronization between the different sources, etc., but this solution allows a certain amount of flexibility to automakers to do exactly what they want with the incident data.

Example of a radar operating at 77 GHz: a radar head requires an integrated circuit to create the emitted signals and harvest the returning signals. This is the front end of the system. An example of a circuit of this type is shown in Figure 4.11.

Refined data

In the case of refined data, the signal is preprocessed at the sensor itself (or near to it) by its provider (automakers, subcontractors, OEMs, etc.). Then, only the refined data are transmitted, at a lower datarate, over a connection, which may be USB or CAN(-FD) or CAN XL. Thus, by definition, this solution is strongly linked to the philosophy adopted by the OEM who has supplied the sensor, and their choice of electronics and proprietary integrated software layers, which the automaker must deal with or negotiate for modifications.

Example of a radar (cont.): Figure 4.12 shows an integrated circuit for signal processing, which applies important preprocessing.

The technical, political, and economic debate as to the choice of these two solutions will continue to rage for many years to come, given that the industrial makeup of the automobile industry is being entirely reshaped by numerous incoming players

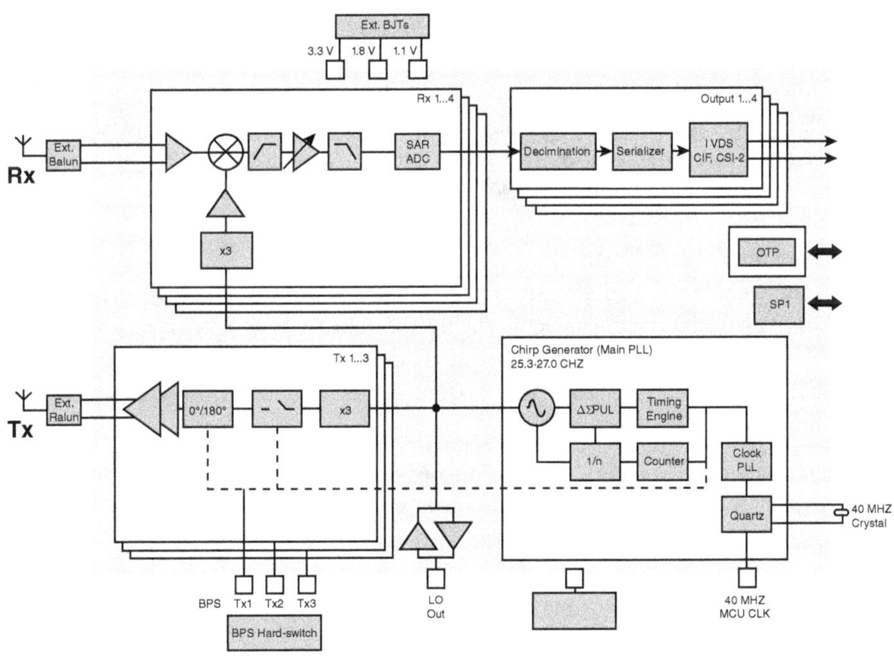

Figure 4.11 Example: block diagram of the TEF 810X from NXP.

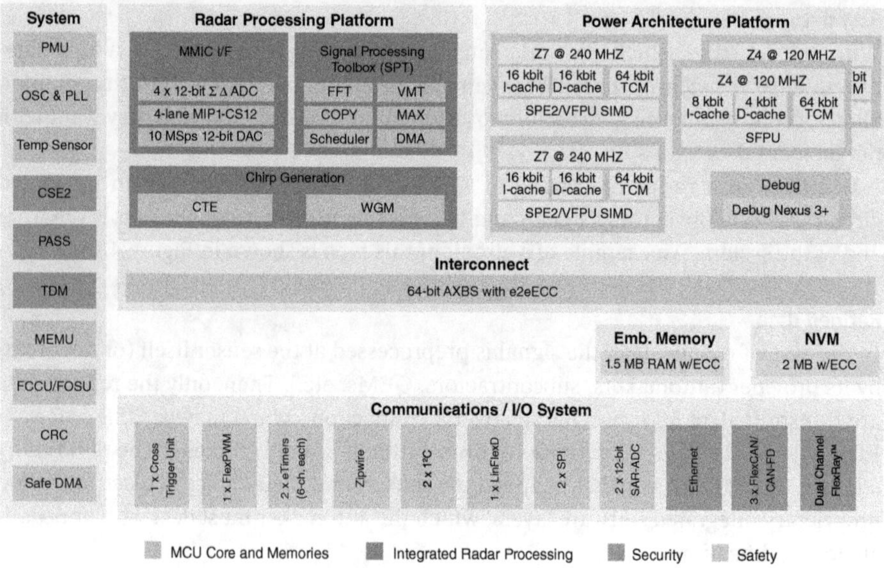

Figure 4.12 Example: block diagram of the S32R27 from NXP.

(startups, new OEMs, etc.). To conclude this broad overview: every automaker, every OEM, every SME, startup, and research institution has its own special formula (the best, the most reliable, the least expensive, the most groundbreaking, of course) and jealously guards its automotive data fusion algorithms and patents for ADASs and autonomous vehicles, which are/will be the nerve centers in the war of these top-of-the-range twenty-first-century automotive applications.

Returning now to the technology, and to illustrate the discussion, we shall describe one of the classic examples of data fusion algorithms integrated into a vehicle's ECU. Its job is to merge the measurements of characteristics and manage image-tracking algorithms, to detect and identify objects in the vehicle's vicinity.

Generic example of data fusion

The example chosen is a combination of systems of lidar sensors and algorithms corresponding to the data of the object in the line of sight, to track that object and classify it in real time, on board a vehicle. This system helps design, produce, and refine ADASs and HADAS applications that are extremely reliable, even in complex traffic environments:

- Automatic/emergency braking (compulsory);
- Anticollision alert system;
- Adaptive cruise control (ACC);
- Other active safety systems;
- For comfort systems and to study the driver's behavior at the steering wheel (levels 3 and 4).

This example draws inspiration from German OEM Ibeo's "ibeoHAD ibeo LUX Fusion System", which has (see Figure 4.13):

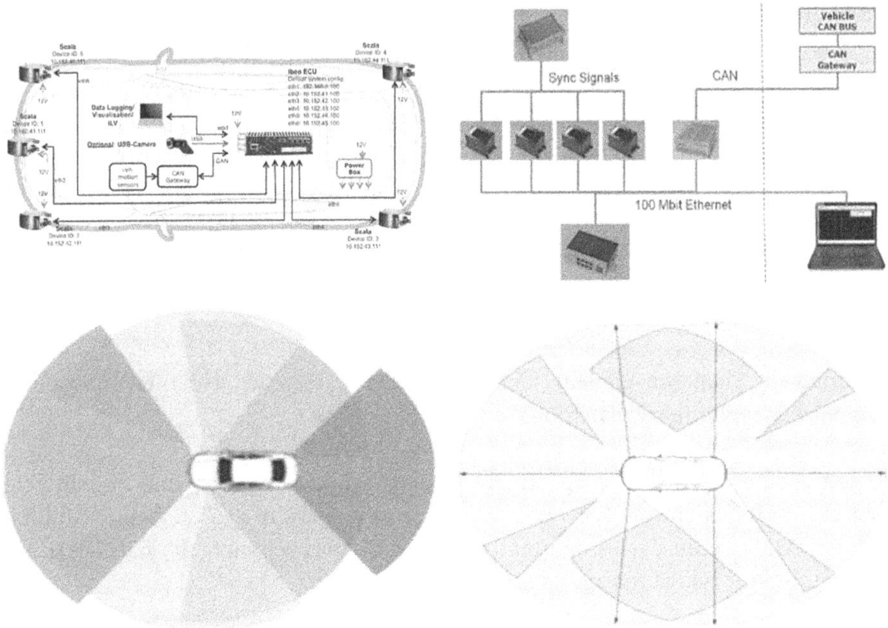

Figure 4.13 Complete overview of the envisaged solution.

- Five lidars, feeding their measurements and data to the ECU by Ethernet (Ethernet ports 2–6) on the switch of the central processing unit. Each lidar requires the use of a converter to transform the sensor output signal into an Ethernet standard signal;
- A USB camera, connected to the ECU. The camera image can be merged with the raw data, and the data stream about the object serves as feedback for the operator;
- The ECU receiving data about the vehicle's motion (e.g. its speed) through a communication gateway in the conventional in-vehicle network (CAN);
- A piece of software, built into the ECU, designed to handle data from up to six lidars in the vehicle, merge the data from those sensors, and produce a full 360° view of the vehicle's environment. Synchronization of the clocks both within and without the system means that time-based data fusion is possible.

Operational principle behind data fusion

There are dozens of principles that can be used for data fusion – some of them free/open and some proprietary. Each company proclaims the advantages of the solution it has chosen for automobiles, particularly in the area of autonomous vehicles. As this book is not intended to be an encyclopedia, we have selected one that is fairly generic, and will draw on this example in the coming sections.

The object detection process may partly take place at the level of each individual sensor (camera, radar, lidar) and partly in the data fusion software. The raw data produced by lidar scanning and the vehicle's motions, amongst other things, serve as input for preprocessing, in the following steps, including a certain degree of intelligence, mainly to recognize the shape of objects.

Preprocessing During preprocessing, points analyzed as being very unlikely to be objects (e.g. rain or the ground) are filtered and marked accordingly.

Segmentation During segmentation, after scanning, clusters or point clouds that could belong to an object are identified. The words "could belong" imply probability.

Feature extraction In feature extraction, the characteristics from individual scans are identified (with more detailed shape recognition). Then come four steps that are carried out by the fusion software, which applies numerous algorithms either sequentially or in parallel.

Association Then, the segments identified earlier are matched with their corresponding objects (preliminary matching).

Object update With object updating, between each sensor sweep and the next, the objects' positions and motions are updated on the basis of an Interacting Multiple Model (IMM), because it is important, of course, to take account of the vehicle's simultaneous motion whilst the analysis is being performed.

Track maintenance In track maintenance, the system checks the credibility of all the information currently being processed. This vector analysis is based on the object's orientation (and, therefore, the direction) and the value of the difference (thus, its magnitude) with respect to its previous position (thus, the point of application). If certain segments are found to belong to the same object, or else are found not to, a logical sequence is applied to either merge those segments or to split them, on the basis of their behavior or appearance. For example: if two objects continuously follow one another, at the same speed and with the same distance between them, we can conclude with a high degree of probability that the two objects are, for example, a truck towing a trailer. Thus, they can be merged into a single computational object.

Classification In classification, the principle of "*track before detect*" is applied. This means that an object is tracked by successive sweeps as many times as is necessary to clearly identify that object (final matching). In this process, moving objects such as cars, trucks, bikes, and pedestrians are assigned a label based on their features. Figure 4.14 shows an overview of these steps.

To summarize, the main functions and technical characteristics of a data fusion unit are frequently:

- The capacity to simultaneously fuse data from 6–8 sensors;
- An object detection range of 200 m or more for vehicles, and 100 m or more for pedestrians;
- The ability to detect multiple echoes, so as to perform well whatever the weather;
- Preprocessing of data (classification of road surfaces, etc.);
- Detection and monitoring of other road users;
- Detailed information about the positions, movements, and shapes of road users surrounding the vehicle, and also static objects in the background;
- Object classification and monitoring;

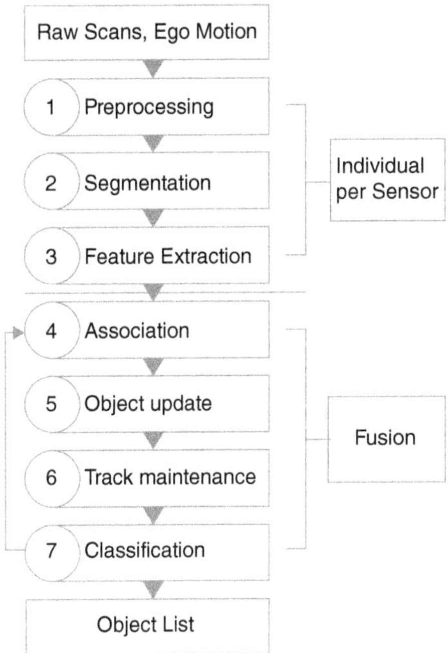

Figure 4.14 Example of the sequence of steps in data fusion.

- Classification and monitoring of road users (available categories include cars, trucks, bikes, and pedestrians);
- Estimation of the uncertainty of the measurements obtained and also the probability of existence for all classified objects (including estimations of the FAR [false acceptance rate] and FRR [false rejection rate]);
- An interface for incorporating the data obtained with other information sources/sensors, such as the radars, lidars, and cameras.

All the above must fit in as small a physical volume as possible, consuming very little energy and not giving off heat – nothing new there!

An example of the visual output from data fusion is given in Figure 4.15.

4.3.2 AI (artificial intelligence) – e-learning – deep learning

Artificial intelligence (AI) has actually been in use for nearly 30 years in the automotive industry. It can be defined partly by reference to human intelligence and partly as a problem-solving methodology using complex combinatory logics without any known algorithms. In all cases, AI must be explicable and robust in order to generate confidence and allow new ecosystems to develop. In our field, it is important to be aware both of the opportunities AI offers and of its limitations.

There are two main AI techniques used in this domain:

- Machine learning, which operates in the same way as a neural network, by understanding concepts and rules;

Figure 4.15 Example of the visual output from data fusion.

- Deep learning, which, because of the capacity to handle a vast mass of data and the increase in computational power, is a technique that combines a huge number of neural layers.

One important aspect of AI is data processing, including: *big data*, which combines a massive amount of heterogeneous information to be processed in real time; *data science*, which is capable of acting on the basis of those data, using extremely powerful computers; and *machine learning*. Although, when we examine them in depth, these three areas are highly different, for the sake of brevity, we have combined the discussion (below) of AI, e-learning, and deep learning. In the domain of autonomous vehicles, AI is mainly applied in:

- Perception;
- Understanding of the scene;
- Decision-making;
- Navigation;
- Interaction with the vehicle occupants;
- Driver monitoring;
- Establishment of transport services (robo-taxicabs, car-sharing, and so on).

In years to come, in level-4 and level-5 autonomous vehicles, AI will have an increasingly important role to play. Every day, a vehicle's AI will be forced to learn all the ordinary (or extraordinary) subtle rules (not formally written or "a little" prohibited by Highway Codes) that drivers all over the world know and apply effortlessly today – otherwise, the traffic system will become hopelessly congested (see the extracts from the Highway Code in Section 2.3). Thus, with basic knowledge and an associated intelligence supplied and implemented in the same way in all vehicles coming off the production line, the AI of every vehicle must harvest data and enrich its knowledge (handling big data that are cold, hot, or sometimes red-hot!) and learn (e-learning), day after day, to tend (constantly) towards the operation of a vehicle that is (almost) perfect for its daily users.

In order to be able, amongst other things, to drive day or night, on the outskirts of Paris or on the Antwerp ring road, for example, in rainy conditions, and be able to react to other vehicles cutting in, in unpredictable ways, vehicle AI is generally based on a mixture of two approaches:

- Rule-based technology (unfailing compliance with the Highway Code, etc.), on the assumption that everyone follows these rules;
- An e-learning-based approach.

In addition, to build up a store of data, it is necessary to have the demonstrator vehicle drive thousands – or even millions – of kilometers in multiple countries, such as France, Spain, Germany, the Netherlands, the UK, the Czech Republic, the USA, and Japan, to teach the vehicle how to behave in all types of environments. In the UK, traffic drives on the left. In the State of Nevada, in the USA, small reflective dots (Botts' Dots) are used for road markings. In Asia, different-colored paints are used. The vehicle must take all of this information on board, in order to be able to operate autonomously in any circumstances, and feed back the information to be managed and distributed to other users by 5G connectivity or another form of V2X in connected vehicles.

Thus, *Homo sapiens* v. *AI* is well under way! Clearly, in this context, we inevitably see the conventional rules of "morals" and "ethics" shifted to a degree, so the question remains: who is to oversee the AI and adjudicate on good or bad behavior, its GDPR compliance, its handling of personal data, etc.? Who will decide whether the fateful decision-making criterion is 1 or 0? That is the big question (see Sections 2.8 and 2.9). Similarly, in an emergency situation, who will have the final decision: the human or the AI? Do our own neurons still have a future in participation and/or cooperation with the AI running a vehicle? What would happen if the machine's "brain" were to have a stroke? Etc. Thus, only when these questions have actually been answered will a true level-5 vehicle ("driverless" and "without steering wheels or pedals") on the open road become more than a dream. Being positive, we can say that with the rapid improvement of ADASs and automated driving technologies, autonomous vehicles are on the way to becoming a reality.

For example, the following already exist:

- AI processing with artificial vision for the cameras of level-3 autonomous vehicles (conditional automation) and level-4 (advanced automation), with low energy consumption and compliant with the NCAP3 (new car assessment program);
- Optimization of processing architectures for artificial vision, with advanced detection capabilities based on an image-recognition engine and hardware accelerators with algorithms including dense optical flow, dense stereo disparity, and object classification. The integrated AI dedicated to the convolutional neural network (CNN) speeds up the deep-learning process;
- Open-source platform solutions that allow OEMs and equipment suppliers the option of developing camera solutions with their own distinguishing features, and benefit from evolving solutions.

4.3.3 E-learning – machine learning – deep learning

There are entire volumes dedicated to these subjects, conference proceedings (see the bibliography), and so on, to which we invite readers to refer for in-depth information. In this book, we shall briefly explain a number of points on the everyday reality of machine learning, which are particularly important in relation to autonomous vehicles. Let us start at the very beginning. In principle, an e-learning system will learn good driving behavior, but absorb everything, so could also learn less good conduct! With that fact acknowledged, we can proceed.

A good machine-learning model must, of course, perform well, but it must also be resilient, unbiased and quantifiable, and it must be a delicate balance between technology and methodology. The hype in the media about artificial intelligence may have given many false hopes. The widespread adoption of machine-learning and deep-learning models suggested that the technology was likely to spread rapidly into all thinking models. However, early feedback typically indicated a shortfall, in real-world vehicles' performances, in relation to what the prototypes promised to deliver. Let us take a look at a number of obstacles often proffered as explanations of why these promises were not kept, and point to some best practices that could help overcome these obstacles.

Methodology

There are two points that must be made in relation to the methodology to develop AI systems:

- Artificial intelligence cannot be developed using a conventional project method, or even agile methodology, because the system must have its own rules and constraints;
- A machine-learning model needs to fail a certain number of times in order to learn. That flexibility and openness to failure is not traditionally found in development projects.

Definition of requirements

Whether in terms of classification (see Section 4.3.1) or prediction, it is sensible to clearly define the problem at hand and ensure that the AI is able to solve it. Thus, the different teams must be involved from the very outset, but also during the training of the model, in order to assess the quality of the learning data.

Consider the example of a model designed to determine whether a department is properly discharging its duties. The AI cannot, unassisted, guess whether the particular dataset indicates a problem of communication with an original client, or an unachieved objective. It is up to those humans who *do* know to teach the system the correct pattern, by validating or rejecting the responses associated with a certain context. Such training is absolutely essential, and if the AI is required to score extremely highly in terms of correct responses, then a particularly lengthy training process will be required at the early stages.

Data qualification

Another potential pitfall relates to the quality of the data fed to the model, and regular updates. Before releasing and rolling out the AI, we must be absolutely sure that it has been trained with appropriate data for the requirements of the technology – as with the principle of *garbage in, garbage out* (GIGO), wrong or inadequately prepared input data will lead to aberrant output. In conclusion, we must dispel the myth that one can simply plug an AI in to any dataset and obtain meaningful results. Lengthy preparatory work is required, to ensure the integrity of the (refined) data, before recording some of those data so that the machine can learn from them. In addition, machine-learning models cannot handle any and all data formats. Thus, we need to adapt the data to the chosen algorithm, rather than the other way around. With that in mind, let us now look at the choice of algorithms.

Choice of algorithm

In the field of automotive AIs, there are at least 60 families of algorithms available, each of which has an average of 10 sub-groups. Certain algorithms are reserved for specific uses: convolutional neural networks for image recognition, recurrent neural networks for natural language processing, reinforcement learning for strategy situations, and so on. Also, it is often sensible to avoid reinventing the wheel, but instead use an existing system.

Take the example of a neural network that has learned to distinguish vehicles from pedestrians. The system could be reused to classify vehicle types (trucks, light cars, etc.) by simply altering the last layers of the algorithm. The main benefit to this approach is a major reduction in the volume of learning data needed (rather than millions of examples, a few thousand might suffice).

A mistake that is often made in AI is to choose a complex algorithm in the hope of obtaining maximum precision in the results. However, additional complexity in an algorithm is often detrimental to its robustness – that is, its ability to produce meaningful results in the longer term. It is more sensible to adapt the complexity of the algorithm to that of the input data. If we use too complex an algorithm for a simple dataset, it will produce incorrect results. This is known as overfitting – the model drowns in details and loses its ability to generalize. To prevent this, we must either change the model or stop the learning process sufficiently early. In conclusion, the more complex the algorithm, the longer the learning phase will need to be.

Infrastructure design

The next step is to choose the infrastructure that is to house the model. It must be robust, stable, and regularly fed with data. The AI needs a constant flow of live information. A model produced with a stock dataset will never evolve unless it is fed constantly with fresh data. Thus, the platform on which the project is built must be able to cope with this requirement.

As they need to harvest, dynamically store, and process data, most AIs are extremely hungry in terms of CPU resources – and electrical power. In addition, certain models have several milliseconds latency. To deal with these challenges, the underlying infrastructure is generally built on a big-data-type model, with a Hadoop software stack architecture in combination with Apache Spark or Flink. This type of platform could also be linked to a traditional SQL database, deployed in the Cloud or internally, and fed by an ETL (extract, transform, load) data stream.

Testing and validation of the model

Then comes testing the AI. The first step is to choose and define a sample dataset to validate the model's results, risks and credibility. The process will use cross-validation techniques, but also statistical methods. One of the objectives is to eliminate skewing in the model, which may arise if a sufficient number of representative examples has not been used to train it. The majority of the dataset used to validate the AI unit will, inevitably, be labeled by humans. However, in principle, they cannot be 100% objective (for example, there may be a certain skew relating to ethical choices in the case of accidents). Consequently, we need to cross-reference the validation of the data in order to limit bias.

Consider the example of an AI whose output is biased (in principle, this should be a rare phenomenon). Consider that an accident will occur on the ground with a 1% error rate. If the algorithm predicts no errors, it will be said to be 99% accurate. In itself, this is an excellent result, but in practice, it is utterly useless! To obtain an accurate view of the phenomenon, we need to re-balance the dataset by oversampling the data. That is, we need to overrepresent cases of error and also homogeneously reduce the examples of non-error that the system sees.

Monitoring the model

When the algorithm goes into production, this does not mark the end of the project; the model then needs to be carefully monitored. Over the course of a vehicle's life (over 10 years), an AI unit trained in a certain context, released at a time *t*, must be capable of evolving constantly to deal with changing operational paradigms, driving conditions, and regulations that will have changed in the meantime. Consequently, while they would originally have been valid, the results may no longer be accurate. Therefore, it is absolutely necessary to carry out regular checks, measure the prediction error rate, and adjust in terms of variables and new input data.

Real-world examples

To conclude this section, let us take a look at two representative actors in the production of automotive AI (among many others): NVIDIA and Tesla. Their products are designed to manage automated driving systems.

NVDIA

This system (shown in Figure 4.16) has an architecture that is capable of multiple configurations:

- A system on chip (SoC) "DRIVE™ AGX Xavier", which delivers raw performance of 30 TOPs (tera-operations per second), consuming around 30 watts of power. This SoC includes six different types of processors in order to handle redundant and heterogeneous deep-learning algorithms;
- A "DRIVE™ AGX Pegasus" system, using two Xavier™ SoC processors and two TensorCore GPUs. The Pegasus can process autonomous vehicle driving in real time at 320 TOPs for deep learning. It can take input from cameras, radar, lidar, and is capable of sensor fusion, surround vision, vehicle localization, and safe and robust path planning.

Figure 4.16 NVIDIA DRIVE Orin™ SoC (system-on-a-chip), which delivers 254 TOPS (trillion operations per second) and is the central computer for intelligent vehicles. Image courtesy of NVIDIA.

Tesla

At an “Autonomous day,” Tesla presented its solution, developed alongside Samsung (see Figure 4.17).

It consists of two integrated circuits, measuring 37.5 × 37.5 mm, using 14 nm CMOS technology. It has a 260 mm^2 silicon surface, with 12 layers of metal, 6 million transistors, 250 million ports, all in a BGA casing with 2116 connection micro-bearings (see Figure 4.18).[1]

It includes:

- A 2.5 Gpixel/s camera processor;
- A 1 Gpixel/s image processor;
- A GPU that can handle 600 GFLOPS;
- A 36 TOPS neural processor (×2) operating at 2 GHz;
- An ARM12 main processor operating at 2.2 GHz;
- An H.265 video encoder;
- A Tesla-proprietary security system;
- The whole system operates at 50 TOPS with total redundancy; and
- Of course, it consumes energy and dissipates (a great deal of) heat: up to 100 W, which needs to be evacuated (see Figure 4.19).

Using these platforms, automakers or OEMs are able to speed up production of autonomous vehicles. In addition, they enable companies to construct their own applications,

Figure 4.17 Example of the Tesla/Samsung automobile AI unit.

Figure 4.18 Photo of one of the two main integrated circuits of the AI.

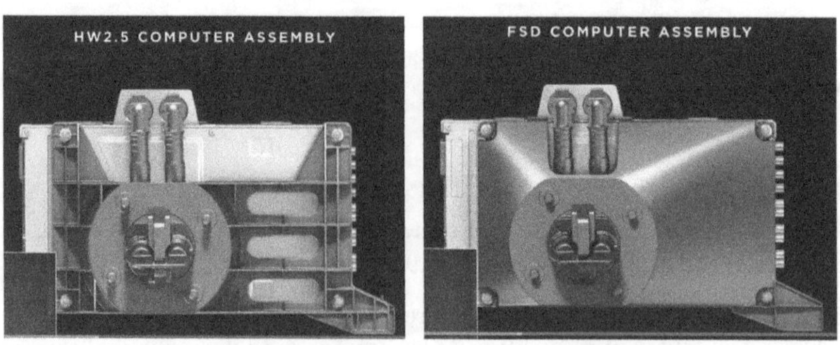

Figure 4.19 Photos of the heat dissipators needed for the integrated circuits.

constantly improve them through over-the-air updates and/or add new functions over time. With deep neural networks, these solutions offer a broad range of software options necessary for autonomous driving:

- Data harvesting;
- Sophisticated perception of obstacles and access paths;
- Classification of objects in the surrounding environment;
- Tracking from one image to the next;
- Driver monitoring;
- Monitoring and vision both inside and outside of the vehicle;
- Identification of road markings;
- Detection of "drivable" spaces;
- With a data logging tool, harvesting time-and-date-stamped data from multiple sensors for the purposes of training, testing and validation, etc.

Such software can also be used inside the vehicle:

- A driver-facing camera can monitor the driver's facial expressions to detect signs of fatigue, or tell whether they are paying attention to the road;
- Visual display capabilities allow passengers inside the vehicle to assess the accuracy of the vehicle's perception, and thus build up a degree of trust between humans and machine.

With this flexibility, companies can define the best hardware configuration for their software across the whole autonomous vehicle portfolio (ADASs for driverless vehicles), all with a unique architecture.

4.4 Hardware architectures of vehicles

Considering the technical conclusions of the previous chapters (estimations and definitions of the desirable and necessary datarates, etc., in relation to the intended functionalities that we have defined), we shall now focus on hardware architectures and topologies of communication networks and their performances in autonomous and connected vehicles.

Obviously, the architectures are sometimes interdependent on the types of engines used (combustion, electric, hybrid, etc.); on where it is possible mechanically to place the different elements of the ECU; on the electromagnetic pollution and radiation that the components produce or that could interfere with them; on the types of strands in the cables being used; on the weight of those cables; and on a host of other factors besides. In addition, some of these architectures necessitate redundancy in the networking, to rectify certain structural and functional shortcomings in the vehicle security (e.g. the structure *X-by-Wire*).

For information purposes, before proceeding with the discussion, it would be useful to bear in mind certain orders of magnitude for some of the classic parameters in a combustion vehicle. Whilst these numbers may not all be 100% accurate, neither are they completely inaccurate. Thus:

- Cabling is the third most costly element in the vehicle (the engine being the most costly, followed by the chassis);

- Cabling is also the third most weighty element (again, the engine and chassis, respectively, are in first and second place);
- The weight of cabling used often runs to around 50 kg, which accounts for the energy consumption;
- On average, 1 km of cabling is used in a vehicle, though it may be 3 km or even more in some cases, with up to 1500 different cables and 3000 contacts in the system as a whole;
- Cabling alone is responsible for 50% of the cost of labor involved in building a vehicle.

Now the scene is set (see Figures 4.20 and 4.21) and the show is about to begin!

Let us begin by examining the different network topologies in the automotive domain.

4.4.1 Network-specific topologies

There are four main types of topology used for communication networks in automobiles: bus, star, ring, and point-to-point.

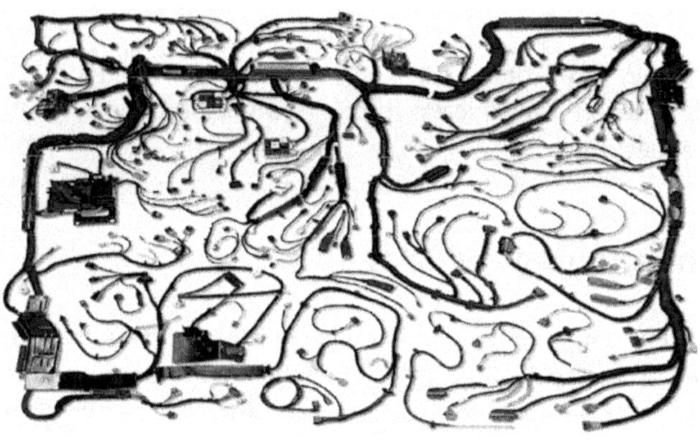

Figure 4.20 Example of all the cables needed for a vehicle to work.

Figure 4.21 Example of how the cables are laid out in a vehicle.

Bus

Bus topology is very widely used (see Figure 4.22). This technique is appropriate for supporting multiplexed communication protocols. Its three main standard-bearers are LIN, CAN with its multiple variants (Low Speed, High Speed, FD, and XL), and FlexRay (see Section 4.5 for details).

The principles of acknowledgement used in some of these protocols (CAN, CAN FD, CAN XL) lead to a direct relation between "digital datarate" and "coverable distance" for a network. Thus, for a given datarate, the network topologies functionally limit the possible length of the network. In addition, the EMC performances of the physical layer chosen for the application – often differential unshielded twisted pairs – and their diameters (thus, the weight of the cables, the connectors used, and so forth) limit datarates to some 10 Mbits/s in order to abide by the RF pollution regulations in force.

Ring

Ring topology (see Figure 4.23) is/has been widely employed in the field of multimedia communication (audio, video, and telematics) and management of multimedia in the cabin of a vehicle. This topology is mainly used for the protocol MOST and its applications. Over the years, the datarate of which MOST is capable has gradually increased from 10 to 150 Mbits/s, using either copper pairs of wires or fiber-optics (FOP).

Given that Ethernet is in the process of supplanting or replacing MOST in many applications, we shall not go into great detail about this protocol. We invite readers wishing to know more to our other publications.

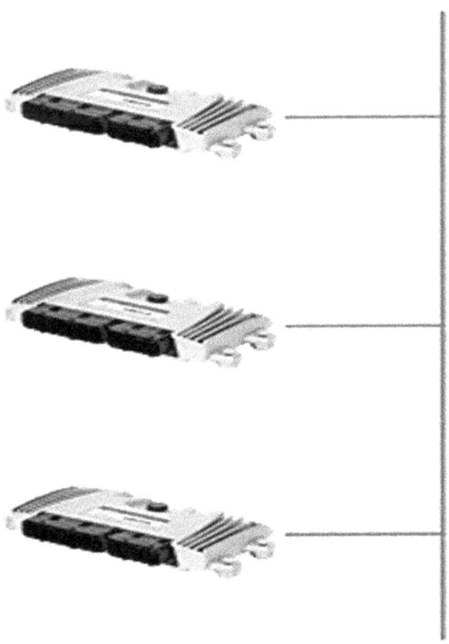

Figure 4.22 Bus topology.

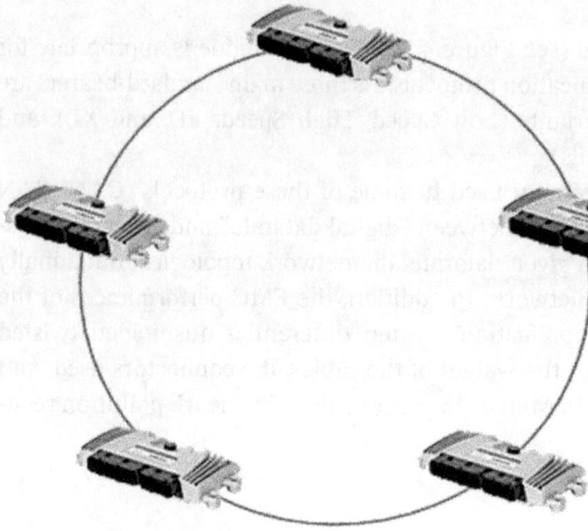

Figure 4.23 Ring topology.

Star

FlexRay is a notable example of a network which, in addition to bus topology, is also able to use star topology (see Figure 4.24). We shall briefly revisit FlexRay a little later on in this book. Readers interested to learn more can consult a specific publication.[2]

In this protocol, the acknowledgments are directly integrated into the content of the data frames, whether in static or dynamic slots, and in no way affect the relation between the digital datarate and the coverable distance. Here, too, the EMC performances of the physical layer chosen for the applications (differential UTPs of different diameters [and therefore differing weights], the connectors used, etc.) limit the datarates to around 10 Mbits/s to comply with the applicable RF pollution regulations.

Point-to-point

Let us linger for a moment over point-to-point topology. In principle, this topology can be applied partially in a bus or star arrangement. In this book, we shall specifically examine Ethernet protocol, mainly in its "point-to-point switched" star topology (see Figure 4.25) – meaning that point-to-point connections between participants in a network are formed in different timeslots. This principle solves the issues of clashes and collisions, but does nothing to solve the problem of signal propagation time as a function of the distance.

4.4.2 Automotive hardware architectures

Put simply, automobiles (using thermal, electric or hybrid propulsion) have been working properly for many years. This is partly thanks to the existing protocols: LIN, CAN, FlexRay, and Ethernet. What is missing, currently, is the certainty that they will always work properly; (almost) the same level of comfort in the car as we

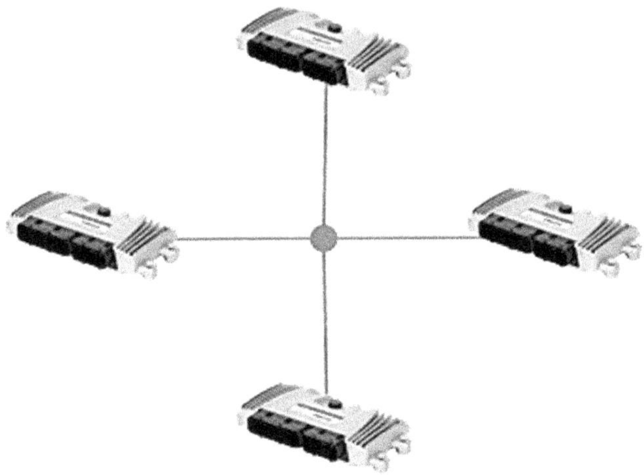

Figure 4.24 Star topology.

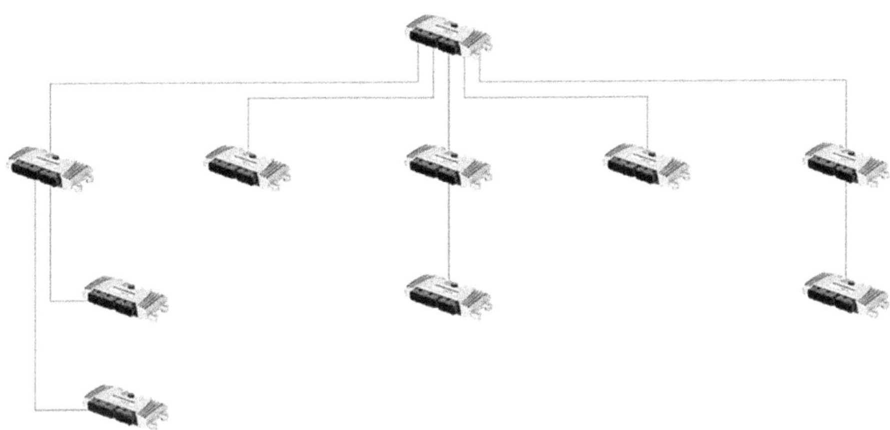

Figure 4.25 Point-to-point switched star topology.

enjoy in our homes; that they can be "green" and we can be safely and reliably transported in any circumstances, increasingly autonomously, and for a (relatively) reasonable price. Simple, is it not? All you need to do is manage the multiple types of sensors (relating to the engine, the suspension, the acceleration, the ambience, and so forth), all the intevrnal and external cameras, radars, lidars, etc., the audio and video systems, telephone connection, computer system, etc., and, of course, the necessary datarates for all these systems (see Chapter 1); encrypted language; all the highly sophisticated ADASs, with artificial intelligence, which have learned their numerous lessons over a long period of time; and the infotainment systems.

Here, then, are the commonly accepted conditions and applications for the introduction of these networks.

Technical requests

A plethora of new technologies have emerged in recent years. The hardware architectures for a vehicle can be broadly divided into three categories:

- What happens under the hood;
- What happens outside the vehicle; and
- What happens within the cabin.

Also, broadly, the whole system can be divided in terms of time, into three generations.

Generation 1 – diagnostics over IP (DoIP)

Let us begin by remembering a few events from 1995 to 2000, which should be borne in mind when discussing this topic.

For many years, a dedicated assembly line produced one vehicle per minute (i.e. 60 vehicles an hour, which equates to around 350,000 vehicles each year). Then, at the end of the production line, only around 40 seconds were allowed to load, read, and verify the software contents in the memories of the various ECUs at different points in the vehicles (currently, this represents hundreds of megabytes of data, or even gigabytes in the case of Flash and eeprom ECUs). The cycles of recording and deletion of a memory element take less than a millisecond per byte. This computes to $10^6 \times 10^{-3} = 1000$ s = around 10 minutes per Mbyte. Of course, multiple memory elements can be recorded at once, but it still takes time! Often, portions of the codes are preloaded in advance (either as the vehicle passes along the production line or by the OEMs supplying the ECUs), but it is still necessary to apply the finishing touches to the vehicles, as quickly as possible. The performances and datarates of the existing IVNs, such as CAN and FlexRay, are not always sufficient. Hence, for a number of years (e.g. since 2002 at BMW), the automotive industry has been applying high-datarate protocols (at least 100 Mbit/s), such as Ethernet 100Base-TX, with a category-5 cable (see Chapter 5). These protocols are used for the test equipment, for on-board diagnostics (OBD), and for activating the flash memories of the ECUs. This helps to save on production costs and after-sales service. In addition, for years, ISO 13400 (Diagnostic Communication of Internet Protocol – DoIP) and ISO 14229 – x (Unified Diagnostic Services – UDS) have been in use in the road vehicle industry, providing stable standards for the current situation in relation to issues of diagnostics.

In addition, overall vehicle topology used to be fairly similar to that shown in Figure 4.26, with each network operating in its own little world.

Figure 4.27 shows a real case of a generation-1 vehicle.

Generation 2 – advanced driver assistance systems (ADAS) and infotainment

The second generation – spanning from around 2015 to the vehicles at the time of publication (in the early 2020s) – includes the launch and the gradual adoption of Ethernet-type in-vehicle networks. Thus, it heralds the beginning of assured communication and shuttling between all types of data through the different systems in the car, and the dawn of safety and infotainment functions in consumer vehicles, in the form of:

- Advanced driver assistance systems (ADASs);
- Information systems;

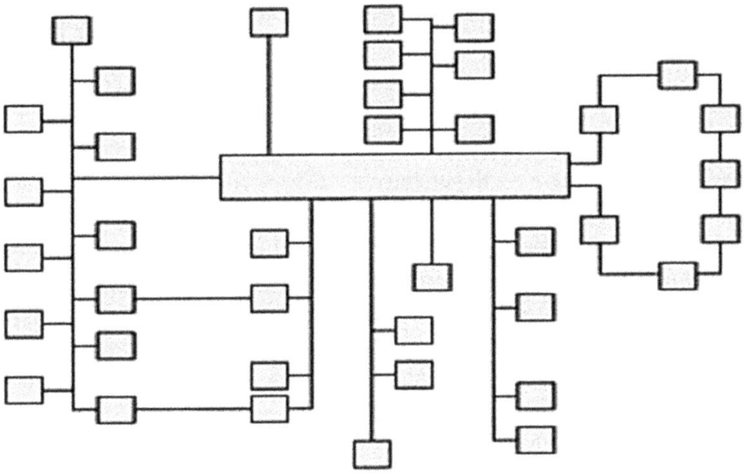

Figure 4.26 Overall topology of a vehicle from around 2010.

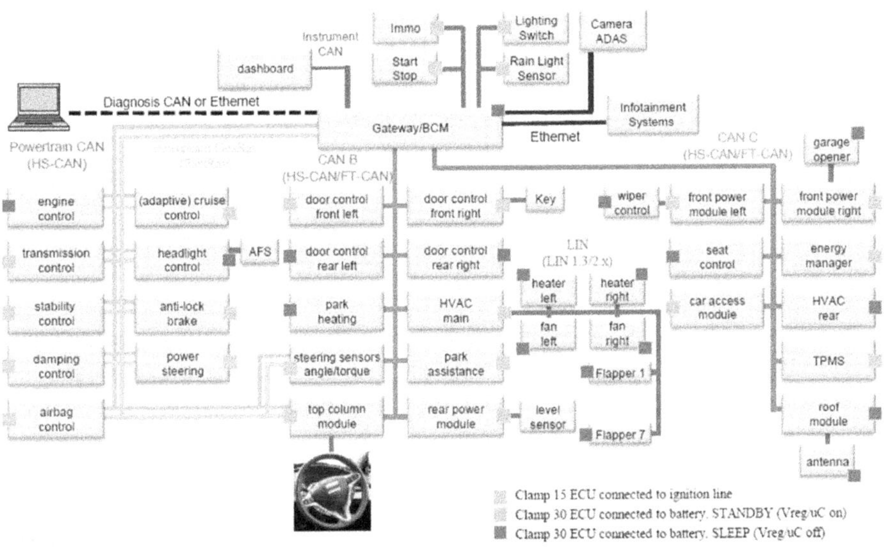

Figure 4.27 Overall topology of a vehicle from around 2015.

- Entertainment systems;
- Systems enhancing the vehicle's and passengers' safety (e.g. blind spot systems);
- Smartphone applications to keep us constantly connected.

Also, Ethernet is an open standard for Local Area Networks (LAN) and defines the bottom two layers, 1 and 2, in the seminal OSI model. Over the past few decades, the IEEE 802 standardization committee has created Ethernet specifications. meaning that multiple physical layers can be employed, achieving datarates between 10 Mbit/s and 10 Gbit/s. As we shall see later in Chapter 5, IEEE 802.3u (100Base-TX), a standard that is widely used both in industry and in the consumer market, was initially selected for automobile diagnostics over IP, as described in ISO 13400-3, Diagnostic communication over internet protocol (DoIP) – Part 3: Wired vehicle interface based on IEEE 802.3.

Outside the vehicle: driver assistance Reversing obstacle detection systems are becoming an almost standard feature on all vehicles (using ultrasound, cameras, etc.), and rear-view solutions are practically essential on all models. However, such systems often use cameras linked to centralized processing by a standard for transmission of electrical signals (video data) at a high frequency (typically hundreds of MHz), operating on LVDS (low voltage differential signaling) lines. This technology works properly with individual cameras, but can cause problems with multiple cameras. Therefore, it is essential to provide greater bandwidth and deliver lesser (and known) latency in the signal processing. It is clear, in this light, that a system based on LVDS is inappropriate and inflexible, in terms of performances and in terms of cost for cables, connectors, and cable harnesses (see Section 4.5.6).

Advanced driver assistance systems (ADAS) are made up of numerous video cameras and/or infrared sensors (up to 15) (see Sections 3.2 and 3.3) or ultrasound sensors. There are abundant communication networks and video signals sent by the cameras, which can be fused with short/long-range radar and lidar data and other sensors to prevent all kinds of accidents. For information, there is already standardization of communication protocols and physical layers, in ISO 17215 – "Video communication interface for cameras (VCIC)," to assist the driver and take over control of the vehicle if necessary (review Section 3.3 on ADASs).

Note
As we shall see later on, a "switched Ethernet" network is capable of linking multiple video cameras to a central unit, for the purposes of synchronization and for subsequent signal processing. In addition, the cameras can use Energy-Efficient Ethernet (IEEE 802.3az), which allows for low-power idle (LPI) mode, and a wake-up function, to save energy when the devices are not in use. Furthermore, Power over Ethernet (PoE) solutions can be used to reduce the size of the cable bundle.

Inside the vehicle: infotainment The solutions used, even today, in infotainment applications are based mainly on proprietary technologies, which are patented and have little or no capacity to evolve (e.g. the ring-shaped MOST network). Each year, automakers devise new ways to connect functions and applications from users' favorite devices to their cars (music, hands-free systems, eyes-free systems, etc., to say nothing of media streaming services, cloud services, and so forth).

An "automotive" Ethernet solution also serves these applications, in a more flexible and economical way, using audio video bridging (AVB) – IEEE 802.1, developed by the audio video bridging working group at the IEEE AVnu Alliance (see Section 5.4). In addition to the transport layer of the IEEE 1722 protocol, which allows time-sensitive audio and/or video feeds to be sent over AVB Ethernet networks and allows different stations to be interoperable, the protocols IEEE 802.1 AS, QAT, QAV, and BA respectively cover the functions of address timing and synchronization, streaming reservation, and transfer and queuing of AVB systems (see details in Section 5.4). It should be noted that synchronized streaming of video and audio data with guaranteed latency is already a possibility, with existing AVB Gen. Ethernet 1 components.

AVB is a set of open standards designed especially for the transport of audio and video streams. The IEEE developed AVB because Ethernet, in its original form, was ill suited to the needs of audio and video networking. The main AVB protocols are (see Section 5.4 for details):

- Precision time protocol (IEEE 802.1AS);
- Traffic shaping (IEEE 802.1Q-2012 Clause 34 – FQTSS);
- Stream reservation protocol (IEEE 802.1Q-2012 Clause 35 – SRP);
- AVB configuration protocol (IEEE 1722.1 – DECC); and
- The formats for audio and video (IEEE 1722 – AVTP and IEEE 1733).

Having examined this near-prehistoric aspect of driver assistance, let us now look at the solutions of the future.

Generation 3 – the "Ethernet backbone" network

In generation 1, the IVN architecture was primarily a heterogeneous network, because of the historical nature of such systems (see Figure 4.28) and the hierarchy of datarates in the LIN, CAN, and FlexRay protocols, with a bus functional topology. Their inexpensive physical layers (UTP) and the more expensive MOST (fiber-optics/wired with ring topology) pose problems in terms of the costs of the physical layers for applications requiring a high datarate.

In generation 2, the Ethernet network is still confined to sub-networks for certain applications such as management of the ADAS and infotainment.

A third-generation IVN system, designed with a clean slate (so not inheriting anything from the past), would allow us to build an architecture in which Ethernet can gradually become the backbone of in-vehicle communication. A concrete example of Ethernet in the world of automobile today (and the not-too-distant future) is illustrated

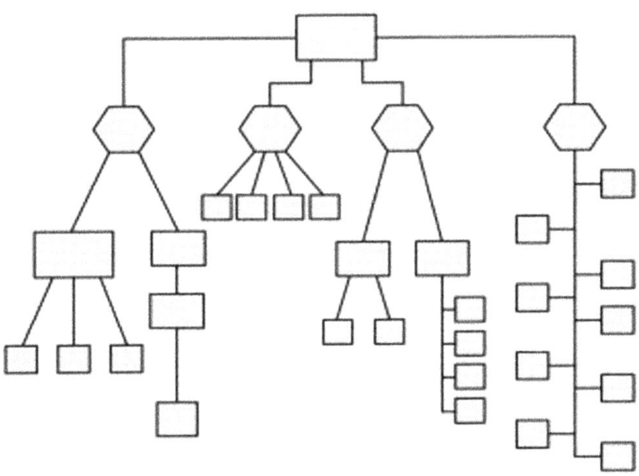

Figure 4.28 A heterogeneous network, as formerly existed.

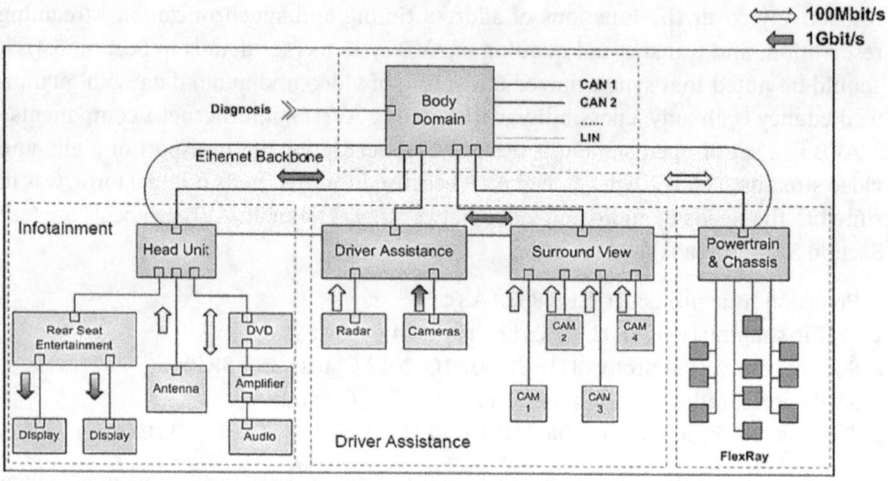

Figure 4.29 Generic example of Ethernet-based network architecture.

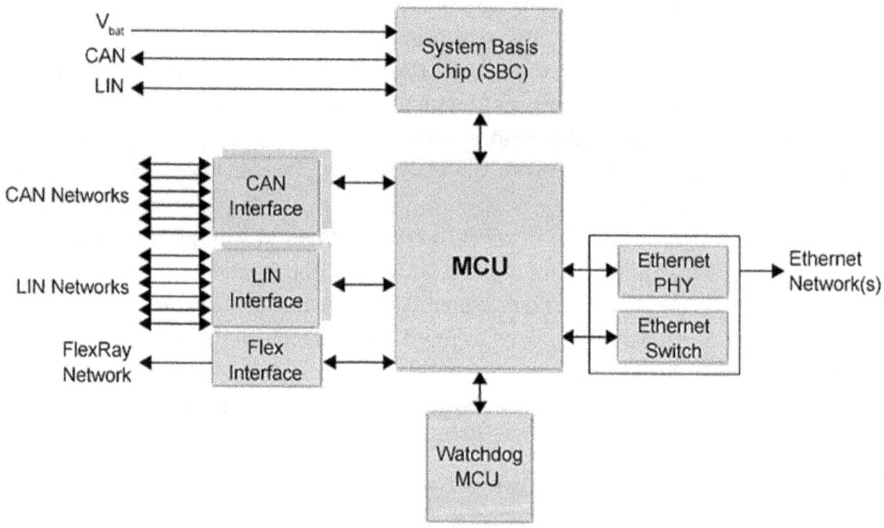

Figure 4.30 Concrete example of Ethernet-based network architecture.

in Figures 4.29 and 4.30. Note that, in these figures, the Ethernet datarate is not limited to 100 Mbit/s, but can actually go as high as 10 Gbit/s.

The numerous reasons to choose Ethernet

The design of such a network is a new paradigm in bidirectional communication between ECUs, connection of the networks in different domains, and transport of different types of data (control data, streaming data, etc.).

- It is possible to organize communication networks in a highly hierarchical fashion, with the main domain controllers (often switches) interlinked by a supervising Ethernet network. In this case, the computers are structured in a hierarchical architecture, where the application domains are linked by a data highway. Ethernet provides all the requisite conditions for such a holistic approach, considering the object to be a component in a bigger whole. Thus, it can be used as a backbone bus topology to link the different application domains, and subnetworks that simply need higher bandwidths;

- The domain controllers in the subnetworks can also communicate using Ethernet, through switches, and using bridges to transmit information between different levels of the network;
- These “switched Ethernet” networks (based on switches and bridges) rely on point-to-point communication, where the available bandwidth is used more efficiently than in networks such as CAN or FlexRay. This concept of switching is used to surpass the boundaries of the domain without wasting time on packaging and sorting the transmitted messages, as would be required with a complex gateway. In addition, this structure offers a solution that can evolve, because each switch port can generally communicate at a rate of 10 Mbit/s, 100 Mbit/s, or 1 Gbit/s, with no change to the upper layers of the protocols (e.g. based on AUTOSAR or similar technology – see Chapter 5);
- The paradigm shift is also sensitive to the way in which a message is transmitted to its destination outside of its domain. Whilst in other complex networks, gateways are needed to perform this function, known and mature IP routing, including switches and routers, is available for backbone networks;
- The IP routing used is completely independent of the subnetworks, and means that a single form of addressing can be used for the whole of the IVN. In addition, IP allows the vehicle infrastructure to connect directly to an Internet network – a trend attributable to the fact that, today, end users want the same level of access to Internet services in their cars as they enjoy at home;
- Another feature of this architecture is that, in principle, there should be only one basic network technology (Ethernet), serving as the foundation for the whole system. The technology therefore needs to be adapted to the different types of communication data, such as diagnostics, video, audio streaming and highly reliable control data. AVB Ethernet and TT Ethernet can provide the different levels of quality of service (QoS), combined with real-time performances and other activities necessary to demonstrate secure coexistence of these different types of data communication on the same network (see Section 5.4);
- A vehicle using hardwired Ethernet offers major advantages over the normally dominant cable technologies, because it is relatively inexpensive, uses a smaller weight of cabling, and provides faster data communication;
- In years to come, access to enormous quantities of data will provide added value for vehicle owners, decision-makers, and repair specialists. Combining these data with the day-to-day behavior of the vehicle an the computational power available in the Cloud will make our cars more reliable, more secure, and more efficient;
- In addition to the data link layer (DLL) and the physical layer (PHY) in automobiles, the upper layers need to be taken into account, as shown in Figures 4.31 and 4.32.

Let us now examine the details of the types of networks used in autonomous vehicles.

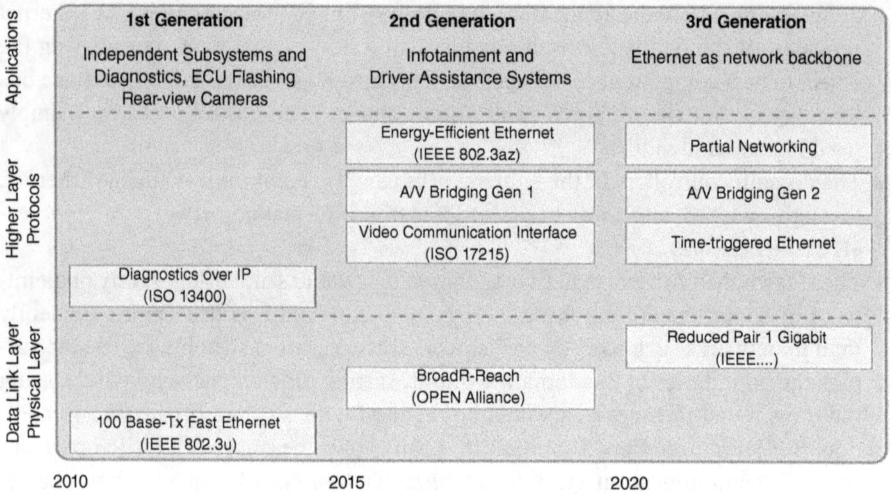

Figure 4.31 Evolution of the content of the different layers in the OSI model between 2010 and 2020.

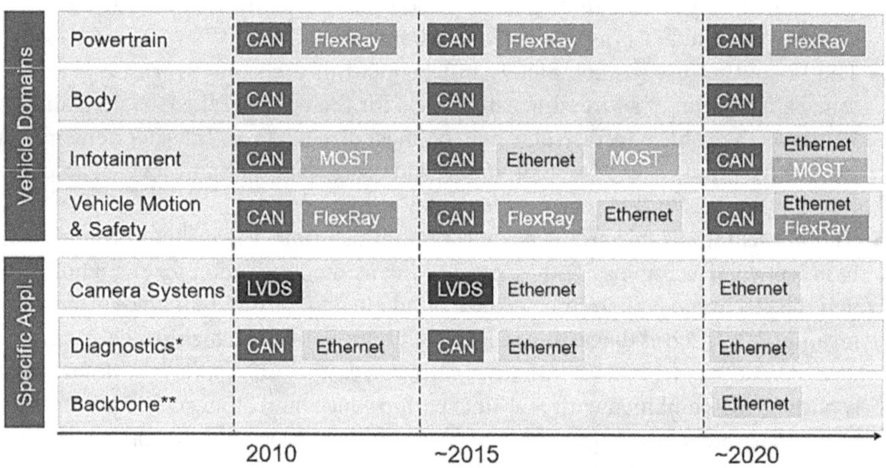

Figure 4.32 Evolution of the content of the application layers between 2010 and 2020 (*Source*: Bosch).

4.5 Types of networks and description of the protocols used in autonomous vehicles

In this section, we shall briefly recap some of the legacy protocols that have been on the market in this area for a number of years. For obvious reasons, this will include things that are already known, so for the sake of succinctness in this book, readers are referred to earlier publications for the details. Here, we shall cover only the novel points or those that are essential to know about when dealing with autonomous and/or connected vehicles.

4.5.1 LIN

LIN (local interconnect network) was developed in 1999. A consortium was set up in March 2000, including the automakers Audi, BMW, Daimler Chrysler, Volkswagen, Volvo Car Corporation and Motorola Inc., and Volcano Communication Technologies AB. The final specification LIN rev. 2.2A has been available since December 2010. It should also be noted that an ISO standard based on this protocol – ISO 9141 – followed, but ISO 17987 Parts 1–7 have been issued. LIN is designed mainly to support the control of mechatronic elements, which are present in automotive applications, but, of course, it can be applied in numerous other fields.

LIN provides a multiplexed communication system. Its level and associated performances are plainly inferior to what we expect from CAN. It is founded on the concept of a (sub-) network, in which there is only one master and a finite number of slave nodes. As the master alone is responsible for managing the network, the communication system is deterministic, because it is entirely dependent on time-sequencing linked to the master node's task management. Thus, LIN's main and original purpose is to provide a "sub-bus" to the CAN network, with reduced capabilities and at a lesser cost. In other words, it provides an economical solution when the required performances are not too great. Thus, it can be used when the bitrate and bandwidth of the network are reduced and the reliability and robustness that CAN offers are not necessary. Note that there are no problems with conflicting and arbitration, etc., for the simple reason that the system works with a single master node and multiple slaves. For example, in an automobile, numerous nodes/participants in a network can use LIN for:

- Controlling the roof (opening or closing it, inclining it, and so on);
- Rain detection, automatically switching on the vehicle lights;
- Controls and functions of the seats;
- Functions at the head of the steering column, controls on the steering wheel, etc.;
- Doors, window controls, rear-view mirrors (positioning, defrosting, etc.);
- Wiper controls;
- Interior lighting controls, etc.

A great many semiconductor producers offer integrated circuits for LIN that have multiple functional options.[3]

4.5.2 SENT

SENT (*Single Edge Nibble Transmission*), specifications for which were published in January 2010 by the Society of Automotive Engineers, entitled J2716, is a serial bus protocol for single-direction, point-to-point communications, used to transmit signal data from smart sensors to an ECU (data transport applications for temperature sensors, high-resolution pressure readings, accelerator position, ventilation volume, etc.). SENT is designed for economic applications where security is critical in automobiles, and it helps replace analog signaling between sensors and microcontrollers. As it is simpler and less expensive, it represents an alternative to CAN Low Speed and LIN.

It should be noted that PSI5, a protocol designed by Autoliv, Bosch, and Continental in the eponymous consortium, is primarily used for safety applications – notably for the airbag systems – so it is not really a competitor, but complements the other protocols. Its position is shown in Figure 4.33.

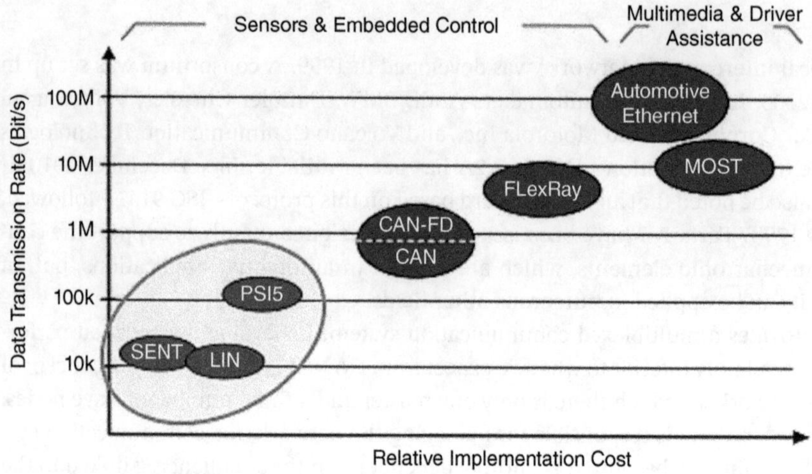

Figure 4.33 Relative positions of the different communication protocols used in the automobile industry.

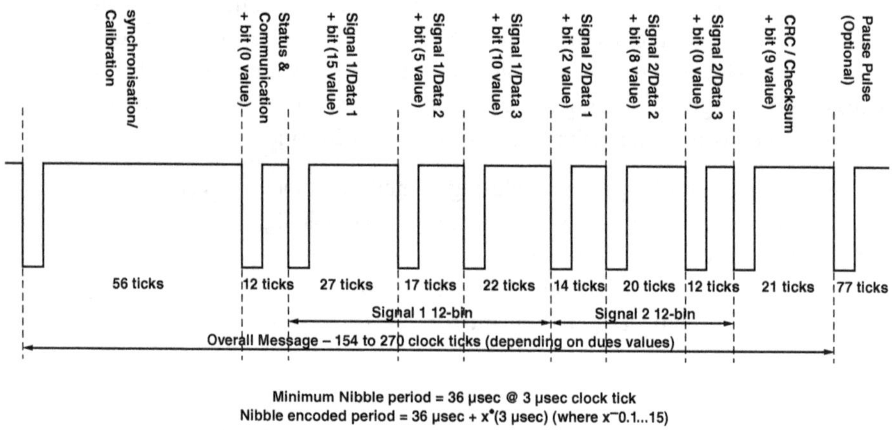

Figure 4.34 Composition of a SENT message.

To operate, the unidirectional SENT requires an asynchronous voltage interface and three wires:

- One line for the signal (low state < 0.5V, high state > 4.1V);
- One line for the power supply (5V);
- One mass line.

SENT message

SENT carries data over four bits (nibbles/quartets). A SENT message is composed as follows (see Figure 4.34):

- A calibration/synchronization pulse (56 clock ticks);
- A pulse corresponding to a nibble (4 bits) containing status information and communication (between 12 and 27 clock ticks);

- A sequence of between 1 and 6 data nibbles (each between 12 and 27 clock ticks), representing the values of the signals that need to be communicated. The number of nibbles is fixed for each application, but can differ from one application to another (if, for example, two 12-bit values are transmitted, 6 nibbles will be communication and 24 bits [6 nibbles] will be signaling data, which represents two channels of 3 nibbles each – e.g. temperature and pressure measurements);
- A nibble for the error detection code (between 12 and 27 clock ticks);
- An optional pause pulse – if used, this compensates for the variable duration of the messages.

A SENT message is usually encoded on 32 bits (8 sequential nibbles). Optionally, the data may be transferred in messages of only 20 bits (5 nibbles), composed of a 12-bit measurement (3 nibbles), a 4-bit (1 nibble) cyclic redundancy check, and a 4-bit (1 nibble) field for status/communication. In accordance with SAE J2716:

- A basic unit of time for SENT communication (clock period being one unit of time [UT], representing the nominal transmitter clock value) may be between 3 and 10 μs (a UT of 3 μs is considered nominal so that fuller descriptions use the same basis);
- The maximum allowable clock variation is 20% of the nominal unit of time, which means that low-cost RC oscillators can be used in the detection device.

SENT is a unidirectional communication standard in which sensor data are sent independently, with no intervention by the data receiver (e.g. the MCU). The total transmission time depends on the data values transmitted and the variation of the clock on the transmitter (in this case, the sensor). The next SENT transmission begins immediately after the previous transmission ends (the last falling edge of the CRC in the SENT transmission is also the first falling edge in the next SENT transmission). A signal transmitted by the sensor consists of a series of pulses, where the distance between consecutive falling edges defines the 4-bit data nibble representing values from 0 to 15.

Explanation of values of data nibbles

A sequence containing between one and six 3-bit nibbles of data (each lasting between 12 and 27 clock ticks), representing the signal values, needs to be communicated (see Figure 4.35):

- 1 nibble = 4 bits = thus 16 different values for the nibble;
- 12 to 27 ticks = 16 values of pulse widths.

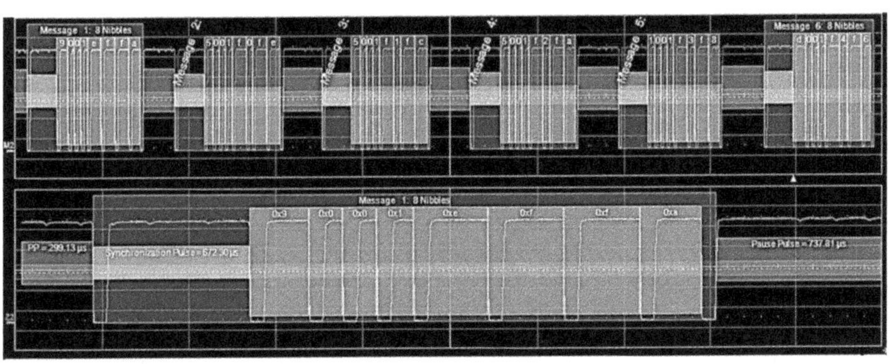

Figure 4.35 Sequence of one to six 4-bit nibbles of data.

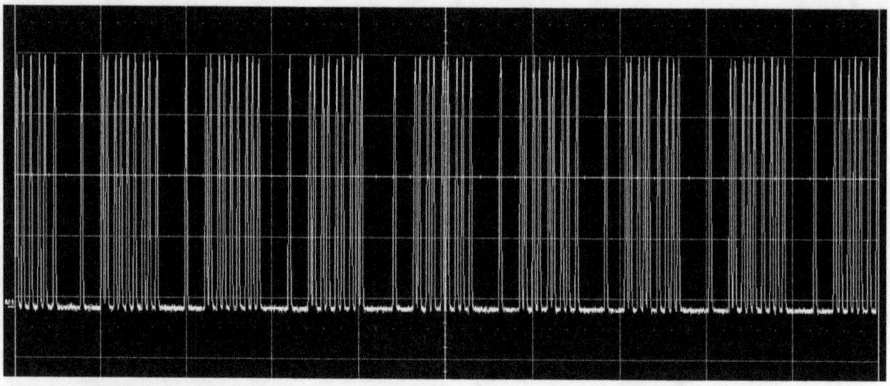

Figure 4.36 Examples of SENT communications with a 13-bit data frame.

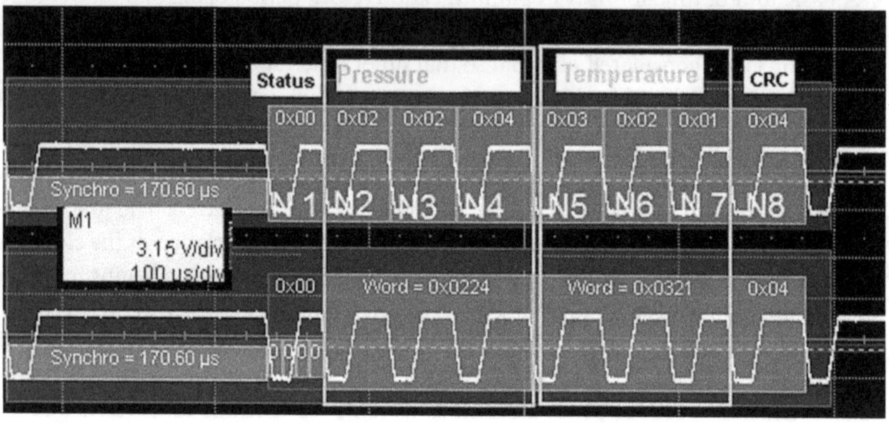

Figure 4.37 Grouping of nibbles conveying pressure and temperature values.

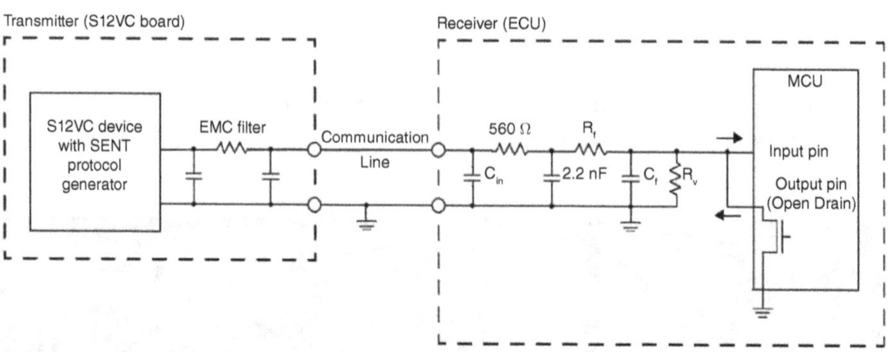

Figure 4.38 Physical layer in SENT.

Examples of the communications are shown in Figures 4.36 and 4.37:

Physical layer

To conclude this section, Figure 4.38 illustrates the physical layer in SENT.

4.5.3 CAN

When it was devised by R. Bosch in 1998, it was decided that CAN – Controller Area Network – should be able to serve all communication applications on board an automobile: in other words, to transport and multiplex various types of messages, ranging from the fastest to the slowest.

Because it was created by an automaker, CAN is designed to work in environments with a great deal of pollution, mainly from electromagnetic interference. In addition to the reliability of transmission that is offered by a high-performance error-detection mechanism, CAN has multi-master functions in order to increase the likelihood of quick recovery once an error is detected. Furthermore, in the event of a bus access conflict, the principle of bit-wise arbitration used allows for non-destructive arbitration when multiple stations are trying to begin transmitting simultaneously. What this means is that in the event of competition, one of the participating stations (whose message has higher priority) will always be granted access to the medium and will then complete the communication on its own. Thanks to this technique, none of the communication capacity (network bandwidth) goes to waste in the event of medium access conflicts. This type of non-destructive arbitration and ranked messages also means that the system can, in real time, easily abide by the response times required for system controls in which the bus datarates are not particularly fast. The drawback inherent in this bit-wise arbitration technique lies in the fact that the maximum network length is directly linked to the chosen datarate: if a high datarate is required, the network length can only be short. Finally, to minimize electromagnetic noise that is due mainly to the fast edges of electronic signals communicating over the bus, the datarate needs to be as low as possible.

ISO standardization

Over the years, the original CAN documents from R. Bosch have been enriched and submitted to the ISO for the development of international standards serving as a point of reference for anyone wishing to adopt the protocol. At present, the main standards refer to applications in automobiles. They are listed in Figure 4.39.

These documents conform, as closely as possible, to the division of the OSI communication model (see Figure 4.40 for the official ISO text).

CAN High Speed and CAN Low Speed Protocols

For details about how CAN High and Low Speeds work, please see our earlier book.[4]

Let us now move on to look at CAN FD.

ISO 11 898	
1	Data link layer and physical signaling (CAN and CAN FD protocols)
2	High-speed medium access unit (< or = 1 Mbits/s)
3	Low-speed (< 125 kbit/s) fault-tolerant, medium-dependent interface
4	Time-triggered communication (TTCAN)
5	High-speed medium access unit with low-power mode
6	High-speed medium access unit with selective wake-up functionality

Figure 4.39 Summary of ISO 11898.

ISO 11 898-x – Road vehicles – Interchange of digital information. This is the generic reference to the CAN standard, which is made up of six documents.

The ISO 11898 series provides specifications for the PHY and DLL (levels 1 and 2 of the OSI model) form CAN technology, for real-time distributed control and multiplexing for use in road vehicles.

ISO 11 898-1	**2015**	***Data link layer and physical signaling*** (including CAN FD) Specifies the data link layer (DLL) and physical signaling of the controller area network (CAN). This document provides the characteristics for setting up an interchange of digital information between modules implementing the CAN DLL with detailed specification of the logical link control (LLC) sublayer and medium access control (MAC) sublayer.
ISO 11 898-2	**2003**	***High-speed medium accessb unit*** Specifies the high-speed (transmission rates of up to 1 Mbit/s), medium access unit (MAU), and some medium dependent interface (MDI) features (according to ISO 8802-3), which include the physical layer of the controller area network. ISO 11898-2 uses a two-wire balanced signaling scheme. It is the most used physical layer in vehicle powertrain applications and industrial control networks.
ISO 11 898-3	**2006**	***Low-speed fault-tolerant medium-dependent interface*** Specifies low-speed, fault-tolerant, medium-dependent interface for setting up an interchange of digital information between electronic control units of road vehicles equipped with the CAN at transmission rates above 40 kBit/s up to 125 kBit/s.
ISO 11 898-4	**2004**	***Time-triggered CAN*** Specifies time-triggered communication in the CAN (TTCAN). It is applicable to setting up a time-triggered interchange of digital information between electronic control units (ECU) of road vehicles equipped with CAN, and specifies the frame synchronization entity that coordinates the operation of both logical link and media access controls in accordance with ISO 11898-1, to provide the time-triggered communication schedule.
ISO 11 898-5	**2007**	***High-speed medium access unit with low-power mode*** Specifies the CAN physical layer for transmission rates up to 1 Mbit/s for use within road vehicles. It describes the medium access unit functions as well as some medium dependent interface features according to ISO 8802-2. This represents an extension of ISO 11898-2, dealing with new functionality for systems requiring low-power consumption features while there is no active bus communication.
ISO 11 898-6	**2013**	***High-speed medium access unit with selective wake-up functionality*** Specifies the CAN physical layer for transmission rates up to 1 Mbit/s for use within road vehicles. It describes the medium access unit functions as well as some medium dependent interface features according to ISO 8802-2. This represents an extension of ISO 11898-2 and ISO 11898-5, specifying a selective wake-up mechanism using configurable CAN frames.

Figure 4.40 Standards governing the lower layers of the OSI model for CAN.

Tests		
ISO 16845-1	2004	Provides the methodology and abstract test suite necessary for checking the conformance of any CAN implementation of the CAN specified in ISO 11898-1.
ISO 16845-2	2014	Establishes test cases and test requirements to realize a test plan verifying if the CAN transceiver with implemented selective wake-up functions conforms to the specified functionalities. The kind of testing defined in ISO 16845-2:2014 is named as conformance testing.

Figure 4.40 (*Continued*)

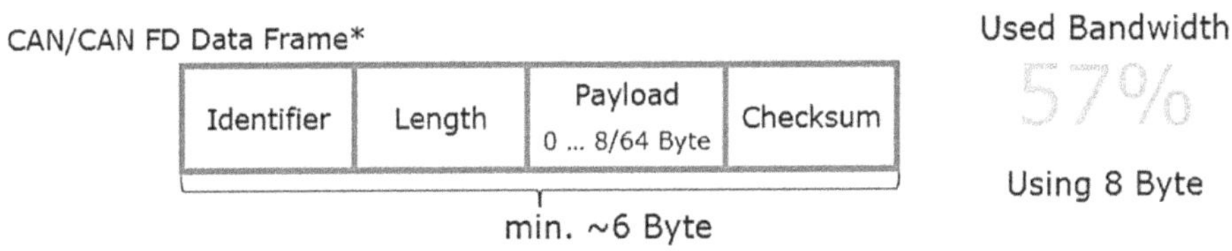

Figure 4.41 Payload/bus occupation in commonplace applications.

CAN FD

Officially published in 1986, "CAN 2.0", which is described by the ISO 11898 series of standards (1 to 6), included structural and inherited limitations. At the beginning of the new millennium, users' expectations began to approach these limits. The main two reasons for the intrinsic limitations of the CAN protocol are:

- The maximum value of its datarate (1 Mbits/s);
- The consequent possible distance over which it can perform properly.

Let us briefly examine both of these points:

- CAN messages contain an overhead part (including the header, the acknowledgement, etc.) that is quite considerable (≥ 50 %). This means that only around 40–50% of the frame format is used to carry the message payload – for example:
  - CAN 111 bits/message for 64 bits of data
  - extended CAN 131 bits/message for 64 bits of data

  At present, the majority of in-vehicle CAN networks have reached the maximum values of data transfer, with payloads/bus occupancy rates ranging from 50 to 95% (see Figure 4.41);
- the second limitation of CAN is closely connected to one of its most significant advantages: its non-destructive arbitration over network access and its specific acknowledgement mechanism. Indeed, at a given bitrate, the maximum distance that can be covered by a CAN network is limited because of the mechanism used during generation of the ACK bit (acknowledging correct transmission of the message). This acknowledgement mechanism means that the total signal propagation time on the bus (return journey) between the furthest two nodes must be less than the duration of a bit. In the case of a CAN bitrate of 1 Mbit/s, this means a maximum distance of around 40 m; the distance also depends on the physical characteristics of the cabling used in the vehicle. For information, most conventional in-vehicle networks have datarates ≤ 500 kbit/s.

The rising complexity of automobile systems means that the payloads are rapidly approaching, and sometimes surpassing, the limit of bandwidth that a CAN 2.0 (High Speed) network is able to support. This is not acceptable with a non-deterministic technology. This problem has been solved either by increasing the number of CAN networks present in a vehicle or by switching to other protocols; the latter solution, though, entails a great deal of effort in system design and replacement (which is often costly) of the hardware and software already in place.

Possible solutions

Of course, as is always the case, multiple concepts/solutions capable of handling higher datarates have been developed and been suggested to replace CAN and are certainly feasible, either by improving what already exists or by developing new generations of bus systems and a host controller interface (similar or different to what is currently in place). We shall now briefly examine these two options and the consequences they wreak.

Back to the drawing board Let us begin with the most radical of solutions. We discard everything that is currently in place, looking a long way into the future (to around 2025–2030), and decide to radically change the system and the technology. This solution means starting from scratch, with a completely blank slate in relation to current technology. In addition, this hypothesis in unlikely to come to pass in the world of industry, for a number of very prosaic reasons:

- It is necessary to continue to make money, to survive, during what would assuredly be a lengthy transition;
- The development, industrialization, validation, etc., would take an extremely long time;
- In the long term, developers would need to ensure synchronization and matching of multiple schedules between incoming data, and the availability of certain motor components, electrical systems, etc., and also deal with the emergence of new technologies, new competing models, etc., on the market;
- They must ensure that the use of higher operational datarates has little or no effect on the vehicle's power consumption, in relation to current solutions (for information, 100 W of electricity consumed by an ECU is equivalent to the consumption of 0.1 liter of gasoline per 100 km, or 2.5 g of CO_2/km); and
- Finally, enormous sums of money would need to be invested to bring the ideas to fruition.

Smooth transition In principle, a smooth transition to a higher datarate and a few additional improvements should cost nothing. If it were to be costly, we can rethink the whole architecture and revert to, say, Ethernet, or any other solution. Thus, the aim is to gradually migrate from old systems (such as CAN) to new ones (such as CAN FD). In this case, all the nodes in the (new) networks must be equipped with (new) compatible controllers. However, this often means that they must also be able to communicate using the older standard.

In addition, this gradual transition and smooth migration mean that, for a certain (fairly lengthy) period of time, there will be a mix of the old and new technologies. This also means we must look at all the implications and the impact of old and new networks on one another, and also juggle with so-called partial network segments. Where the signal is passed from one type of network to the other, the transition runs through gateways, and the impact of one network on the other can be quite significant (often, the actions/interactions of a CPU have an effect on the messaging of two or three other adjacent networks/sub-networks). Consequently, we need to re-validate numerous CPUs, representing a considerable investment of both time and money, and, once again, we glimpse the prospect of temporary asynchronism between the time taken to develop a new vehicle and the time taken to develop systems and computer solutions.

These solutions are obviously not as advanced as those mentioned in the foregoing sections, but this technical leap will be less expensive and less risky, the result will remain compatible with existing technology, and will allow us time – time to see how new high-datarate technologies evolve on the ground.

Solving the problem

Many automakers have chosen the less risky strategy of migration, and have chosen to put back the date of their full-Ethernet "big bang" somewhat. During this migration on the market, high-datarate communication is generally limited to specific use cases, such as downloading boot-up and/or maintenance software. During these operations, other nodes that cannot support the migration are kept in sleep mode or on standby. In addition, if the new (faster) forms of communication are also able to limit the data fields to a length of eight octets, as before, there will be no need to alter the application program beyond the initial configuration of the controller.

CAN FD

To bridge the gap in performances with other protocols, R. Bosch, the company that developed CAN, published a version 1.0 of the specification "CAN with flexible data-rate, or CAN FD" in 2011. The specification was presented publicly in April 2012, at the Thirteenth international CAN Conference (iCC), focusing on CAN in Automation (CiA).

The transition to CAN FD represents more of an evolution than a revolution. We can briefly list the properties of CAN FD. Its development was based upon:

- The communication protocol CAN 2.0, described by ISO 11898-1;
- Maintaining the physical layer of CAN (High Speed), and remaining closely compatible and interoperable with ISO 11898-2;
- Tacit (backwards) compatibility with that standard;
- The requirement to increase the network's datarate (in terms of the number of octets that can be transported and the digital datarate);
- The need to support two different bitrates in the same message;
- Preservation of the arbitration phases at the same bitrate as CAN 2.0;
- Consequently, only the protocol controllers may need to be reinforced with CAN FD;
- Given the desire for similarity, the aim of preventing or reducing the need for major software modifications;

- The desire to keep the hardware unchanged as far as possible;
- Compatibility with the EMC regulations.

Given these aims, the structure of CAN FD has led to an approach consisting of notably increasing the digital datarate for the payload, which is between the network access and arbitration phase and the acknowledgement phase. This means it is possible to:

- Keep the start-of-frame structure essentially the same;
- Use CAN's arbitration method;
- Increase the data bandwidth by making a few significant changes to the frame format;
- Include a few additional bits in the CRC field, which open up new possibilities;
- Increase the effective datarate (> 1 Mbits/s up to around 8 Mbits/s), solely in the data field of the frame, by reducing the bit time after arbitration;
- Improve the payload signaling header, so that a larger payload than before can be sent (> 8 octets and up to 64 octets per frame) and thereby reduce the header-to-payload ratio, in view of the longer payload field;
- Improve the CRC sequence, so we can obtain longer frames, whilst preserving the same Hamming distance as that of the existing CAN protocol;
- Preserve the end-of-frame structure compatible after return to a longer bit time, identical to that at the start, at the transmission of the CRC Delimiter, before the receivers send their ACK bits, as before, by the same mechanism.

We shall now examine the structure of a CAN FD frame, as we have done in earlier works for standard CAN.

ISO standardization of CAN FD

When the ISO was drawing up a standard for it, the original "CAN FD" protocol, designed by R. Bosch, was slightly modified, in relation to:

- Its error detection capacity (in particular, a 3-bit stuff-bit counter and a parity bit were added);
- The way in which the CRC is computed.

Therefore, Bosch's original CAN FD is no longer precisely the same as the ISO's CAN FD. These improvements mean that "ISO CAN FD" is incompatible with the original Bosch CAN FD. Where relevant, to avoid misunderstandings and confusion, the CAN in Automation (CiA) working group (and we authors) recommend using the terms "ISO CAN FD" for the ISO version and "non-ISO CAN FD" for the Bosch version.

The "ISO CAN FD" frame is an improvement to the existing CAN frame, which it replaces. To unify the terminology and prevent inconsistency, the ISO has amended the existing standards ISO 11898-1 and -2 (CAN High Speed physical layer), and -5 and -6 have been merged to form ISO 11898-2:2015. In addition, all products complying with ISO 11898-1 and 11898-2:2015 must be labeled "ISO CAN FD" and clearly indicate whether they are "ISO CAN FD" or "non-ISO CAN FD".

Description of the ISO CAN FD frame

It should be noted that, for brevity's sake, hereinafter we shall simply write CAN FD, in the knowledge that we are talking about "ISO CAN FD".

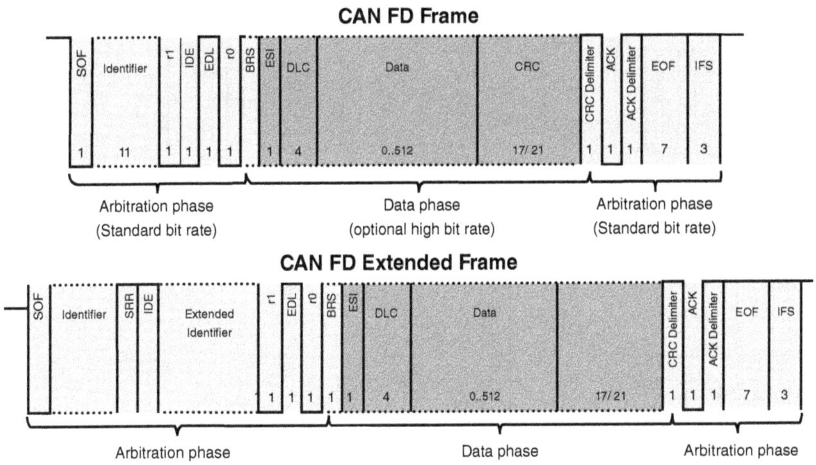

Figure 4.42 General layout of a CAN FD frame.

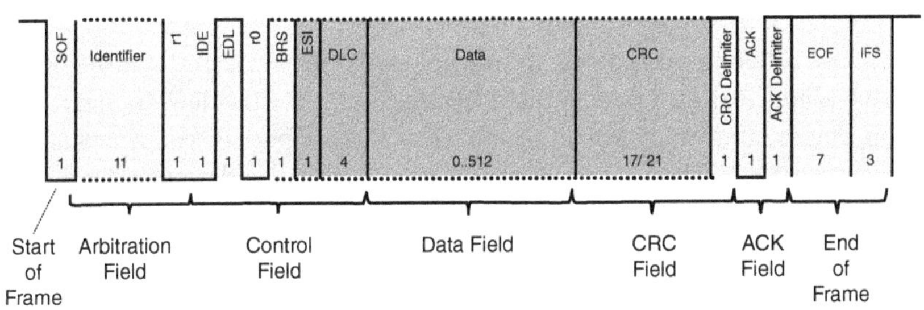

Figure 4.43 The seven fields in a CAN FD frame.

As Figure 4.42 shows, a CAN FD frame is made up of three phases: two arbitration phases (one at the start and one at the end of the frame), both at the standard CAN datarate, and a third data phase in the center of the frame, which may be at a higher datarate. All three strongly resemble those used in standard CAN, but with certain exceptions, of course.

In addition, as shown in Figure 4.43, we find the seven fields used in a conventional CAN frame: namely, SOF (start of frame), Arbitration, Control, Data, CRC, ACK, and EOF (end of frame).

SOF The start of frame is entirely unchanged (see Figure 4.44).

Arbitration The identifier fields, which are 11 or 29 bits long, and serve mainly for arbitration between participants and for network access, are unchanged, depending on whether we are operating in standard CAN or extended CAN. The slight differences between CAN and CAN FD now begin to show (after arbitration) (see Figure 4.45), because the format of the new frame uses certain bits reserved for CAN and new bits. Therefore, the three CAN bits between the identifier and the data length

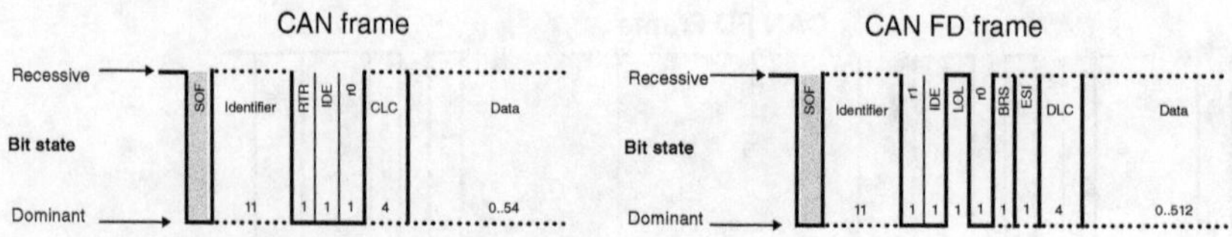

Figure 4.44 Comparison of SOF fields.

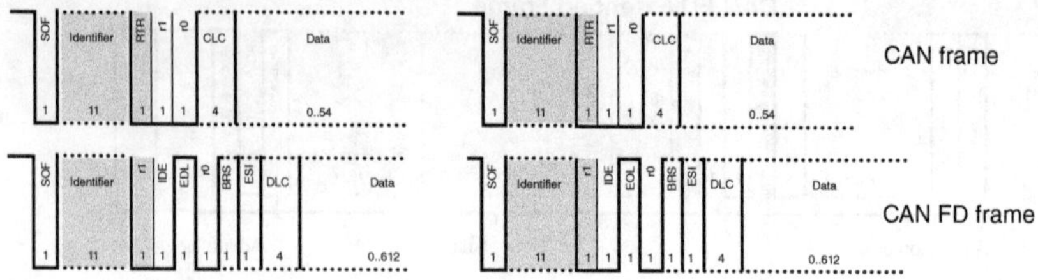

Figure 4.45 Comparison of arbitration fields.

code (DLC) have different names and are expanded. By means of this mixture of old and new bits, a node can distinguish the different formats of incident frames when receiving messages, allowing CAN FD controllers to take part in communications using conventional CAN. This also facilitates the establishment and progressive migration of nodes from CAN to CAN FD.

Let us look at the structure of these new bits in CAN FD:

- The **RTR bit** has been removed, because the remote frame function is rarely used. When necessary, these frames are transmitted as ordinary CAN remote frames, without enhanced speed;
- The **"r1" bit** is still dominant and installed instead of the RTR bit;
- The EDL bit (extended data length) has been added instead of the reserve bit r0 from CAN:
  - EDL = dominant indicates a standard CAN format frame;
  - EDL = recessive indicates a CAN FD format frame (with new DLC and CRC).
- The bit r0 (reserved bits): "r0" is transmitted as dominant and it is again reserved for future variants of the protocol.
- The BRS (bit rate switch) bit:
  - BRS = dominant indicates that the bitrate does not change;
  - BRS = recessive indicates that during the data phase, the bitrate is raised, but says nothing about the value of the new datarate chosen, which is limited only by the features of the low-pass filters of the line drivers.

  As of the sample point of the BRS bit, as shown by Figure 4.46, the CAN FD controller reduces the internal value of its *time quantum* and the nominal duration of the CAN bit *nominal bit time* that has been in place up to that point is ready to change to a new, shorter value called the *data bit rate* (see below), in accordance with the CAN FD datarate chosen by the controller.
- The **ESI** (error state indicator) bit:
  - ESI = recessive indicates that the transmitter node is in an *error passive* state;
  - ESI = dominant indicates that the transmitter node is in an *error active* state.

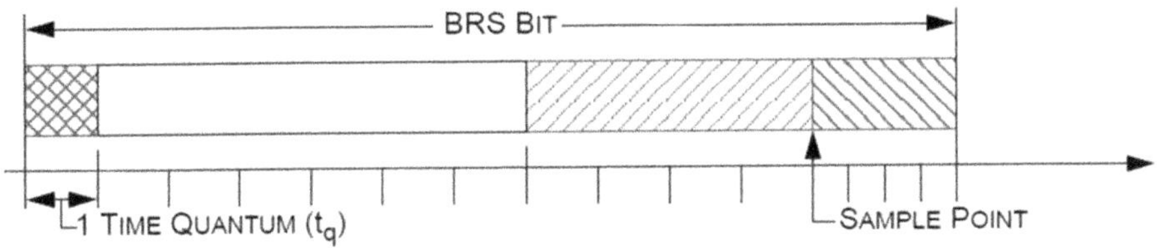

Figure 4.46 Definition of a sample point.

CAN standard			**CAN FD**		
EDL bit (Extended Data Length)	Number of octets (payload)	DLC binary values	Number of octets (payload)	EDL bit (Extended Data Length)	
Dominant	00	0000	00	Recessive	
	...	...	...		
	07	0111	07		
	08	1 000	08		CAN FD mod1
	08	1001	12		Often called CAN FD mod2
	08	1010	16		
	08	1011	20		
	08	1100	24		
	08	1101	32		
	08	1110	48		
	08	1111	64		

Figure 4.47 Values of an extended data length bit.

Data length code (DLC) and data field The data length code (DLC) field, which comes next, is made up of four bits, representing 16 possible values. These bits are used for both CAN and CAN FD formats. To ensure compatibility with the current incarnation of CAN, the first nine values of the DLC, codes 0 to 8 (both inclusive), are used to indicate a data length of 0 to 8 octets inclusive. In addition, standard CAN (non-FD) ignores the three least-significant bits (LSBs) beyond DLC = 8. This means that binary codes 9–15 also indicate a payload of 8 octets.

On the other hand, in CAN FD, for data lengths greater than or equal to 8 octets, the DLC field uses all binary values achievable with four bits, from 1001 to 1111, to define stepwise discrete data lengths as presented in Figure 4.47.

Data field In CAN FD, the data field is either 0 to 8 octets, as it is in standard CAN standard, or discrete values of 12, 16, 20, 24, 32, 48, or 64 octets that are transported in accordance with the principle of *MSB first* (most significant bit first) (see Figure 4.48).

Another tricky point in relation to the data field arises due to the change of digital datarate, which was only slight beforehand, when the BSR bit is executed. In principle, the position of the sample point is different in the two configurations of the

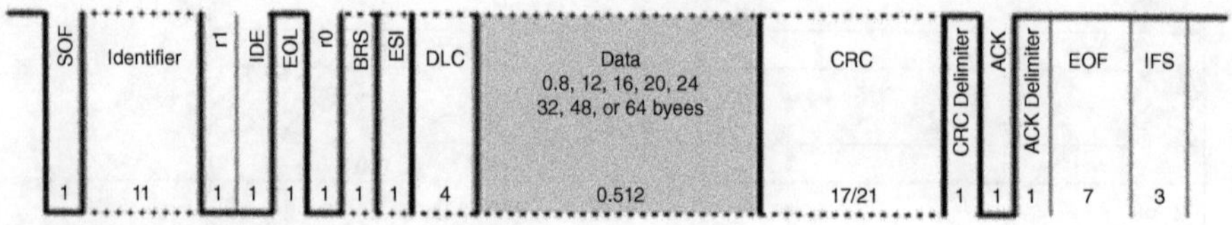

Figure 4.48 Data field.

Chosen values			Bit time parameters		
		Prescaler m	PROP_SEG	PHASE_SEG1	PHASE_SEG2
CAN	Nominal bit time	2	6	4	4
CAN FD	Data bit time	1	1	4	4

Figure 4.49 Bit time parameters.

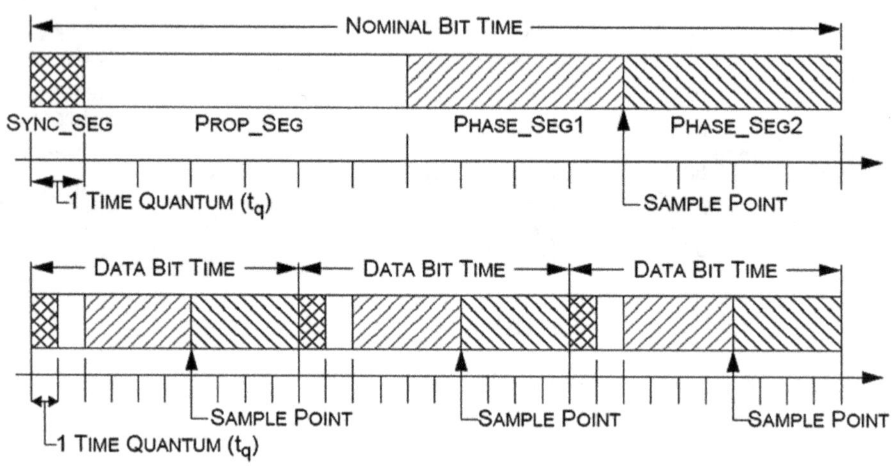

Figure 4.50 Comparisons of the position of the sample point between the two datarates.

timing bit (normal bit time in CAN and data bit rate in CAN FD), because the bit time is reduced. To restore a reasonable proportion to set the sample point, the length reserved for the prop_seg must be reduced if the data bit rate is set too fast. In addition, when the bitrate is reversed when the BRS bit is executed (or, conversely, when the system returns to the slower bitrate when the CRC delimiter bit is executed), the bitrate must be switched immediately after the sample point, making the length of these two bits momentarily "intermediary". The sum of the length of these two bits must be the same as the sum of one bit of the nominal bit time plus one bit of the data bit time. When the datarate is switched because an error condition has been detected, the switch time can be shifted to after the sample point, by the length of the information processing time. Figures 4.49 and 4.50 illustrate the above remarks, showing an example of switching when a particular division of the bit time has been chosen.

Non-technical but useful remarks CAN FD offers three main benefits:

- The ability to communicate with longer data fields than before (adjustable stepwise from 8 to 64 octets, representing a maximum of 512 bits), in order to switch the EDL bit to a recessive state;
- The ability to transfer those data as quickly as possible (rate not currently defined), in order to switch the BRS bit to a recessive state; and
- Finally, by the same device, the ability to improve the payload/overhead ratio and the network occupation time.

In CAN FD, we can certainly continue to transport packets of only 0–8 octets (granted, faster than before), but to do so would seem somewhat wasteful, and does not offer much of a gain in terms of network occupancy time, given the low payload/overhead ratio. All of this means that with CAN FD, it makes sense to transport the maximum octets per frame possible (the maximum being 64). As if at random, the latest applications are often linked to transmissions of "voluminous" files, which are often divided or fragmented into pieces of 64 octets each – i.e. often catering for the requirement to download files when personalizing or setting up a vehicle.

Up until now, we have not mentioned the new value of the "fast" datarate during the data phase. We shall now see, over the course of a brief discussion, where the (physical) limits in terms of maximum network datarate lie. However, if our stated goal is to avoid slowing down the messaging over the network, and for CAN FD to retain the same maximum duration for the data field as in a standard CAN frame, then we need to be able to transmit 64 octets of a CAN FD frame in the same space of time as the former 8 octets making up a CAN frame. This means a payload 8 times greater, so the bitrate must be increased by a factor of 8 during the data phase. In other words, the network must be capable of a momentary bitrate of 8 Mbits/s (see the bottom section of Figure 4.51).

CRC of the frame and CRC delimiter

CRC field It is necessary to take certain additional measures, notably in relation to those due to the possibility of new payload lengths (more bits to be transported

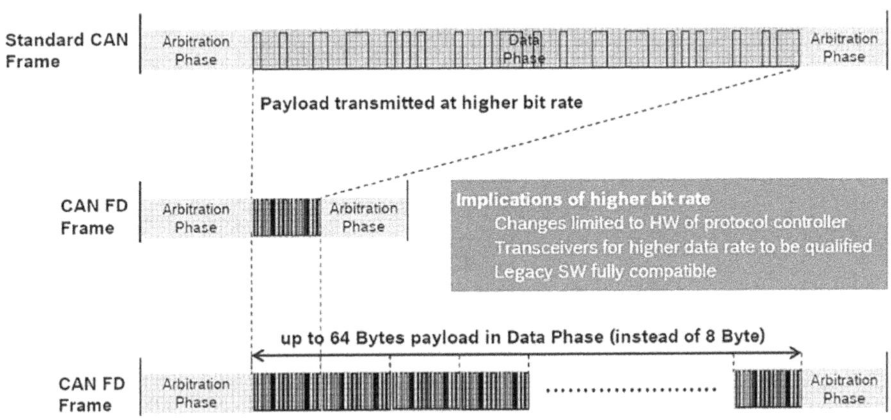

Figure 4.51 Comparison of CAN and CAN FD.

– review sections on DLC). Thus, the polynomial used for calculation (BCH) of the CRC must be modified and the CRC field length adjusted (see Figures 4.52 and 4.53) as a result, to maintain a Hamming distance (6) identical to that of a standard CAN frame.

CRC delimiter bit

The CRC delimiter bit is a very special bit, coming just before the ACK bit which – for reasons of the return journey of the signal, depending on the length of the existing network – must return to the normal CAN for the duration of the CRC delimiter bit. We must revert from fast mode to normal mode, and quickly recalibrate the sample point on the normal bit duration. Thus, there is internal juggling that is very specific to the CAN FD processor. It is beyond the remit of this book, but we strongly encourage interested readers to consult Bosch's website[5] for numerous details (see Figure 4.54).

Some remarks about the CRC and the CRC delimiter:

- Unlike CAN, which does not take account of stuff bits in calculating the CRC, CAN FD takes account of these (new) stuff bits;

	Data fields in the frame	CRC	Polynomials used to calculate the CRC
CAN	0 to 8 octets inclusive	15-bit	g15 = × 15 + × 14 + × 10 + × 8 + × 7 + × 4 + × 3 + 1
CAN FD	0 to < or = 16 octets	17-bit	g17 = × 17 + × 16 + × 14 + × 13 + × 11 + × 6 + × 4 + × 3 + × 1 + 1
	17 to 64 octets inclusive	21-bit	g21 = × 21 + × 20 + × 13 + × 11 + × 7 + × 4 + × 3 + 1

Figure 4.52 Polynomials used to calculate the CRC in CAN and CAN FD.

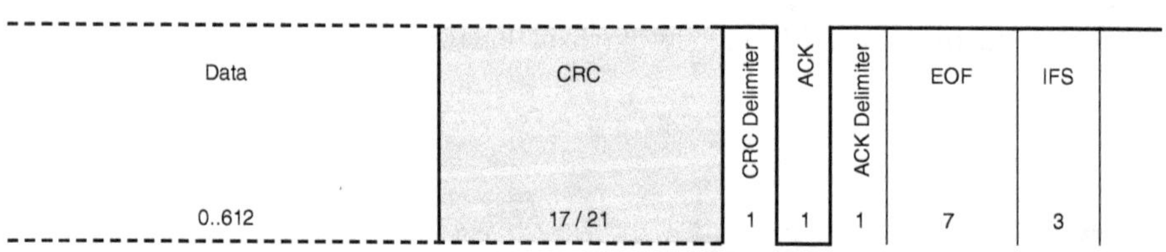

Figure 4.53 CRC field.

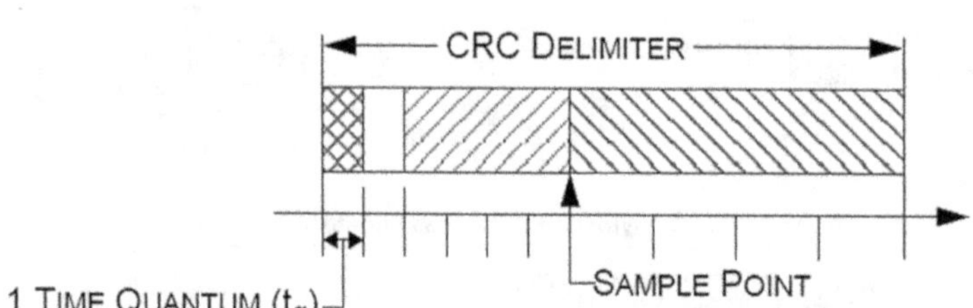

Figure 4.54 CRC delimiter bit.

- The CRC delimiter is transmitted as a single bit. However, because of phase shifts, among other things, the receiver can accept CRC delimiters of a duration of two bits;
- The specific moment when the data phase in CAN FD ends is situated at the same time as the sample point of the first bit in the CRC delimiter.

Chronogram of a CAN FD frame Finally, Figure 4.55 shows a couple of examples of chronograms for CAN FD frames.

Comparison of performances between CAN and CAN FD To conclude, the table below (Figure 4.56) compares the average datarates of CAN FD that we can expect, on the basis of the data payloads transported and bitrates in the arbitration and FD phases.

Reality

The stated purpose of CAN FD is to keep "everything the same" – notably, the same bundles/strands (wires, twists, crosstalk, etc.) except perhaps changing the line drivers. Thus, a number of problems arise.

Bit integrity The minimum bit duration is reduced from 1 µs (for a maximum datarate in standard CAN of 1 Mbit/s) to 1/8 of a µs = 125 ns (for example, at a datarate of

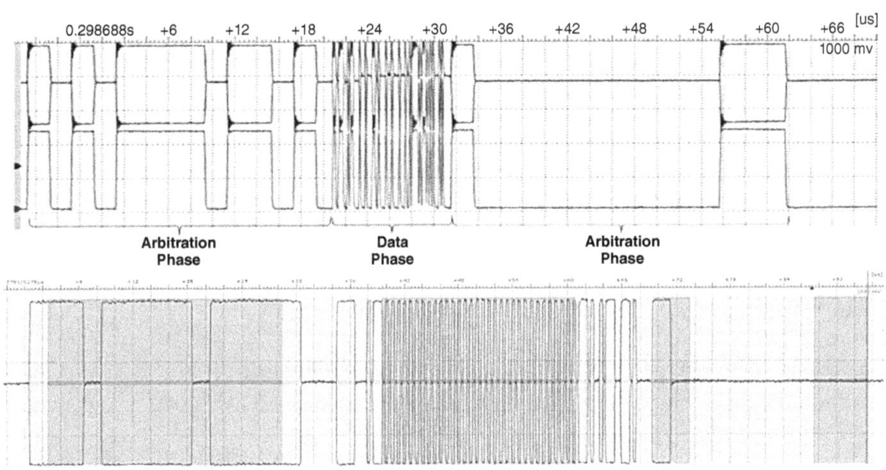

Figure 4.55 Examples of chronograms for CAN FD frames.

	Number of octets of data	Bitrate			
		Arbitration	FD option	Average	Frame duration
CAN	8	1 Mbit/s	-	-	111 µs
CAN FD	8	1 Mbit/s	4 Mbit/s	2.3 Mbit/s	50.75 µs
CAN FD	8	1 Mbit/s	8 Mbit/s	2.9 Mbit/s	39.875 µs
CAN FD	64	1 Mbit/s	4 Mbit/s	3.5 Mbit/s	163.75 µs
CAN FD	64	1 Mbit/s	8 Mbit/s	5.9 Mbit/s	96.375 µs

Figure 4.56 Average datarates of CAN FD as a function of the data transported and the bitrates.

8 Mbit/s in CAN FD). All other things being equal (the wires' payload capacity, network lengths, and topologies, etc.), the rise time and fall time are (fairly considerably) altered, so the electrical signals will be altered, and therefore so will the durations of the reconstructed bits. The temporal ranges of bit sampling and the positions of the sample points for measurement of the bits will be affected. In addition, it is a short step from there to saying that the eye pattern of the signal is disrupted, and thus that the BER (bit error rate) is also reduced (independently of the modification of the value of the CRC, which is constructed only to correct the new number of bits in the data field, but which has a significant impact on the signal integrity). In short, all these elements must be carefully checked.

Sample point Standard CAN defines the bit time, bit value, and different bit segments. In addition, we have discussed where to position the sample point in order to read the value of the bit. Unfortunately, by speeding up the digital datarate, it is not possible to keep this ensemble completely the same, because certain aspects and time parameters can be compressed (bit time, because this is a matter of choice), some are not easily compressible (partially, the resynchronization segments for a given time quantum), or simply cannot be compressed (the signal return journey time over a network of a given length), etc. The bit shape, which is theoretical itself (see Figure 4.57), is significantly affected and deformed by the intrinsic properties of the medium used, the rise and fall times of the electrical signals, and the relative asymmetries as a function of the nominal values and the hysteresis of the "upper" and "lower" measurement thresholds, and edge decisions indicating the presence of bits (see Figure 4.58, showing NXP-Freescale, at 500 kbits/s, 2 Mbits/s, and 8 Mbits/s), which often leads to major dissymmetry (between +10% and –20%) between the durations of the resulting recessive and dominant bits.

This means we need to review the copy in the signal processing part of the CAN FD (micro-)controller, one value of the bitrate at a time (2, 4, 6, 8 Mbits/s) in relation to the new temporal positions of the sample points, and thus in relation to the latent errors this can cause.

Radiation With CAN FD, independently of the signal integrity, it is also important to mention the problem of RF pollution that can be engendered by the electrical signals,

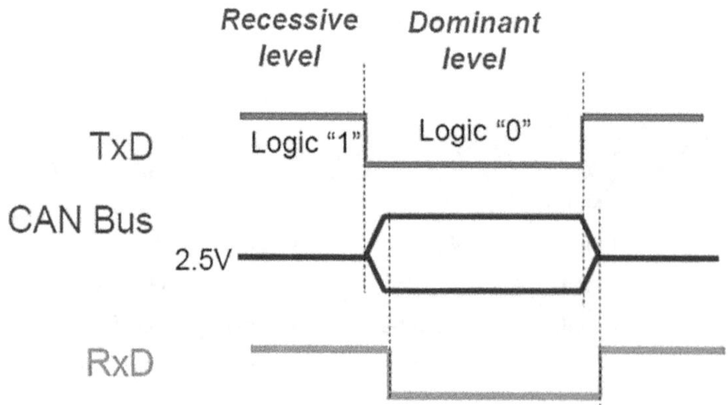

Figure 4.57 Bit deformation.

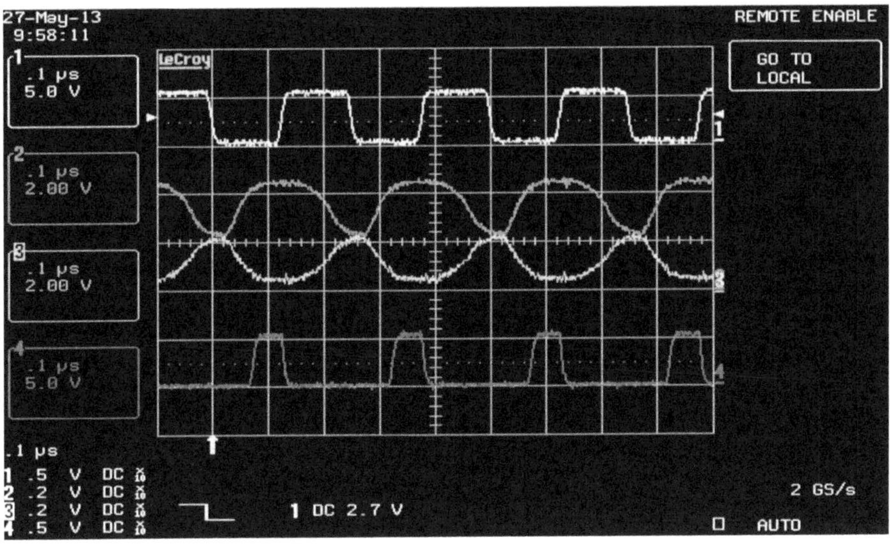

Figure 4.58 Concrete examples of bit deformation.

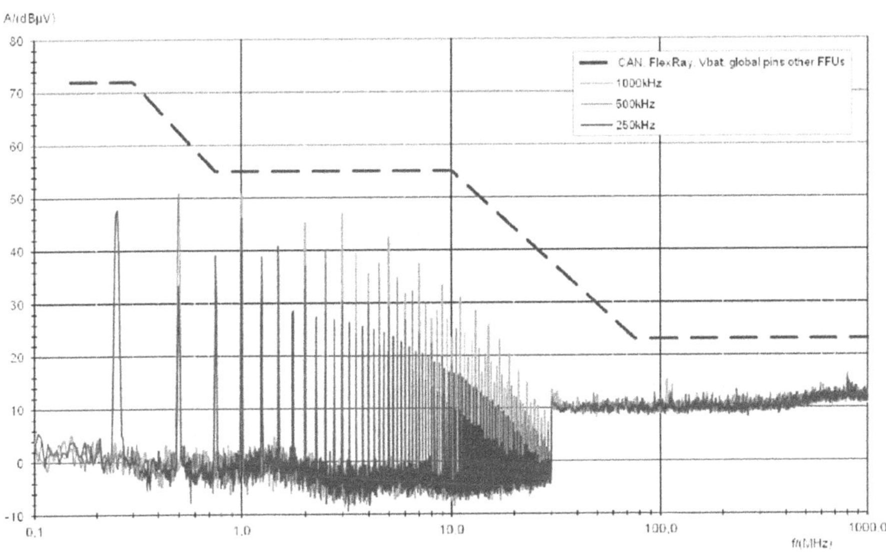

Figure 4.59 Position of radiation in relation to the mask (example TJA 1044).

either in driven or radiated mode, because again, on the existing support (the current cabling), the bitrate (frequency) changes, and the signal shapes evolve. In addition, the maximum datarate for standard CAN (1 Mbit/s) was chosen because, just at that datarate, there began to be tangents to the RF pollution masks, in view of the shape and amplitude of signals for CAN High Speed – ISO 11898-2 – in an unfiltered network (see Figure 4.59). Very frequently, it was necessary to add some type of external low-pass filter (a choke, a capacitor, etc.) to comply with the standards.

Similarly, whilst FlexRay has managed to climb to 10 Mbits/s with the same RF pollution mask, it is mainly because the swing of the signal amplitude on the differential twisted pair is lesser. However, the possible datarates for CAN FD are between 1 and 10 Mbit/s, but with a voltage swing that is greater in CAN FD than it is in FlexRay. The question is then the limit to which we can raise the bitrate: 5, 6, or 8 Mbits/s. To what degree can we round the signals to comply with the masks without compromising signal integrity and without significantly increasing the cost of the solution?

Impact of network topology In order to increase the bitrate of CAN FD and prevent problems due to the response time limit in the ACK frame, alternative solutions have been examined in terms of the network topologies – star or tree structures – in which the arbitration (or message routing) is achieved by an active star. With this idea in mind, the CiA demonstrated that the bus topologies produce better results than star and/or mixed topologies.

Passive, active, and partial networks, and coexistence

Let us now take a look at the concepts of "passive," "active," and "partial networks" and "coexistence of CAN and CAN FD" on the same networks (Figure 4.60).

Components for CAN FD

We shall now offer a brief, and non-exhaustive, overview of integrated circuits (as at mid-2020), products/families of microcontrollers, transceivers, and line drivers for the physical layers of the multiplexed networks CAN and CAN FD.

Line drivers/transceivers In order to present the physical details of CAN and CAN FD networks, remaining impartial, we have selected examples of transceivers from NXP's range of products, but there are many others available on the market.

Transceivers for CAN FD active First, let us mention the successors of the classic CAN drivers, such as TJA 1044 and 1057 from NXP (guaranteed up to 1 Mbits/s). It is the components TJA 1044GT and TJA 1057GT that are variants of the ancestor circuits for active CAN FD applications. The loop delay symmetry is guaranteed up to 2 Mbits/s across the full range of uses in automobiles (temperature measurement, etc.). In addition, the propagation delay is improved, tested, and guaranteed, so longer cables can be used.

Partial Networking Transceivers for CAN FD passive The TJA1145 FD transceiver is an improvement and a pin-for-pin replacement of the current "CAN Partial Networking" driver TJA1145 A. This circuit is designed for use in CAN FD applications with partial networks, in accordance with ISO 11898-6 (so with selective wake-up functions). What, then, is the idea of this improvement? It is quite simple, and becomes apparent when we look at the notion of "CAN FD passive."

In principle, a CAN FD protocol controller is also able to communicate using standard CAN. Thus, it is possible to use CAN FD controllers in specific operational modes – for example, to download software at the end of the assembly line or update it during vehicle maintenance, while other controllers that do not support CAN FD are placed, and kept, on standby. At present, if we have classic TJA 1145 A transceivers in a CAN network, they are placed on standby and we can take advantage of the semi-operative state of this portion of the network to carry out some work on the rest of the network

Network type	Features	Comments
CAN PN (CAN Partial Network)	– On the same network, for each timeslot, certain nodes are placed on standby or in sleep mode, while the nodes that are still awake operate (if possible) without disturbing those that are asleep. – Then, all or some of the sleeping nodes are awakened to participate in the network once more, and other nodes can be placed in sleep mode.	"Partial Networking" transceivers, as defined by ISO 11898-6 to enable selective waking of the computers on the same CAN network to save energy, are also impacted by CAN FD: they must be able to correctly decode the frames so as not to mistakenly awaken a non-CAN FD node – for example, in mixed networks, where CAN FD is used for faster reprogramming of the computers.
CAN FD active	A node that can transmit/receive and correctly interpret CAN FD frames – and therefore, by default, do the same for standard CAN frames.	We do not say that a circuit is CAN FD active or CAN FD passive if a node on the network is only literate in CAN. It may be that a CAN FD active node only works on a network equipped exclusively with CAN FD active nodes – for example: TJA 1144 GT.
CAN FD passive	A node that is unable to communicate in CAN FD (that is, a standard CAN node), but above all, which is not perturbed by the transport of CAN FD frames, and does not cause problems for the network when CAN FD communication comes through, as though it were not there at all, and/or were completely transparent.	Example: TJA 1145 FD.
Mixture and coexistence between CAN and CAN FD	– For reasons of application, when CAN and CAN FD communications are using the same network, it is preferable to opt for mutual non-aggression. – The old infrastructures do not interfere with the new. – The new infrastructures do not interfere with the old. – A CAN FD circuit does not interfere with older components, which would not understand CAN FD. – Such is the case with ISO 11898-6 circuits.	May be configured to recognize FD frames as valid frames, but will not be awoken by FD frames that are received (FD passive).

Figure 4.60 Definitions of partial, active, and passive networks and coexistence of these solutions.

in CAN FD with the nodes that are still awake, the classic TJA 1145 A transceivers, sleeping with one eye open (so to speak), will witness the passing of CAN FD frames that they do not recognize, and thus will generate network errors (Figure 4.61).

To prevent these unwanted effects, the TJA 1145 FD was designed. In a new registry, it holds a special control bit, which can validate "CAN FD passive" operation, so certain nodes can ignore the passage of CAN FD messages through the network. Thus, when CAN FD communication is needed between active nodes, CAN FD traffic on the network can be tolerated without being disturbed by the inactive nodes in silent mode. The partial network in question is then placed into sleep and/or low-power mode. This is where the name "CAN FD passive" comes from.

TJA 1145 FD transceivers that are in sleep mode, configured to ignore CAN FD traffic, can only be awoken by a wake-up message sent in normal CAN format, when the CAN FD communication has completed. The set of registries enabling CAN FD frames to be ignored is described by ISO 11898-6. "CAN FD passive" is an important function, because it ensures the coexistence of traditional CAN nodes with CAN FD nodes in the same network (e.g. by simply replacing the TJA 1145 of the existing nodes with TJA 1145 FD), and facilitates migration between the CAN FD circuits at 2 Mbits/s and faster speeds, helping to take full advantage of the high datarates offered by CAN FD (Figure 4.62).

Transceivers for CAN FD active and passive Derived from the existing TJA1145 FD range of products, the TJA1146 operate in both *active and passive* CAN FD modes.

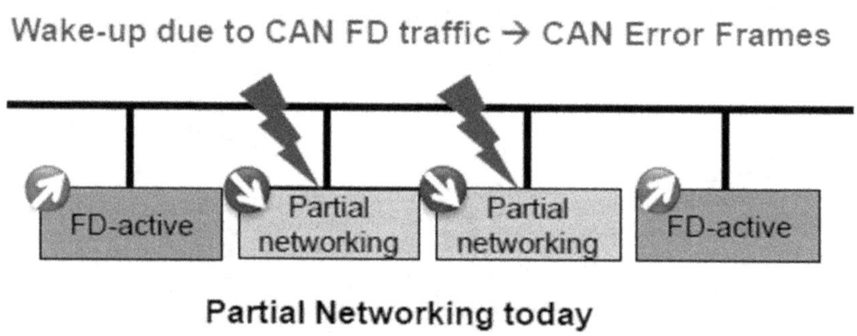

Figure 4.61 Problems of nodes awakening due to the passing of CAN FD frames.

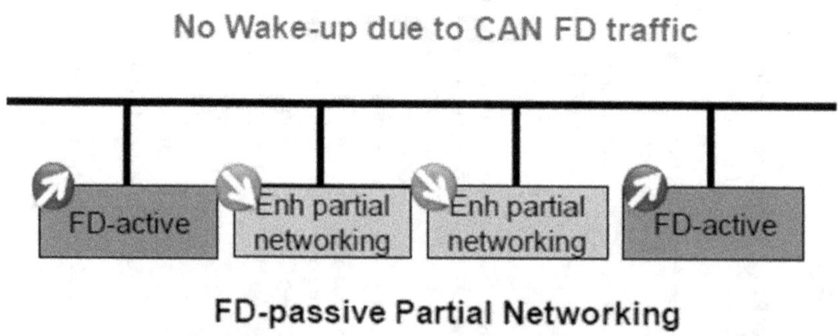

Figure 4.62 No problems of awakening due to the passage of CAN FD frames.

CAN and CAN FD networks and their innumerable topological variants, from one user to another, are highly complex to study, model, simulate, measure, analyze the results and performances, compare the reality with the simulations, evaluate the sources and cases of error, consolidate, etc. – all the more so when one considers that all of this must be done in a laboratory environment as well as in the field. Additionally, and more generally, to provide new electrical parameters for the use cases of CAN FD ≥ 2 Mbits/s (loopback delay, loop delay symmetry), we must investigate the factors limiting the maximum bitrate, the transceiver technologies, cables, the different topologies, the EMC emission constraints, immunity, the use of filters with or without a choke, etc. In addition, we must characterize the performances of CAN FD itself by assessing the possibilities of integrated circuits with partial networking architectures in accordance with ISO11898-6, in CAN FD passive mode, low and very low consumption (sleep mode), with high datarates and across a wide temperature range, define what is guaranteed by design, what needs to be measured as standard, what needs to be measured over an extended range, how reliable the component is, its life span, write and contractually guarantee the content of the data sheets, etc.

Partial networks, consumption and ecology Given the high price of fuel, reducing vehicles' fuel consumption is one of the main areas of innovation in the automotive industry, and participation in that effort is one of the priorities for semiconductor suppliers. This is the reason for the introduction of partial networking technology, which allows more precise control of the activities of the vehicle communication networks by intelligently deactivating the ECUs that are not needed. Thus, it becomes possible to reduce vehicle fuel consumption and CO_2 emissions without sacrificing performance or user experience. The first such solutions implemented offered a 3% reduction in CO_2 emissions, equivalent to 0.11 liters of fuel per 100 km. This is one of the numerous steps necessary to improve vehicles and enhance their efficiency, and to increase fuel efficiency. Ultimately, the potential for growth or maintenance of the automotive combustion engine will depend largely on whether we shall one day manage to achieve these potential savings and consume a maximum of 2 liters per 100 km.

Security of CAN networks without encryption The closer a vehicle comes to being truly autonomous, the greater is the concern over the possibility of a cyberattack. To prevent this danger, CAN High Speed transceivers in the TJA115x range provide security controls in the data link layer, guarding against spoofing, tampering, and flooding attacks (Figure 4.63). These components can replace the existing CAN High Speed transceivers.

If no cybersecurity incident is detected, the TJA115x family behaves like TJA104x transceivers, the only difference being that the TJA115x has self-polarization, for example. If a cybersecurity incident is detected, the frame circulating on the bus is invalidated, by being tagged with an active error flag at the end of the frame. This happens before the frame can be stored in any receiver buffer memory. In the event of an incident caused by the local host, the transceiver temporarily disconnects the local host from the network.

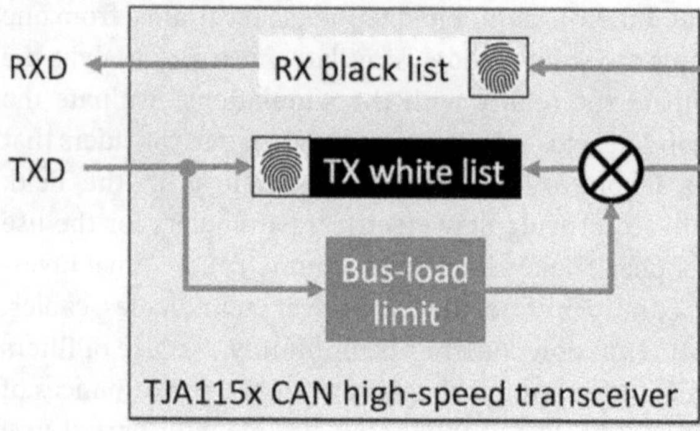

Figure 4.63 Example of a circuit implementing multiple types of security – TJA115x from NXP.

It is possible to replace the standard high-speed transceiver without needing to modify the software of the host controller, but a few minor modifications must be made to configure the base parameters, such as the CAN identifiers or the filtration parameters. The configuration may either remain open for more secure updates, or be locked. The TJA115x has the capability to log and report security incidents on the network to the local host. These CAN transceivers may be used either in a conventional CAN network or in CAN FD networks. They protect their own configuration updates. The proposed design is implemented entirely at a hardware level, and operates independently of and isolated from the host controller. This means that it provides an inherent level of security, and is specially designed to have a minimal impact on the system configuration to compensate for the absence of the sender ID in the CAN network. It can be added into a network progressively (one ECU at a time), without impacting the other elements, or having an effect on the message latency, the bus workload, or increasing the processor's load. The protective mechanism is implemented against tampering, so each time the target ECU receives a protected frame, it checks that the expected sender has sent it. In addition, because the security mechanisms do not need initialization (of the individual computers) or synchronization (of multiple computers on a bus), the network is immediately protected as soon as they are implemented. Denial-of-service incidents can be countered by limiting access to a pre-set authorized band. These bandwidth limitations are also known as "inhibit times."

Conformance and tests

In order for implementations to comply with the ISO standards, a program of conformance tests (the CAN-FD reference model) has been developed. It pertains to the validation/certification of *transceivers* operating at 4 and 8 Mbits/s. In addition, the SAE has developed a network layer and the applications J1939-22 and the physical layer J1939-17, connected for the HGV industry.

CAN XL[6]

As described above, the CAN protocol was first published more than 35 years ago. Twenty-five years later, an improvement called CAN FD began to be discussed. After its successful release, standards and specifications were drawn up for the corresponding physical layer, and CAN FD transceivers and micro-controllers became available.

The next step in the journey came in 2018, with the advent of the next level of CAN: CAN XL:

In December 2018, initiated by Volkswagen and under the leadership of Holger Zeltwanger, Director General of CiA (*CAN in automation*), under the presidency of Dr. Arthur Mutter (Bosch), the CiA SIG "Next generation CAN" – a group of experts drawn from CiA and various automakers – established the "CAN XL Special Interest Group (SIG)." This group is working on specifications for CAN XL. With several OEMs, semiconductors suppliers, and development tool manufacturers, etc., the group has established three task forces (TFs):

- CAN XL physical layer TF;
- CAN XL higher layer TF;
- CAN XL security TF.

In March 2020, at the Seventeenth International CAN Conference in Baden-Baden Germany, CAN XL was officially introduced.

CAN XL – available specifications

The specifications available as at September 2020 are presented in Figure 4.64.

- After the release of these specifications, the procedure of international standardization with the ISO will be started.
- CiA organizes "plugfests" to test the interoperability of CAN XL protocol controllers, but also of the CAN SIC XL transceivers in network environments, as soon as prototypes or engineering samples are available.

CAN XL specifications (status in September 2020)	
CiA 610 series	
CiA 610-1	Data link layer and physical signaling requirements
CiA 610-2	Data link layer and physical signaling conformance test plan
CiA 610-3	Physical media attachment sub-layer requirements
CiA 610-4	Physical media attachment sub-layer conformance test plan
CiA 610-6	Media independent CAN interface conformance test plan
CiA 610-7	Higher-layer function requirements
CiA 610-8	Higher-layer function conformance test plan
CiA 611 series	
CiA 611-1	SDT value specifications
CiA 601-4	CAN SIC (signal improvement capability) version 2.0.0

Figure 4.64 CAN XL available specifications – September 2020.

General philosophy of CAN

For nearly 40 years now, the set of protocols CAN Classic, CAN FS, and CAN XL has been operating on the basis of a general philosophy founded on the following pillars:

- In-depth knowledge of the automotive market, with its ins and outs;
- Applications for vehicles of all levels, from the bottom to the very top of the range;
- The numerous and diverse topological implementations of the networks/buses;
- Excellent knowledge of signal processing in the physical layer;
- The stringent technical constraints of networks;
- The stringent financial constraints of the market;
- Compatibility, interoperability, and flexibility with existing networks and new ones.

CAN XL is a development and improvement of the well-established CAN Classic and CAN FD protocols (Figure 4.65). Over the course of development, payload has gradually increased, from 8 bytes in CAN up to 64 bytes in CAN FD, and CAN XL represents a major increase, capable of carrying a 2048-byte payload. Of course, in order to achieve acceptable transmitting times with such a big payload, higher bitrates are needed. For a transmitting time below 2 milliseconds with a 2048-byte payload, a bitrate of 10 Mbit/s or higher in the data phase is necessary.

CAN XL – General view	
Overview	For the purposes of flexibility, compatibility, and interoperability, CAN XL operates largely on the same principle as Classical CAN and CAN FD protocols The CAN XL frame is divided into "arbitration phase" and "data phase." CAN XL uses low transmission speeds of between 500 kbit/s and 1 Mbit/s in the arbitration phase (arbitration and ACK fields). A rate of 500 kbit/s in the arbitration phase is set, to ensure the same distances between ECUs in networks like CAN FD.
Arbitration	CAN XL separates the arbitration and addressing functions. CAN XL only supports 11-bit identifiers. The medium access method is CSMA/CR (carrier sense multiple access/ collision resolution), which resolves competing write access through bit arbitration. CAN XL obeys a strict priority system that allows the more important frame to be transmitted with no delays.
Data phase	In the data phase, CAN XL bitrates are scalable over a wide range, from 2 Mbit/s to 10 Mbit/s. Bitrate switching between the "arbitration phase" and "data phase" is compulsory with CAN XL. Scalable payload length, from 1 byte to 2048 bytes. (It features some embedded layer management information for higher-layer protocols. If necessary, this enables communication systems to package Ethernet frames in CAN XL frames, and/or to use IP communication via CAN XL).
Security	CAN XL features a high level of data transmission reliability. With a Hamming distance of 6 for headers and frames, and also format checks, it actually outperforms FlexRay and exceeds the capabilities of Ethernet CRCs.
Compatibility	Backwards compatibility with CAN FD. It is highly scalable in relation to applications but also to the supported bitrate, as CAN XL can be used with many different transceivers.

Figure 4.65 CAN XL – General view.

CAN XL – data link layer

The following are the key features of the CAN XL data link layer protocol CiA 610-1:

- Large data field of up to 2048 bytes, in order to execute TCP/IP protocols within only one CAN data frame;
- Datarate of 10 Mbit/s;
- Higher-layer management information;
- As simple a data link layer and frame format as possible, with only 11 bits for the ID;
- Improved reliability because of two CRC fields – one for the header and the other for the total frame content;
- Similarly to Ethernet, the CAN standard (ISO 11898 series) specifies a DLL including two sub-layers (LLC and MAC);
- Logical Link Control (LLC): sub-layer between the OSI network layer and the medium access control sub-layer;
- Medium Access Control (MAC): responsible for moving frames from the LLC sub-layer to the PMA (physical medium attachment) sub-layer and protects the transmission by means of stuff bits, CRC fields, etc.
- CAN XL retains backwards compatibility with CAN and CAN FD.

LLC sub-layer The LLC frame (see figure 4.66) contains five Fields and all content needed for all CAN frame formats and types, including the selection of a specific CAN frame format. In the interaction between the LLC and MAC, the content of those parts of the LLC frame that are not used for the selected CAN frame format shall be ignored.

The LLC frame supports all three CAN protocol generations: Classic CAN, CAN FD, and CAN XL.

MAC frame in XL format The MAC sub-layer comprises the functions and rules related to encapsulation/decapsulation of the transmitted/received data, error detection, signaling, and management of medium access.

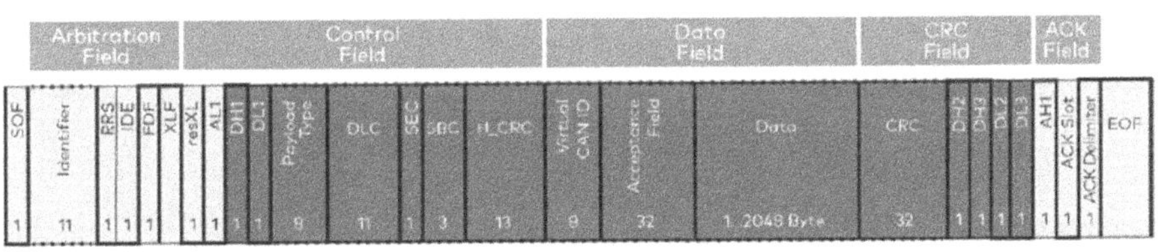

grey: arbitration phase nominal, bitrate compatible with CAN, CAN FD
red: data phase, bitrate from x up to 10 Mbit/s
- IDE: identifier extension
- FDF: flexible datarate field
- XLF: extra large field
- SEC: DLL security indication
- DLC: data length code
- VCID: virtual CAN network ID
- AF: acceptance field

Figure 4.66 LLC frame format as specified in CiA 610-1.

There is only one CAN XL MAC frame format, called "CAN XL Frame Format" – .xlff. The frame has a variable length and can hold between 1 byte and 2048 bytes in the data field, while the data length can change in one-byte increments:

- On transmission, an LLC frame is converted into a MAC frame;
- On reception, a MAC frame is converted into an LLC frame.

MAC frames in XL format are composed of seven different bit fields, as shown in Figure 4.67 (the fields marked in green are automatically added by the MAC sub-layer and the grey fields are provided by the LLC frame).

CAN XL Frame

Let us now look again at the whole of a CAN XL frame and give some explanations Phase after Phase (see Figure 4.68).

In the Arbitration field The 11-bit **Identifier bit** – priority ID sub-field – in the Arbitration Field provides the uniquely assigned priority of the CAN XL data frame.

In the Control field In this long control field, including "Control + CRC + ACK:"

- SDT – Service data unit type

SDT Service data unit type is a feature that is usable for higher-layer protocols. It is embedded (OSI) layer management information as described in ISO 7498-4:1998 and is similar to the EtherType field in the Ethernet frame.

CiA 611-1 specifies the SDT values and the corresponding usage to unfold the power of this field:

  - ◦ Content-based addressing (i.e. use of message IDs);
  - ◦ Node addressing;
  - ◦ Node tunneling of Ethernet frames;
  - ◦ Classical CAN and CAN FD data frames.
- Protocol Type

SOF	Arbitration	Control	Data (field)	CRC	ACK	EOF
1	15 bit	81 bit	1 to 2048 byte	36 bit	6 bit	7 bit

Figure 4.67 CAN XL MAC frame fields.

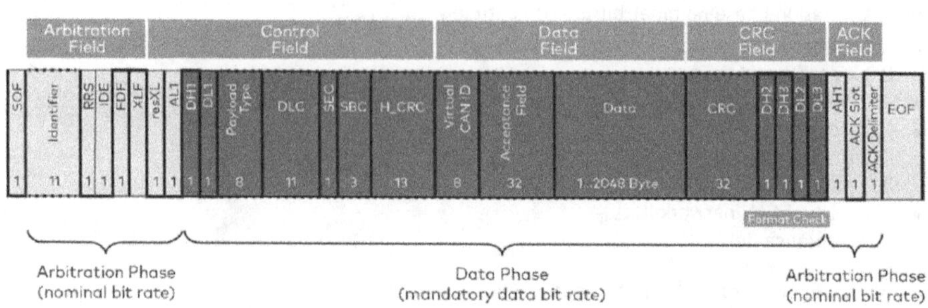

Figure 4.68 Example of CAN XL frame.

The 8-bit Protocol Type field generally indicates what kind of data are located in the payload. The 8-bit Payload type indicates the next OSI layer protocol. This can be used, for example, to propagate data through a software stack. It can also be used to specify addressing information and for filtering.

- Data Length Code – DLC
  DLC from one bit to 11 bits (i.e. data length 2048 bytes, from 0 to 2047 bytes) facilitates the use of CAN XL as a data link layer for TCP/IP-based applications or to transport complete Ethernet frames.
- Optional DLL security

CAN XL TF security specifies the CADsec data link layer security protocol. The CADsec protocol features a header with cipher control information, the CAN secure channel ID, and a freshness value. The 16-byte trailer contains the authentication tag.

The 1-bit **SEC** field indicates whether this CAN XL data frame uses the CADsec protocol.

In the Data field Virtual CAN network ID (VCID)

The 8-bit **Virtual CAN network ID (VCID)** allows up to 256 logical networks to be run on a single CAN XL physical network segment. This means that multiple protocols can be used in parallel on the same physical CAN network. This field is also an embedded (OSI) layer, managing information, as described in ISO 7498-4:1998.

- Acceptance

The addressing 32-bit **Acceptance** sub-field can contain the node address or indications of content, such as message ID.

Protections and CRC fields To improve the security of transportation of the data frame, CAN XL requires protection against the following errors:

- Bit inversion
- Bit drop or Bit insertion
- Burst errors

To do so, CAN XL defines four protection mechanisms:

- Two CRC (cyclic redundancy check) fields:
  - 13-bit Preamble/Header CRC (PCRC or HCRC) field in the
    Control field
    - Includes SOF, arbitration field, payload type, DLC, SBC and dynamic stuff bits;
    - Excludes static bits: resXL, ADS, and fixed stuff bits;
    - If a H_CRC error is detected, a receiver ignores the rest of a frame.
  - 32-bit frame CRC (FCRC) field in the
    CRC field
    - Includes SOF, arbitration field, payload type, DLC, SBC, H_CRC and the Data Field;
    - Excludes static bits: resXL, ADS, and fixed stuff bits.
  - The CRCs are cascaded, which means FCRC protects the whole frame, including the PCRC.

Both CRCs are able to detect any five randomly distributed bit-errors. This corresponds to a Hamming distance of 6.

- Offering better protection than FlexRay or Ethernet CRCs
- Stuff Bit Count
- Protecting dynamic stuff bits in the CAN XL Header
- Format checks

ACK field, CRC field, and EOF field The CRC field, ACK field, and EOF field are illustrated in Figure 4.69.

CAN XL and transceivers

CAN XL is highly scalable in terms of bitrate and the medium access unit (MAU) physical sub-layer (implemented in transceiver chips or system base chips), and will operate with all transceivers:

- Up to 1 Mbit/s
  CAN XL controllers can be used with **Classic CAN High Speed**.
- Up to 2 Mbit/s
  CAN XL controllers can be used with **CAN FD**.
- Up to 5–8 Mbit/s
  CAN XL controllers can be used with **CAN FD-SiC** and **CAN SIC** (**signal improvement capability** specified in CiA 601-4 version 2.0.0) transceivers using the AUI (attachment unit interface) as specified in ISO 11898-2:2016).
- 10 Mbit/s and beyond

CAN XL controllers can be used with **CAN SIC XL** transceivers for support.

Higher-layer protocols

Standardization of higher-layer protocols is essential for interoperability of devices with CAN XL connectivity. The CAN XL higher layer describes the specifications of SDU types, Multi-PDU concept (similar to the concept known from Autosar), whereby several PDUs can be aggregated and sent as a Multi-PDU inside a single CAN XL MAC frame.

Physical layer in CAN and CAN XL

As explained several times before, the CAN network is a serial bus system/topology, which is able to connect more than two nodes on a network. Therefore, collisions are possible; to manage these collisions, CAN uses the well-known CSMA/CR (Carrier Sense Multiple Access/Collision Resolution).

In the first arbitration phase, one or more nodes can transmit a CAN frame on the network at the same time. The node with the highest priority wins the arbitration. For that reason, a CAN transceiver controls only one level (TxD = 0). This is called the dominant

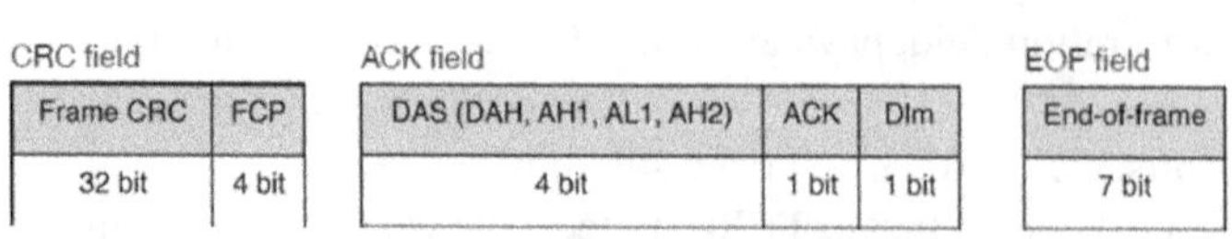

Figure 4.69 CRC field, ACK field, and EOF field.

level and, during the recessive level (TxD = 1), the transceiver output stages are high ohmic and the termination resistors are responsible for the recessive state on the network. This concept allows a transmitting node to overwrite the recessive state on the network with a dominant level without the risk of damaging a transceiver transmitting a recessive level at the same time. With such a concept, collisions on the network can be supported.

However, this concept brings a major disadvantage: the transceiver's output stages driving the CAN line change from high impedance to low impedance, and vice versa, between dominant and recessive levels. Within a point-to-point topology, this impedance change creates reflections (values and polarities) on the network and, on star topologies, the impedance values also change.

- Examples

  On a star point topology with three lines (one line for the incoming wave and two lines for the outgoing wave), the two outgoing lines are parallel. With two times 120 Ω (nominal) impedance, the overall impedance for the wave is 60 Ω. Directly, we have a reflection factor of −0.33 (Figure 4.70).

These reflections are caused with every transition on a wire, irrespective of whether the network levels change from dominant to recessive or vice versa. However, a long time taken for the oscillations to fade ("ringing") limits the maximum bitrate in the data phase because the "sample point" has to be set after the ringing has finished in order to get a reliable sample – and before the next bit. (The sample point is the point where the bit value is sampled and measured – see Figure 4.71 and reference books from Dominique Paret.)

Termination impedance	Reflection factor
30Ω	-0.6
40Ω	-0.5
60Ω	-0.33
120Ω	0
100kΩ (Transceiver input imp.)	+0.99

Figure 4.70 Reflection factor for 120 Ω wire impedance in CAN.

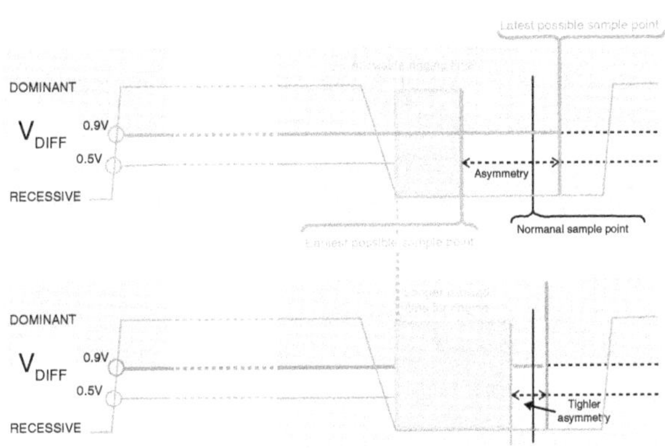

Figure 4.71 Signal symmetry and the nominal sample point.

CAN SIC

To achieve higher bitrates and place the sample point favorably, the length of ringing must abe reduced and the transition from dominant to recessive has to be controlled by the transceiver. This is the concept of the **CAN SIC** transceiver, already in use with CAN FD.

- Note: In the Ethernet PHY, the problem is different because a round-robin approach is implemented that allows the collision-free network access via PLCA (physical layer collision avoidance).

Performances required for CAN SIC and CAN SIC XL transceivers "CAN SIC" already exists and is in use in CAN FD. The main targets for the "CAN SIC XL" transmitter used with CAN XL are:

- In the arbitration phase, support CSMA/CR and minimum 500 kbit/s;
- In the data phase, support minimum 10 Mbit/s;
- Reduce the timing asymmetry of the signal;
- Hardware compatibility, with the same pinning as for CAN FD transceivers.

To cover all these requirements:

- In the arbitration phase, CAN SIC, which is already used with CAN FD, is used;
- In the data phase, to achieve bitrates higher than 5 Mbit/s, a different transmitter concept is needed. An alternating network voltage (differential voltage) concept is chosen, based on the idea of FlexRay. The advantage of FlexRay is that the levels are symmetrical to the ground and the receiver thresholds. The impedances of both levels are close to the wire impedance (less reflection) and the timing asymmetry is very slight.

To achieve higher bitrates, independently of concept (i.e. topology/distance/number of nodes, protocol, etc.), all minimum values of the symmetry parameter in the CAN SIC transceiver are improved and reduced. The main impact stems from the reduction of the minimum limits for transmitted recessive bit width, from –45 ns to –10 ns, and the reduction of the receiver symmetry minimum value, from –45 ns to –20 ns.

To support the CAN SIC transceiver, two different concepts based on CiA 601-4 are available:

- Transmitter (Tx)-based concepts;
- Receiver (Rx)-based concepts.

Transmitter-based concept In the Tx-based concept, the transmitter actively controls the dominant-to-recessive transition and afterwards, up to 500 ns of the subsequent recessive phase. In the case of shorter recessive bits, the transmitter changes from active recessive to dominant directly. If the recessive bit is longer, the transmitter changes from the active recessive to the passive recessive (high ohmic) state, as happens in a standard CAN FD transceiver.

With the CAN SIC transceiver, bitrates of up to 5 Mbit/s in star topologies and 8 Mbit/s in linear topologies can be achieved.

Receiver-based concept In the Rx-based concept, all nodes suppress the recessive signal after the dominant-to-recessive transition, triggered by the internal receiver. The suppression time depends on the product and is optimized for a particular bitrate.

For example, for 2 Mbit/s, the transceiver suppression time is up to 450 ns.

CAN SIC XL

For higher bitrates (10 Mbit/s and above) the CAN SIC XL transceivers **CAN SIC** specified in CiA 610-3 are suitable. New parameters have to be considered with CAN XL. Also, there are new fields – ADS field in the control field and DAS field in the ACK field – in the CAN XL frame (see below).

Now, the CAN SIC XL transceivers have three modes – SIC mode, Fast TX mode, and Fast RX mode – to achieve "fast bits" in the data phase but also allow arbitration in the same frame:

- In **SIC mode**, the transceiver drives dominant and recessive bits, as happens in Classic CAN;
- In **Fast TX mode**, the transceiver drives level-1 and level-0 signals with differential voltage levels of –1 V and + 1 V;
- In **Fast RX mode**, the transceiver does not drive the network.

Additionally, CAN SIC XL transceivers support the medium-independent CAN interface (MICI) that signals mode switching:

- MICI (medium-independent CAN interface)

  To signal the mode switch from the CAN controller to the transceiver, CAN XL controllers and transceivers use MICI, based on a TX-based single-path PWM (pulse-width modulation) symbol. This preserves the two-pin interface (RxD, TxD), for CAN SIC XL transceivers as well.

The CAN SIC XL transceiver has now two modes instead of one, as the CAN and CAN FD transceivers do. The new modes are:

- **Slow mode** (arbitration phase):
  - Slow mode is used in the arbitration phase and based on the CAN SIC transceiver concept. All parameters are compliant with CiA 601-4.
- **Fast mode** (data phase):
  - In fast mode, the transceiver controls both levels;
  - The network levels (Vdiff) alternate between + 1 V (level 0) and –1 V (level 1).

Fast mode In fast mode, the transmitter concept changes completely, compared to the established HS CAN and CAN FD transceiver. The output signal is transmitted as a symmetrical alternating differential signal. The new levels are named:

- Level 0 if TxD0
- Level 1 if TxD1

The receiver threshold is 0 V with a tolerance of ± 100 mV. The output levels are now symmetric in relation to the receiver threshold, which reduces the timing asymmetries of the transmitter and receiver.

The output impedance of the transmitter output stage will be 105 Ω for both levels, and fits the most widely used unshielded twisted pair cables. For the CAN SIC transceiver, the output impedances are different for the dominant and active recessive states, and not subject to a standard specification. Transmitter output stage impedances match the cable impedances, helping to reduce reflection in a network.

All these parameters are specified in CiA 610-3.

The SIC transceiver mode changes The transceiver modes are controlled by the CAN XL controller.

Without a mode change, the transceiver can be used as a CAN SIC transceiver only. This allows the CAN SIC XL transceiver to be used in combination with the CAN FD and/or CAN XL protocol.

To do so, in the CAN XL protocol, two fields are reserved for the transceiver mode for bitrate switching (Figure 4.72):

- The ADS field (arbitration to data switch);
- The DAS field (data to arbitration switch).

ADS and DAS fields In CAN XL, bitrate switching is now mandatory and switching is performed at the AL1/DH1 and the DL3/AH1 edges. The falling edges between DH1/DL1 and DH3/DL2 are used for synchronization.

- **The ADS field** is a **part of the control field.** It is between the arbitration field and the data field.
  The ADS field consists of four bits:
  - The first bit is ADH (A = arbitration bitrate), TxD = 1.

ADH is transmitted in the arbitration bitrate, and during this bit, the CAN SIC XL transceiver will be switched from slow mode (SIC mode) into fast mode (Fast TX or Fast RX mode). The MICI sends PWM symbols of arbitrary value to enact the proper transceiver mode switch. All CAN XL nodes ignore the sampled value of the ADH bit.

The ADH bit is a recessive bit, but at first, the network level remains dominant, and after the mode-switch command from the CAN XL controller, the CAN SIC XL transceiver changes from the dominant to the recessive level with SIC performance. After a defined time the network level changes from the recessive level to level 1. In parallel, the CAN SIC XL transceiver switches the receiver thresholds from slow mode threshold levels to the fast mode threshold.

- DH1, DH2, and DL1 are transmitted with the data bitrate and used in the CAN controller for synchronization:
  - DH1 (D = data bitrate), TxD = 1
  - DH2 (D = data bitrate), TxD = 1
  - DL1 (D = data bitrate), TxD = 0
- The **DAS field is a part of the acknowledge field**. It is between the CRC field and the EOF field.

The DAS field consists of four bits, all transmitted at the arbitration bitrate.

- The first bit is DAH (A = arbitration bitrate), TxD = 1.

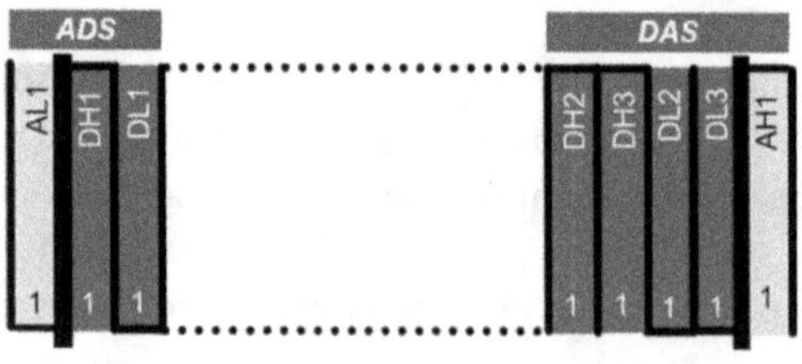

Figure 4.72 ADS and DAS.

This is the point at which the transceiver mode in the CAN XL SIC transceiver is switched back to the SIC mode from the fast to slow mode. At the beginning of the DAH bit, triggered by the CAN XL controller's mode-change command, the transmitter changes from level 0 to active recessive and after the signal improvement time, the transmitter changes from active recessive to passive recessive. In parallel, the receiver thresholds are switched from the fast to slow mode threshold level.

- The edge AH1 to AL1 is used in the CAN XL controller for synchronization in the arbitration bitrate.
  - AH1 (A = arbitration bitrate); TxD = 1
  - AH2 (A = arbitration bitrate); TxD = 1
  - AL1 (A = arbitration bitrate); TxD = 1

Example CAN SIC transceiver: TJA146x family – NXP The TJA146x series from NXP actively improve signal ringing in a network, reducing its effects and helping to operate larger topologies at bitrates significantly faster than before. CAN Signal Improvement technology facilitates network design that moves beyond complex linear CAN FD network topologies at 2 Mbit/s and with reduced specification industrial control of cable harnesses. By incorporating unterminated stubs and star points into the network, the total cable length can be significantly reduced, saving on overall weight and cost. Enabling larger topologies and higher bitrates also brings the potential to combine multiple network branches into a single network, and freedom of design. The TJA146x family also enables us to do away with external components, such as ferrites, whose purpose is to manage signal ringing.

The features of the TJA146x series, even at higher bitrates, are:

- A highly symmetrical transmitter;
- Much tighter bit timing;
- Easy sample point positioning;
- Excellent EMC emission and immunity;
- Robust and reliable intra-network communications;
- Full backwards compatibility with existing CAN transceivers;
- Simple drop-in solution to enhance performance in CAN FD networks.

TJA146x is fully compliant with ISO 11898-2:2016 and CiA 601-4 v2.00 specification for CAN Signal Improvement Capability (SIC) used in CAN XL (Figures 4.73 and 4.74).

Transceiver versus protocol The protocol itself has a minor impact on the choice of the transceiver concept. Two aspects are important:

- The maximum required bitrate;
- The network topology.

When the dual mode of the DM-SIC transceiver needs to be used to achieve bitrates higher than 5 Mbit/s or to improve signal integrity at lower bitrates, then the CAN XL controller is necessary. Only this controller is able to support dual-mode transceiver function.

The CAN SIC XL transceiver may also be used in combination with CAN FD and Classic CAN, but in these combinations, only the slow mode will be supported. Figure 4.75 lists the possible combinations.

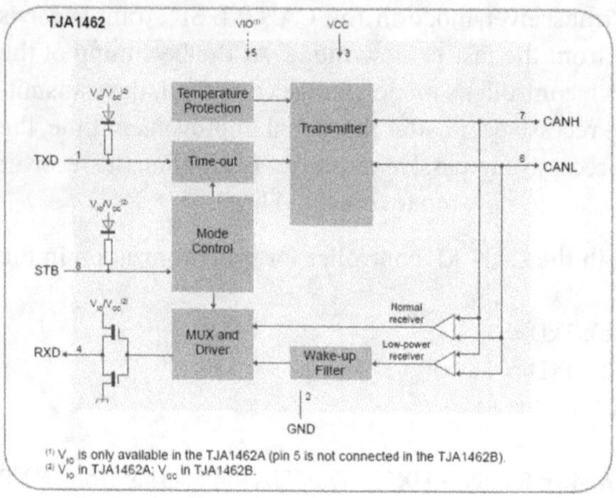

Figure 4.73 CAN SIC and SIC XL Line driver – TJA146x from NXP.

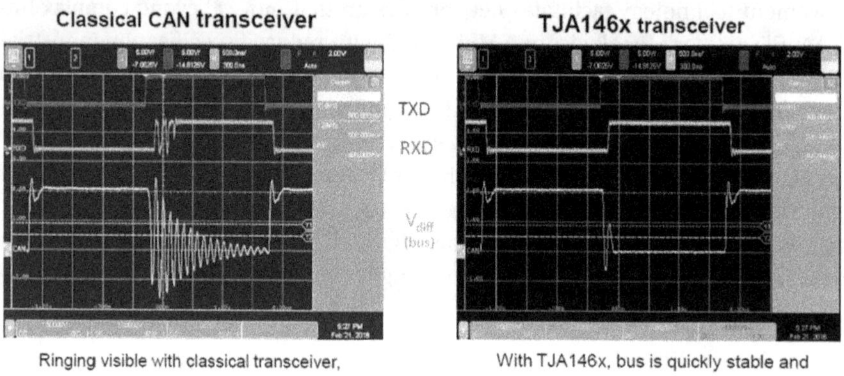

Figure 4.74 Examples with Classic CAN and CAN SIC TJA1146.

Protocols	CAN FD	CAN XL
	HS-CAN transceiver	HS-CAN transceiver
	CAN FD transceiver	CAN FD
	CAN SIC transceiver	CAN SIC transceiver
	CAN SIC XL (slow mode)	CAN SIC XL (dual mode)

Figure 4.75 Possible protocol/line driver combinations.

Relations between bitrate and topologies The maximum bitrate depends on the network topologies – not only on protocol.

The higher the number of stubs, stars, and nodes, the lower are the maximum possible bitrates. Also, the ratio between the stubs of a star topology has an impact on reflection and ringing in the network. However, higher bitrates require a minutely detailed analysis of the network. Thus, with CAN FD and CAN XL, on the one hand, and CAN FD, CAN SIC, and CAN SIC XL transceiver, on the other, there are a broad range of possibilities to make the best choice for your application. The choice can be guided by network simulations.

Applications – signal-based and service-oriented communications

These two types of communications will be discussed in greater detail in Chapter 6.

Signal-based communications – CAN With CAN, for typical control tasks, the signal-based approach is tried and tested, having been in use for almost three decades. Together with the priority ID principle used with CAN, the system ideally satisfies real-time requirements.

A major feature of signal-based communication is the predefined static communication matrix. Signals such as temperatures, pressures, speeds, or revolutions always represent the same fixed parameter, which is mapped onto an established CAN frame and sent to ECUs. In addition, so-called PDUs have been introduced (see details in Chapter 6), which form an intermediate layer and make communication more flexible. On the other hand, nothing stands in the way of upgrading from Classic CAN or CAN FD to CAN XL in service-oriented communication, since a great deal of know-how and development time has already been invested in wire routing and careful design of cable harnesses.

Service-oriented communications – Ethernet Today, in the automotive field, the focus is mainly on advanced driver assistance systems (ADASs), autonomous driving, electric mobility, and continuous Internet or cloud connectivity. High-performance systems such as radars, lasers, video cameras, sensors, and data fusion applications in the vehicle are an indispensable prerequisite for autonomous driving. They generate enormous volumes of data, which are transmitted only during runtime. Such data cannot be mapped statically; instead, the communication system must serialize the data dynamically.

The challenge is how to transmit and process this burgeoning volume of data in real time. To do so does require a dynamic link connection between the data sink (consumer) and data source (provider). Currently, the advent of automotive Ethernet with IP technologies is causing a number of fundamental changes, and service-oriented communication is becoming established in vehicles in parallel to signal-based communication. The ability to transmit dynamic data structures is one of the major advantages of service-oriented communication, which goes hand in hand with Ethernet and IP technology.

Signal and service oriented communications at 10 Mbit/s – CAN XL Applications need data and services, and it does not matter who provides them.

In this context, CAN XL provides the basis for efficient cooperation between IP technology and classic signal-based communication. To reach the lower end of Ethernet networking, Ethernet 10BASE-T1S is reaching down from above in the 10-Mbit/s domain and CAN XL, reaching up from below, is expanding into the 10-Mbit/s domain. Both are converging and are able to provide communications at this bitrate.

Both 10BASE-T1S and CAN XL domains could frequently operate as network branches under a 100BASE-T1 domain (Figures 4.76 and 4.77).

- It is possible to use a switch to couple 10BASE-T1S with 100BASE-T1, with no complications;
- On the other hand, a gateway is required to connect CAN XL branches.

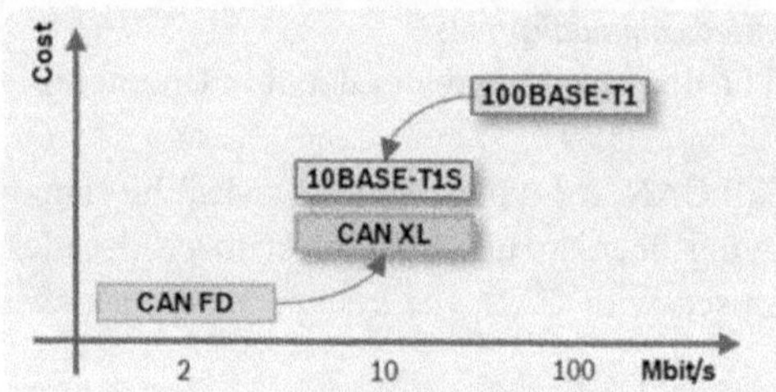

Figure 4.76 Bridging the gap between CAN FD and Ethernet.

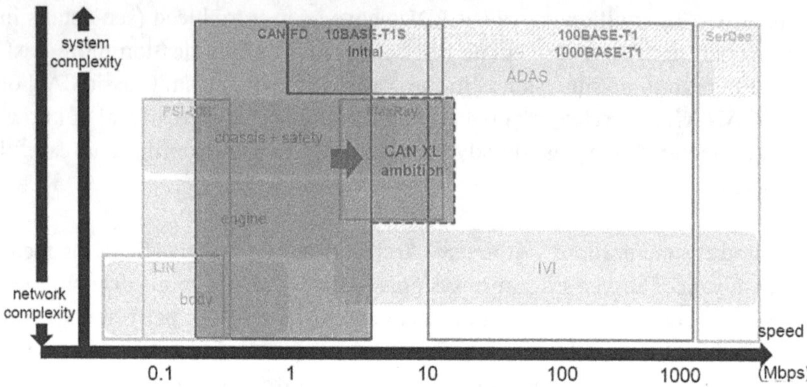

Figure 4.77 Convergences in IVN standards by application.

	CAN	CAN FD	FlexRay	CAN XL10BASE-T1S	ETHERNET
communication	signal-based		time-critical		
	signal-oriented			signal-oriented service-oriented	service-oriented
max Mbit/s	1	8	10	10	
usual Mbit/s	0.5	0.5–5	10	2–10	100–1000
applications	engine management body control		engine management body control chassis control brake system audio applications (microphone, eCall) radar, ultrasound sensors fast end-node links in zonal architecture TCP/IP over CAN		infotainment ADAS telematics connectivity

Figure 4.78 Application fields for CAN, CAN FD, FlexRay, CAN XL, and Ethernet.

Since the fields of application of MOST can also be sufficiently covered by Ethernet, this system will likely be replaced in the medium term. (* MOST, for infotainment applications, covers the 25 Mbit/s to 150 Mbit/s range.)

Given the rise of automotive Ethernet and in view of the growing variety of communication systems, it is sensible to seek to consolidate the solutions, in order to limit complexity and costs (Figure 4.78).

Conclusions

With a maximum data phase bitrate between 2 and 10 Mbit/s, where 100 Mbps Ethernet is excessive and FlexRay too complex, CAN XL features very large payloads (about 2048 bytes maximum) and is suitable for backbone applications in complex network architectures. In contrast to 10-Mbit/s Ethernet, CAN XL is more reliable, more robust, and more cost-effective for complex bus topologies (Figure 4.79).

For networks that do not require deterministic or IP-based communication, CAN XL redefined the market in terms of possibilities. For future applications, it is not just the increased transmission speed that is important. It provides many features that can be used for higher-layer protocols similar to Ethernet. The 8-bit service data unit type in the CAN XL data frame specified in CiA 611-1 indicates the higher-layer communication service (e.g. the network layer) to be used. The CiA SDU type is similar to EtherType in Ethernet data frames. Additionally, the CAN XL data frame provides the 8-bit virtual CAN network identifier (VCID). It enables multiple network applications with the same SDU type to be run on a single network. This was already possible in the past, but the CAN-IDs needed to be assigned uniquely by the system designer.

One of the key advantages of CAN XL is its scalability. CAN XL networks can be based on CAN High Speed transceivers (ISO 11898-2:2016), CAN SIC transceivers (CiA 601-4), and CAN XL SIC transceivers (CiA 610-3). This scalability gives the network designers the greatest possible degree of freedom.

Moving forward, CAN, CAN FD, and CAN XL offer new solutions to real day-to-day problems, and allow players in the industry to achieve greater maturity in solutions for higher-datarate onboard networks such as 100 Mbit/s and 1 Gbit/s point-to-point switched Ethernet, which can be used in the design of autonomous vehicles. In short, the advent of CAN XL marks the beginning of major shifts to come, in the move away from low-datarate Ethernet.

4.5.4 FlexRay

Here, once more, we refer readers to a book dedicated entirely to FlexRay.[7] Nevertheless, there follow a few points that are more specifically applicable to autonomous and connected vehicles.

Reminders

By way of introduction and recap, we shall briefly mention a few points that are applicable to all network architectures.

To begin with, the original version of CAN, designed almost 40 years ago, is still perfect for numerous applications but, over time, some of its limitations have come to light:

- CAN is an "event-triggered" protocol but lacks a means of "real-time" function – put differently, a "time-triggered" approach. To rectify this problem, the first response was to create a time-triggered application overlay called "TTCAN" (ISO 11 898-4);
- CAN's raw datarate is limited to 1 Mbit/s, but its applications in CAN FD mode tend to be orientated toward raw datarates of between 5 and 8 Mbit/s. With this in mind, we need to rethink and reshape many things, because the maximum datarate of 1 Mbit/s practically corresponds to the technical limit below which we do not need to

	Transmission speed 10 Mbit/s		
	CAN XL	**10BASE-T1S**	**10BASE-T1L**
Range	Same as CAN/ CAN FD	"S" stands for short distance, or short range, which covers distances of up to 25 meters, and exists explicitly for automotive applications.	"L" stands for long distance, which covers ranges of up to 1000 meters, and is typically employed in industrial applications.
Topology	Same as CAN/ CAN FD	Network All users are connected to a common Ethernet cable (multi-drop bus topology) by short tap lines ("stubs") measuring max. 10 cm in length.	Switched Ethernet versions
	Ability to use more complex topologies with a star and long stubs	More restrictive network topology – only permits 10 cm max. stubs	
		Proven topologies cannot be replaced on a one-to-one basis with 10BASE-T1S	
	Unshielded twisted pair	Unshielded twisted pair serves as the physical layer for 10BASE-T1S ("T1")	
	Retains the high dominance of classical CAN variants with signal-based communication in numerous vehicles		

Figure 4.79 Conclusions on transmission speed at 10 Mbit/s in the automotive field.

	Transmission speed 10 Mbit/s		
	CAN XL	**10BASE-T1S**	**10BASE-T1L**
Summary and prospects	CAN XL is a CAN variant to which existing Classic CAN and CAN FD networks could easily be upgraded. It offers transmission speeds of up to 10 Mbit/s. It bridges the gap between CAN FD and 100-Mbit Ethernet (100BASE-T1). In appropriate fields of application, CAN XL communication can facilitate smaller and therefore less expensive controllers than Ethernet. With payload lengths of up to 2048 bytes, CAN XL also delivers what will be required in future to transport Ethernet frames and utilize IP communication. Together, CAN XL and 10BASE-T1S could provide a link between signal-based communication on the lower levels and service-oriented communication on the higher-level systems. With appropriate extensions in the various protocol layers, this will open up some advantageous options. With their different approaches, both models offer advantages and disadvantages, and theoretically could exist in parallel to each other. The decision as to which communication system becomes predominant in this area in future depends on considerations of cost as well as technical factors and, last but not least, on backwards compatibility with Classic CAN and CAN FD. Applications for compact and midsize cars: It is precisely this migration path that makes CAN XL interesting for automakers who focus primarily on compact and midsize cars. In this mass market, full autonomous driving (levels 4 and 5) will not become a reality for some time. At best, we will see simple assistance systems that have already been in common use for years, such as anti-lock braking systems. Without radar sensors, high-resolution cameras, and the like, there is no compelling need for an Ethernet-based network; instead, the classic systems will predominate, led of course by CAN. For such vehicles, CAN XL offers the ideal platform for further development on the basis of the existing vehicle architecture. There is no need to redesign the cable harnesses, controllers and protocol stacks. The simpler protocol stack for CAN than that of IP means that smaller and thus less-costly controllers can be used. One goal for CAN XL would be to continue this tradition.		

Figure 4.79 (*Continued*)

talk about line propagation, reflection coefficient, stubs, the Smith chart, etc. Above this value, however, we cannot design physical layers and protocols without speaking of these physical parameters;

- It is difficult, if not impossible, in CAN to create a network architecture/topology that provides redundancy in the physical layers, and thus hope to create systems that are solely controlled by wired connections – this concept is known as *X-by-Wire*. For example, the steering, brakes, suspension, etc., would all be controlled by hardwired connections.

These last two issues were solved by the advent of FlexRay, which is briefly recapped below.

"Event triggered" and "time triggered" aspects

Probabilistic aspect of CAN

By its very construction, CAN transmits communication frames when events occur in a node – in other words, it is an event-triggered system. In principle, at a given time, no node can know whether its message will be immediately transmitted, because of the need for conflict management and arbitration on the basis of the track identifier values. This means that CAN messaging (transmission of messages over the network) has a probabilistic aspect, because it is subject to arbitration, which itself is a function of the respective identifiers of the messages competing to occupy the bus. Consequently, temporarily, the transmission and the associated latency are heavily dependent on the probability of occurrence of the respective identifier values. In addition, the likelihood that this arbitration process will take place is excessively high, when the bus is often occupied and all the nodes that have not been able to access it are waiting for the right time to try again to access the bus. If they all begin their attempts at the same instant, just after the interframe phase, then, automatically, they are all subjected to arbitration! The problem then arises as to when we wish to definitely communicate (either transmit or receive) at a given time, and so be temporally deterministic. There is nothing in CAN, CAN FD, or CAN XL that allows such determinism. Therefore, it is necessary to create systems for which certain actions are triggered deliberately at specific moments. This is known as a "time-triggered" (TT) system – in our case, the FlexRay concept.

Deterministic and "time-triggered" aspect of the applications

TT systems are also referred to as pseudo-real-time systems. When the systems have to function in real time, the major problem arises when we want to be sure of transmitting at receiving at a set time, or within a specific time-slot. Thus, the communication is deterministic. For this purpose, we typically reserve specific time-slots for certain types of information needing to circulate on the network. In principle, the way in which these time-slots are created is completely free and unrestricted. The only tricky point is how to ensure perfect synchronicity among all participants, so that each node can talk or respond in turn, without disrupting the scheduling of its peers. With that in mind, it is generally necessary to cyclically transmit a "reference clock" to the whole network, to reset the ticks of every participant. The approach in FlexRay would not be this exactly, but is close.

FlexRay

It is FlexRay 3.1 that serves as the foundation for ISO 17458. The following are the main salient points in these documents.

Protocol management

Let us begin by stating the philosophy on which FlexRay operates, which is fundamentally different from that of CAN and other protocols. This is necessary in order to fully understand all the subtle points of this protocol.

- It should be noted that in terms of its structure, FlexRay was designed to provide a deterministic communication system with static slots in which there can be no collisions for medium access, meaning that there is no arbitration of the static slots on the transmission channel and collisions cannot occur during normal operation.

In order to make the system flexible and widely applicable, it is necessary to:

- Be able to operate in real time – that is, to communicate at a specific moment, over a known maximum period, and be certain of that being the only transmission occupying the network at that time, so avoiding all risk of collision;
- Where necessary, communicate at a variable datarate, so taking an unknown time to complete the transmission.

For this purpose, FlexRay communicates in communication cycles, within which, network access is governed by two paradigms.[8] Thus, within a cycle, two completely distinct zones must be created:

- The first is time-triggered, known as the "static segment," divided into equal static slots, which are exclusively owned by certain CPUs, allowing those CPUs to transmit their data. These exclusive attributions of slots to CPUs are determined offline – i.e. during the design of the system, in principle eliminating all competition for network access and other edge effects when the system is online (operating normally). Of course, it is the responsibility of the designer or the network manager to correctly choose and define these values. This principle, known as time-division multiple access (TDMA) offers collision avoidance by its very structure. It is enacted synchronously in slots whose duration is clearly defined by the system designer;
- The other is event-triggered, known as the "dynamic segment."

This is almost all there is to it.

Communication cycle Let us briefly examine the general structure of the FlexRay communication cycle, illustrated in Figure 4.80.

FlexRay communication takes place in 64 recurrent communication cycles, each made up of a static segment, a dynamic segment, an optional symbol window and, finally, a slot in which the network is in idle mode (*network idle time* – NIT). We shall briefly discuss these slots.

Communication frame The general content of the FlexRay communication frame is made up of three distinct main segments:

- The header segment;
- The payload segment;
- The end-of-frame segment, known as the trailer.

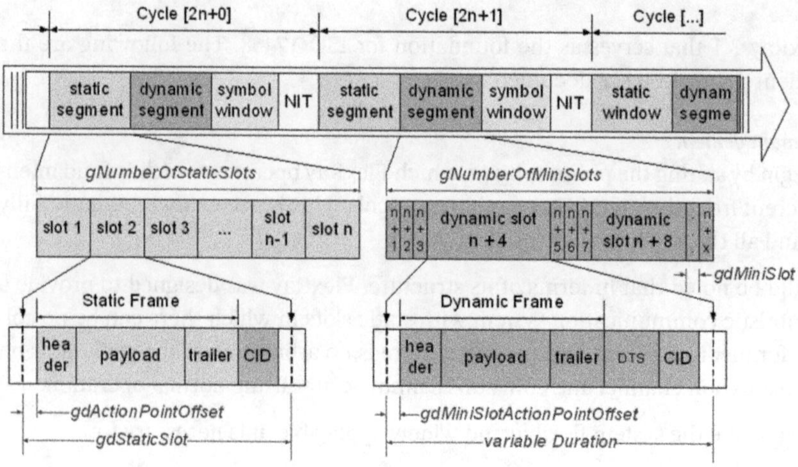

Figure 4.80 General structure of the FlexRay communication cycle.

Medium access

The medium access techniques are one of the key points about this protocol. To begin with, it is the duty of each system designer to define the tasks of each of the nodes in the network (point-to-point connections, centralized tasks, etc., and, usually, distributed tasks) and to precisely define the tasks that need to access the network, deterministically or otherwise. Thus, conceptually, the designer has:

- Firstly, two possibilities on how to access the medium:
  - Either via the static segment for quasi-real-time tasks. If need be, this technique would allow each node to use multiple time slots during the same communication cycle;
  - Or via the dynamic segment for asynchronous (spontaneous) event-triggered tasks, subject to arbitration and dynamically adjustable bandwidth during operation, depending on the operational needs of the whole system.
- Secondly, in a communication cycle, relative shifting of a segment in relation to the next, recalling that, in principle, there should be no encroachment or interference between the static and dynamic segments, that totally different data can be sent on the two channels during the same time slot, and that different nodes can use the same time slots on different channels.

Let us explain how this special configuration works, and the possibilities it opens up. Figure 4.81 shows the static and dynamic segments in a communication frame, with the way in which they are divided being chosen by the designer.

This choice is attributable to the desired bandwidth allocation (in terms of Mbit/s), based on the operational requirements of the intended system. The payload is divided into two segments with appropriate dimensions.

Once this is done, it is necessary to assign time slots to each of the participants in the network that need quasi-real-time medium access. These time slots are irrevocably assigned to the nodes during the static segment. This is shown in Figure 4.82. In this figure, for example, application A (brakes) can communicate during the first time slot, application B (suspension) during the second, application E (clutch control) during the fifth, etc.

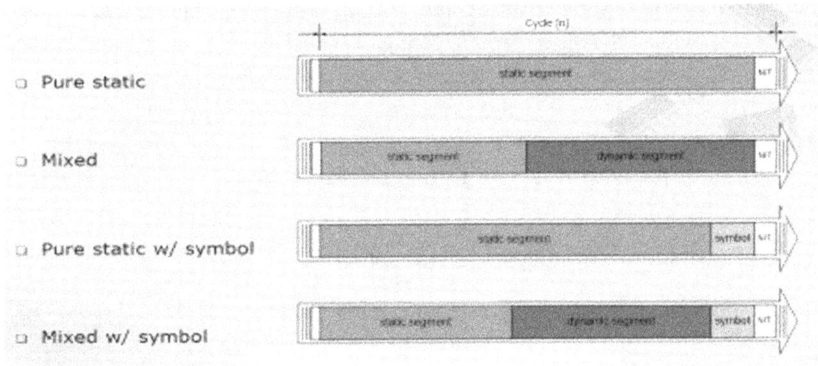

Figure 4.81 Static and dynamic segments, arranged as chosen by the designer.

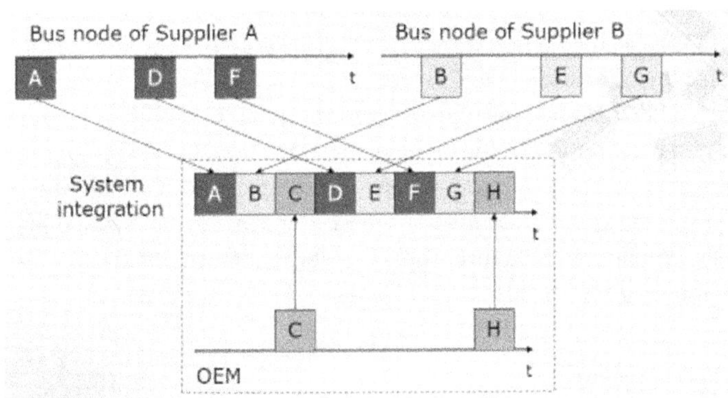

Figure 4.82 Example of division of time slots between different applications.

This method greatly reduces the time and cost of testing, and increases flexibility for development, as variants are easily possible, and the competitive choice of industrial partners to manufacture the system. All of this offers FlexRay an undeniable advantage over other protocols.

Redundancy

For vehicles that have ever fewer mechanical systems, makers are turning to X-by-wire as a solution (braking, steering, suspension, road handling, the body zone, etc.), so for safety's sake, redundancy is needed, and FlexRay offers it with its two connection channels.

The future of FlexRay

In 2020, there were rumors emerging frequently about the future of FlexRay. Some commentators, who are focused on security and functional redundancy, absolutely swear by it. Others predict it will be displaced in a few years' time, by CAN XL or 10 Mbits/s Ethernet on a Single Twisted Pair, able to transport larger amounts of data (see Chapter 5).

In conclusion, it is highly probable that, in the long term, Ethernet and FlexRay will continue to coexist – Ethernet for the sake of greater bandwidth in non-safety critical applications and FlexRay for some years to come, in safety-critical applications where redundancy is required, such as chassis controls (steering, suspension, etc.) and *X-by-wire* (steering, brake controls, accelerator, etc.). It is for this reason that FlexRay is presented here.

FlexRay components

Refer to the authors' previous works for further information.

4.5.5 MOST (media-oriented systems transport)

To recap, the main purpose of MOST was (and remains) to transport audio data in a vehicle. It is also used for video applications in a vehicle, when the digital datarates can be adapted to onboard applications (e.g. MPEG2).[9]

General

Today, in a vehicle, in addition to conventional audio applications, it is necessary to establish links between radios, navigation devices, and the associated systems, including display units (multiple screens in the dashboard, seats, etc.), players, CD changers (audio, video CD, etc.), USB readers, speech recognition systems, mobile telephone systems, the active sound distribution in the vehicle, the HMI, etc. It is also important to have infrastructures in which certain modules can easily be added or removed.

Before briefly recapping the specific features of MOST, in terms of associated digital datarates, Figure 4.83 summarizes the main properties of the majority of audio/video signals that can be sent over a MOST bus.

The concept of MOST

MOST was designed to distribute serial and digital AV and multi-media signals in an automobile. Let us briefly recap the main technical features, starting with the physical layer.

Physical layer and medium

The physical layer, originally designed around a twisted pair (copper wiring), has evolved. Today, MOST also uses plastic optical fiber (POF), to simultaneously benefit from a higher digital datarate (150 Mbit/s, and claimed higher than 1 Gbit/s), a broader range of applications and better immunity to external interference, and to avoid polluting the surroundings with radiation.

Topology

The applicative topological structure of this protocol is a daisy chain (a form of ring topology), as illustrated in Figure 4.84.

– 4 stereo audio channels	(4 . 2) 1.4 Mbit/s	11.2 Mbit/s
– multiplexed video	+ between 2.8 and 11.0 Mbit/s	+ between 2.8 and 11.0 Mbit/s
	+ a reserve of 4.0 Mbit/s	+ 4.0 Mbit/s
Total		= between 18 and 26.2 Mbit/s

Figure 4.83 Example of datarates of AV signals in a vehicle.

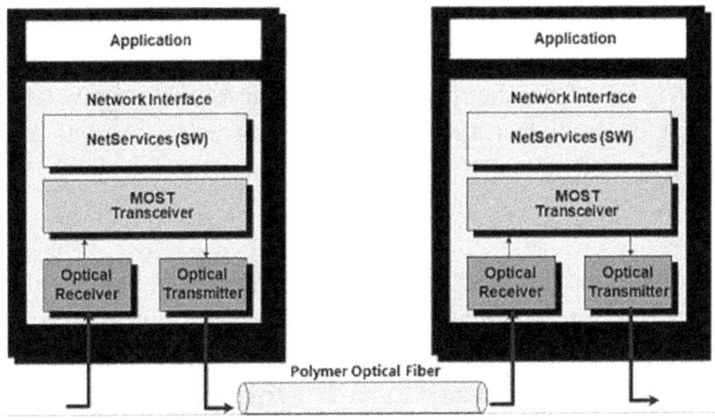

Figure 4.84 Daisy chain topology.

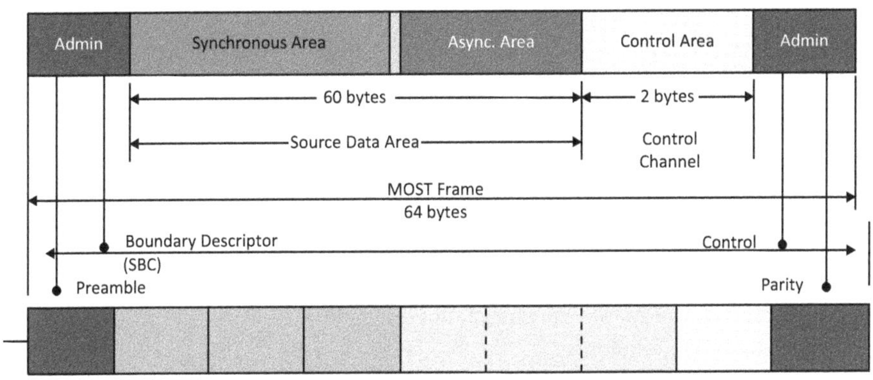

Figure 4.85 MOST communication frame format.

Data transfer and associated datarates

Digital data are transferred in two different modes: synchronous or asynchronous.

Synchronous operation Synchronous mode is the most common mode of operation. In this case, a master sends a reference clock to all other participants in the network. By synchronizing with that clock, the other nodes negate the need for buffers and devices to sample the transmitted information.

Therefore, we merely need to have recurrent frames in the same format in order for each participant in the network to find the octets (channels) of interest to it. This format of a communication frame means numerous functional configurations of nodes can easily be predetermined (see Figure 4.85).

For example, if 6 channels are reserved for the transmission of audio tracks:

6 channels × 8 bits/channel × 44 100 = 2.1168 Mbit/s
and 29 (from channel 12 to 42) to transport video signals:
29 channels × 8 bits/channel × 44 100 kHz = 10.231 200 Mbit/s

Note that this last value is sufficient to multiplex a number of video sources with acceptable quality (resolution) on the same medium.

Asynchronous operation In this mode of operation (without a clock provided by one of the participants), the maximum datarate is 14.4 Mbit/s and the associated dedicated channels operate at 700 kbit/s. Amongst other things, asynchronous mode allows us to transmit short bursts of signals, including those for navigation voice prompts and other driver assistance messages.

The future of MOST

The future of MOST does not look bright. Predictions are being made every day which say that it is on its way out. Certain people believe it will be phased out in favor of Ethernet (100 MB or 1 GB or faster), transporting larger quantities of data (see Section 5.3). In conclusion, it is likely that, in the near future, Ethernet will replace MOST, offering greater bandwidth in non-safety critical applications (infotainment, audio, and video are not too critical, but there may be problems with synchronization between sources). We shall examine this in detail when presenting Ethernet TSN and AVB in Sections 5.4.1 and 5.4.2.

4.5.6 LVDS

LVDS (low-voltage differential signaling), standardized by ANSI/TIA/EIA-644-A, is a standard differential data transmission method conforming to the requirements of high-datarate telematic applications. This standard supports datarates from 100 Mbit/s to 2 Gbits/s, with low energy consumption. In addition, it offers numerous other advantages, including:

- Compatibility with low-voltage power supply;
- Robust signal transmission;
- Low noise generation;
- High noise rejection.

To transmit, the LVDS signal uses two lines operating in differential mode, which leads to significant common-mode rejection and a significant gain in noise tolerance. Because of the differential mode, the signal swing can be reduced to a few hundred millivolts, enabling the system to achieve faster datarates, because the signal rise time is now far shorter.

Commonly used LVDS transceivers operate in dedicated point-to-point configurations (see Figure 4.86).

In this configuration, LVDS is capable of transmitting signals at a high datarate over significant distances, using little energy and producing little noise and pollution. These performances can be achieved with a CAT3 cable and connectors, and/or FR4 hardware. Other topologies/configurations are also possible, but are rarely used.

Figure 4.86 Physical layer of LVDS in point-to-point mode.

	CANCAN FDCAN XL(controller area network)	LIN(local interconnect network)	MOST(media oriented systems transport)	LVDS(low voltage differential signaling)	FlexRay
Origins	Bosch GmbH CiA	Consortium	Proprietary		Consortium
Release date	CAN 1981 CAN FD 2000 CAN XL 2018	2001	2001	1994	2005
Standards	ISO 11898 ...	ISO 17987	None		ISO 10681
Topology	Shared serial bus	Serial bus with master/ slave architecture	Ring	Point-to-point	Serial bus, single or double channel
Datarates	1–1000 kbit/s 8 Mbit/s in FD 10 Mbit/s in XL	20 kbit/s	50 Mbit/s (Cu) 150 Mbits/s (POF)	1–2–4 Gbit/s	5–10 Mbit/s
	Highly reliable	Slow	Very fast	Very fast	Fast and reliable
	Collision management: CSMA/CR	Requires only one shared cable (whereas CAN requires two)			Medium access is divided: TDMA (no repeat in case of error)
Cost	Cheap, and very widely available	Less costly than CAN	Expensive	Inexpensive	Expensive
Applications	Automotive domains: motor, chassis and body	Electronics for small systems (mirrors, electric seats, accessories)	Video connections and infotainment data stream control	Camera connections, video data and screens	Highly dynamic controls and security using X-by-wire (chassis monitoring, active suspension, braking, adaptive cruise control)
Supplier	Multiple	Multiple	Single	Multiple, but generally mutually incompatible	Multiple

Figure 4.87 Overview of the main specifications for the different protocols and networks existing between 2015 and 2020 for automotive applications.

Applications of LDVS

There are numerous applications for LVDS in the automotive field. They include point-to-point transfer of analog signals between cameras (e.g. a reversing camera) and the CPU. In addition, the data stream for digitized or native digital video, HDTV, and color graphics require a great deal of bandwidth (100 Mbit/s or more):

- High-resolution screens in industrial applications;
- Flat screens in the automobile market.

Today, the transport and transfer of data within a system are the main use of LVDS solutions, requiring standard communication protocols, such as IEEE 1394 (FireWire), Fibre Channel, and Gigabit Ethernet.

The future of LVDS

Here, again, there are noises frequently being made (some of them good, some bad) about the future of LVDS. Some foresee it being discontinued in favor of Ethernet at 100 MB or 1 GB or even faster, which is capable of transporting larger amounts of data (see Chapter 5). In conclusion, it is highly likely that some day Ethernet will replace LVDS, in order to take advantage of the greater bandwidth in non-safety critical applications (infotainment, audio, and video are not overly critical, but there may be problems of synchronization between sources). We shall address this matter in detail in the presentation of Ethernet TSN and AVB in Sections 5.4.1 and 5.4.2. This situation may change rapidly in time, with the emergence of Gigabit Ethernet backbone networks, for multiple domains and multiple applications.

4.5.7 Overview of automobile heritage networks

To conclude this chapter, Figure 4.87 offers a brief overview of the main peculiarities of the different networks for automotive applications that have been handed down to the current generation, listing them in order of their release date on the market.

What we now need to do is to carefully examine Ethernet, its story, the mysteries it holds, and its industrial variants that have been adopted in the automotive industry. That is the subject of Chapter 5.

Notes

1 For further information, see the detailed presentation of this integrated circuit at https://www.youtube.com/watch?v=Ucp0TTmvqOE.
2 Dominique Paret, *FlexRay and Its Applications – Real Time Multiplexed Network*, Wiley.
3 For further details, refer to: Dominique Paret, *Multiplexed Networks for Embedded Systems: CAN, LIN, FlexRay, Safe-by-Wire*, Wiley.
4 For further details, refer to: Dominique Paret, *Multiplexed Networks for Embedded Systems: CAN, LIN, FlexRay, Safe-by-Wire*, Wiley.
5 CAN with Flexible Data-Rate Specification – version 1 (April 2012).

6 This section is built from CiA documents and articles written by Florian Hartwich (Robert Bosch), Holger Zeltwanger and Yao (CAN in Automation), Magnus-Maria Hell (Infineon Technologies), and Oliver Garnatz, Peter Decker (Vector Informatik), and Bernd Elend and Tony Adamson (NXP Semiconductors). Many thanks to the team.

7 Dominique Paret, *FlexRay and Its Applications – Real Time Multiplexed Network*, Wiley.

8 Paradigm: The term "paradigm" is often used in relation to FlexRay, because of the numerous variants that exist around the fundamental principles.

9 Dominique Paret, *Multiplexed Networks for Embedded Systems: CAN, LIN, FlexRay, Safe-by-Wire...*, Wiley.

5

Ethernet and Automobiles

This chapter is devoted to the features of automotive Ethernet:

- Firstly, in order to have a clear understanding of how this technology has emerged, we offer the essential overview of industrial Ethernet;
- This overview leads into a discussion of the numerous variants of Ethernet;
- We shall then describe the specifications required for automotive Ethernet;
- We shall examine the consequences of the dawn of Ethernet for new vehicle hardware architectures; and
- Finally, we shall discuss the automotive Ethernet paradigms that are currently popular.

Let us begin by laying the groundwork for this part. For over a decade, there has been fierce competition over how to deliver very high-speed digital data transfer in automobiles:

- In one corner, the long-reigning champion, used mainly in the passenger compartment, for entertainment with a major focus on multi-media (infotainment), the MOST bus (a proprietary solution), is still putting up spirited resistance, with solutions ranging from 15 to 150 Mbits/s capacity, or even 2.5 Gbit/s using fiber-optic technology;
- In the other corner is the challenger for the ultimate title, already having made a name for itself in many other areas, Ethernet – with capacity between 100 and 1000 Mbits/s – which offers the possibility of transmission on non-shielded twisted pairs of copper cables. Ethernet is making its presence felt, not only in the passenger compartment, but also under the hood. It is used by all branches of new applications to connect the 10 Mbit/s FlexRay and the ADAS, safety, and infotainment systems. It is likely that, in time, it will be used also for autonomous vehicle applications!

It is certainly true that, in addition to the existing CAN (high-speed connection at 1 Mbit/s), CAN FD (5–8 Mbit/s), CAN XL (10 Mbit/s), and FlexRay (5–10 Mbit/s) solutions for in-vehicle network (IVN) architecture, the automotive industry has turned its

Autonomous and Connected Vehicles: Network Architectures from Legacy Networks to Automotive Ethernet, First Edition. Dominique Paret and Hassina Rebaine.

attention to Ethernet, which is commonly used for both industrial and consumer applications. Ethernet:

- Is subject to IEEE standards;
- Is flexible and simple;
- Offers high bandwidth;
- Has high component reuse;
- Is interoperable with IP (Internet Protocol);
- Has associated software packages;
- Has available development and testing toolkits.

However, in order to become a true success story in the automotive field, it needs to be optimized in terms of applicability, performance of the physical layers, datarates, scalability, pollution and RF/EMC compliance, energy consumption, robustness, etc. – and, as usual, all this must be achieved for as low a cost as possible.

Ethernet specifications are governed by the "IEEE 802.3" suite of standards, issued by the Institute of Electrical and Electronics Engineers. The generic protocol has innumerable variants, in terms of: the internal make-up of its communication framework; its network access protocols; its numerous datarates; and the concrete designs of its physical layers. We shall come back to this point later on. Is this variability not part of the appeal of Ethernet, though, one might reasonably ask. Obviously, in this book, it will be examined in relation to its possible uses in automobiles, depending on the possible or desired architectures and the intended ranges of applications. It is this automobile/applications aspect that reduces and/or influences the possible choices amongst the myriad possible "802.3xxx" variants.

Before moving on with this chapter, there are two important points to be made.

To begin with, to pre-empt a deluge of (rightful) protests, it should be noted that Ethernet has been in existence for over 40 years (it was invented in 1978!) and the 100 Mbit/s solution and IP technology have been available since the early 1990s. Why, then, has it taken nearly 20 years for the automotive industry to look at Ethernet as a potential next-generation networking solution on board vehicles? In other words, why has the solution not been looked at long before now?

The answer is that, at 100 Mbit/s, there has been no high-performing – and not prohibitively expensive – hardware available that would have been appropriate for use in vehicles. For example:

- Ethernet (at 100 Mbit/s or more) does not meet the EMI/RFI requirements of OEMs operating in the automobile market (it produces too much radiofrequency interference, or "noise", and is also overly sensitive to interference from other equipment in a car);
- Ethernet could not offer sufficiently low latency to achieve the microsecond-scale performance needed for communications requiring a fast response (for example, between the sensors and the control units);
- Ethernet did not have the means to control the allocation of bandwidth to the different data streams, and therefore could not be used to transmit data shared between multiple types of sources;
- Ethernet did not offer any means of time synchronization between the peripheral devices, and thus, the ability to handle data samples from multiple devices simultaneously.

Second, why is it that an inexpensive 100 Mbit/s Ethernet physical layer has not been developed at an earlier stage? Quite simply, there were no high-datarate applications in automobiles to serve at an earlier stage! Rome was not built in a day, and neither were Ethernet solutions in vehicles.

We shall look at all of these points in detail to understand the final choice – typically, the reasoning behind a choice can be put down to the desire to get "the best possible solution for the lowest possible price." Firstly, though, let us examine what designers want to do with this technology – in other words, the desired applications in automobiles.

The stage is set, the lights are down. Curtain up!

5.1 Industrial Ethernet

5.1.1 A little history

The Ethernet standard was devised at the beginning of the 1970s, on the basis of an industrial demand. It was developed by Xerox laboratories, and then evolved into Ethernet II. It is sometimes referred to as DIX, in reference to the association of the Digital Equipment Corporation, Intel, and Xerox. Later on, Ethernet was included in the work on the OSI model in the early 1980s:

- Layer 1 (the elements in the physical layer, broadly defined by IEEE 802.3 standards);
- Layer 2 (the CSMA/CD access method corresponds to MAC (Medium Access Control) in the data link layer).

Since then, Ethernet technology has become totally independent of individual manufacturers, which partly accounts for its popularity. Although progress in datarates has meant cheap physical media can no longer be used, the technology can still be implemented simply. In addition, existing infrastructures are evolving to be able to support multimedia technologies without the need for substantial further investment.

In order to fully understand its applications in the world of automobiles, it is important to review the details of the numerous stages in its industrial history, and the set of documents published by the 802 Committee: GET IEEE 802, which defined four classes of Ethernet datarates:

- 10 Mbits/s Ethernet: the original definition of subcommittee IEEE 802.3;
- 100 Mbits/s Ethernet, or "Fast Ethernet;"
- 1 Gbits/s Ethernet, or "Gigabit Ethernet;"
- 10 Gbits/s Ethernet, or "10 Gigabit Ethernet."

Within each of these categories, there are numerous subdivisions. The most widely used of these are described below.

5.1.2 Principle

Ethernet connections are governed by a set of simple, general principles:

- All nodes are of equal importance in the network; there is no master network controller;
- Transmission may be full-duplex or half-duplex (signals sent in both directions alternately, but not simultaneously). Today, the use of half-duplex transmission is dwindling;

- Data are transmitted via paired copper wires or optical fibers, with a dedicated cable for each direction of communication. This is full-duplex mode, and in an automobile system, a single pair of wires would be used for each connection;
- It is possible to add, connect, or remove a network element without impeding the operation of the whole system. This is highly practical for offering different options on vehicle models.

In relation to the principles stated above, the widespread adoption of the switching mode used simplifies the access method, as it eliminates all the collision-management infrastructure. In closing, the use of these principles has shown that it is much easier to design networks and equipment using Ethernet than with many other technologies.

5.1.3 Ethernet frame in IEEE 802.3

The sections below outline how the Ethernet frame is constructed. They give the description and format of the Ethernet frame under IEEE 802.3 (which is, in fact, structurally very simple), and information about the technology-specific requirements for CSMA/CD (carrier sense multiple access with collision detection).

Access method and specifications of the physical layer

Remark: For further details, please refer to the hundreds of official IEEE documents published on the subject.

In Ethernet protocol, a message/frame is made up of three distinct parts (see Figures 5.1 to 5.4 for the general overview and detailed examination):

- The preamble (made up of 7 + 1 = 8 bytes = 64 bits);
- The Ethernet packet (addresses, length, variable data sizes);
- The control trailer (composed of 4 bytes = 32 bits).

When a frame is received by an online device, that device transmits only the packet part of the frame to the operating system.

Let us briefly examine each of these parts, one by one.

Preamble

The preamble, which constitutes the start of the message, is a fixed binary configuration of (7 + 1 =) 8 bytes = 64 bits, used to help and/or ensure the synchronization of the receiver circuits and thus allow the transmitter/receiver to tolerate slight differences in their respective bitrates. This 8-byte preamble is divided into two fields (see Figure 5.2):

- A 7-byte field whose value is a long succession of (56) bits (NRZ coded), alternating between 1 and 0 – e.g. 1 0 1 0 1 0... – serves as a *clock run in* signal. This signal enables the receiver to (re-)synchronize its clock with the incident signal;
- A 1-byte (8 bit) signaling field, called the *start of frame delimiter* (SFD), whose value is 1010 1011, concludes the preamble.

The next part of the signal is the frame packet.

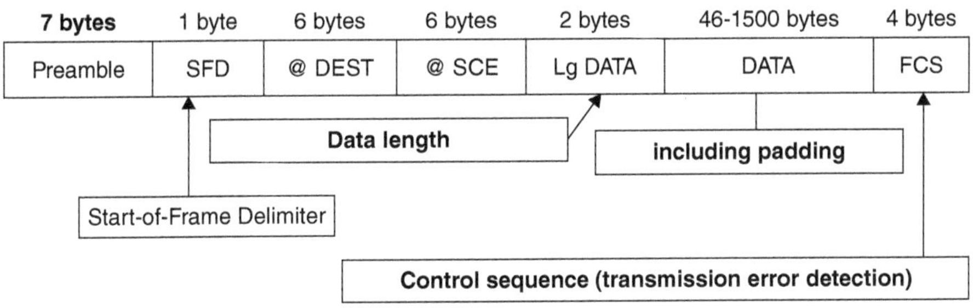

Figure 5.1 Typical format of an Ethernet frame.

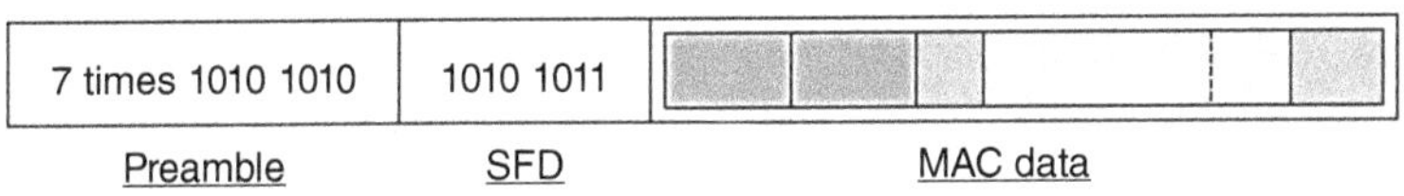

Figure 5.2 Ethernet frame header.

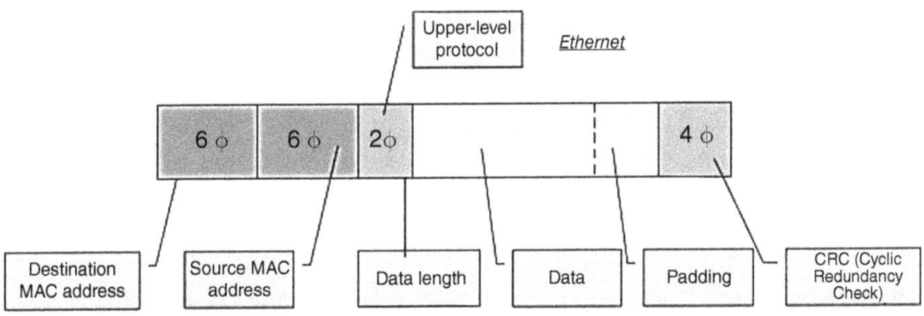

Figure 5.3 Ethernet frame packet.

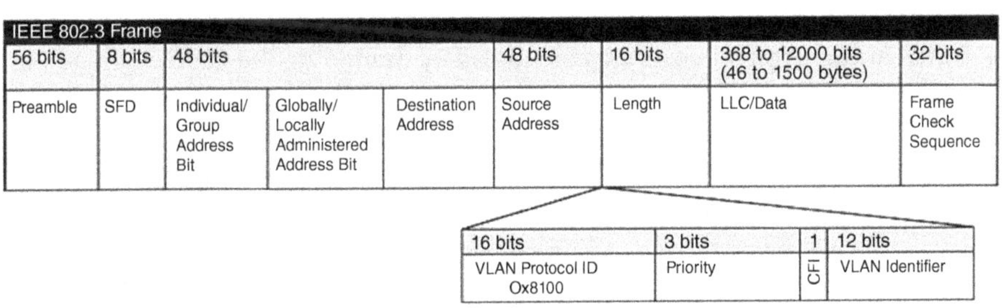

Figure 5.4 Extended frame details.

Ethernet frame packet

An Ethernet frame packet is made up of two distinct parts (see Figure 5.3):

- A header, composed of (6 + 6 + 2 =) 14 bytes;
- Data (which may vary in size).

Header The header field contains information about the Ethernet protocol itself. This header is made up of three distinct fields:

- Destination address, encoded on 6 bytes = 48 bits;
- Source address, also encoded on 6 bytes = 48 bits;
- Message length/type on 2 bytes = 16 bits.

Destination address The destination address, encoded on 6 bytes = 48 bits, indicates the element (the node) to which the data are being sent. Normally, all other elements discard packets whose destination address shows that they are for a different node. Only the node with the corresponding address reads the packet, stores it in an internal buffer memory, and then interrupts the running of its CPU so that its operating system can read and process the buffered packet. However, there are three exceptions to this fundamental rule, whereby the destination address refers only to a single element that is to use the packet:

- All bits in the destination address are at 1, which corresponds to the so-called standard *broadcast* address. The packet thus broadcast is processed by all nodes on the network.
- By mutual agreement, a subset of elements can form a *multicast group* and intercept all packets with a particular destination address that does not correspond to an element on the network. An element may belong to more than one multicast group.
- An element can place its interface(s) in "promiscuous" mode and process all packets sent over the Ethernet network. This is commonly known as "packet sniffing." It serves both to monitor the network and to carry out diagnostics.

Source address No clarification is required. The source address is encoded on 6 bytes.

Data length With respect to the data length:

- The data sent may be between 1000 and 1500 bytes;
- Different frame moments are differentiated by examining the data length and the types of field (Figure 5.4).

Data The data field generally contains data from the network layer. The field is between 1000 and 1500 bytes in length.

Padding

Trailer – FCS (frame control sequence) The "trailer" contains a CRC (*cyclic redundancy check*), built on a polynomial of degree 32, calculated only on the packet, so that the destination element can check the value of the incoming message CRC against the CRC that it generates locally. This helps guard against packet corruption (which is usually due to electronic interference).

We have now briefly described the generic Ethernet frame and its content.

By way of a real-world example, Figure 5.5 shows how a frame is filled.

Let us now examine the numerous variants of Ethernet.

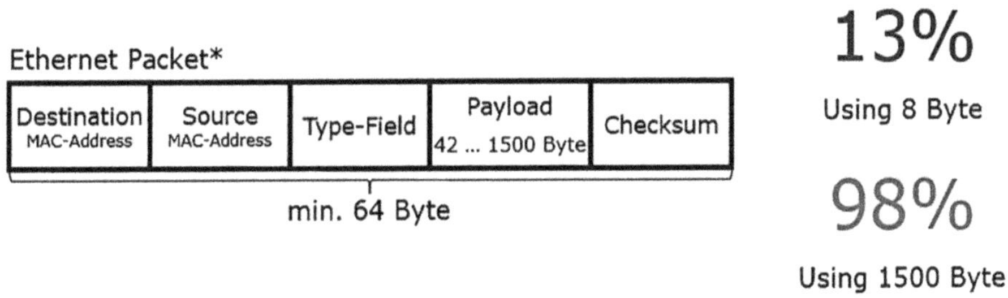

Figure 5.5 Example of frame filling.

5.1.4 Common Ethernet variants

We shall discuss the most widely used Ethernet variants, beginning with half- and full-duplex modes.

Half-duplex

When the first Ethernet standard was published in 1985, all communications were in half-duplex, meaning that a node could not simultaneously transmit and receive. Therefore, these nodes operated in the same collision domain. Access to the medium needed to be controlled at a higher level and synchronized across all existing nodes in the potential collision domain. Because of the danger of collisions, real-time communications cannot be assured.

Full-duplex

Full-duplex Ethernet was standardized in 1998, with the transition from 802.3 to IEEE 802.3x. In full-duplex mode, a node can simultaneously transmit and receive. In principle, a full-duplex link doubles the bandwidth of the network, but in practice, capacity is limited by each node's internal processing time. In full-duplex, a maximum of two nodes can be connected over a single link, and thus each node in the network can have a unique collision domain. This operation completely prevents collisions and negates the need for the traditional CSMA/CD. Typically, either *node-to-switch* or *switch-to-switch* configurations are used. Therefore, this technology is not viable as a real-time solution without using smart, fast switches capable of connecting network segments with unique collision domains for each node. This is known as *switched Ethernet.*

Switched Ethernet

Switched Ethernet represents a major leap forward. To solve problems stemming from collisions, switches have been developed to maximize the available bandwidth, particularly on simple media such as a twisted pair of cables. A switch is a multiple bridge, serving as a point-to-point link between a host element and the switch, which then defines a particular segment with its own collision domain. With switched Ethernet, the elements/nodes only compete for media access within their shared collision domains. The use of a destructive, non-deterministic network access system,

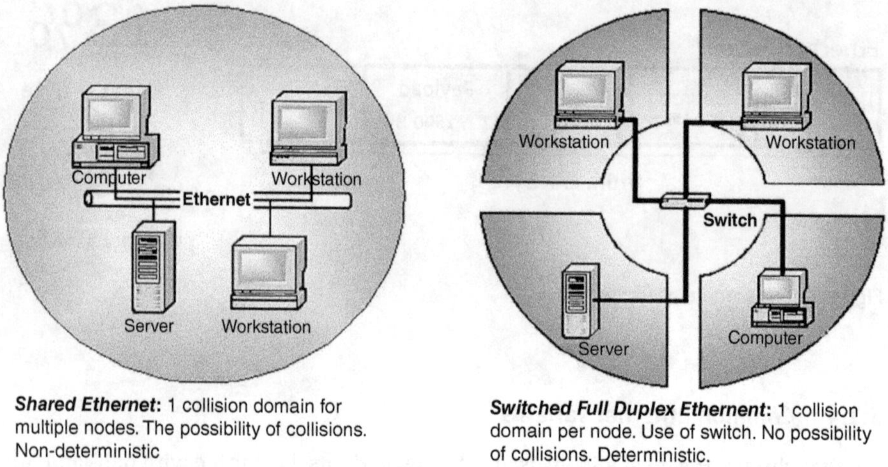

Figure 5.6 Comparison between switched full-duplex and shared Ethernet.

such as CSMA/CD, negates the viability of shared Ethernet as a real-time networking solution.

The most widely used method to prevent collisions, which provides an almost deterministic Ethernet, is to create unique collision domains for each node. This ensures that the node has exclusive use of the medium (bus), eliminating competition for access. This result is obtained by using full-duplex connections and hardware elements such as switches and bridges. They are able to isolate collision domains, segmenting the networks, because each port in the switches or bridges is configured as a single collision domain (see Figure 5.6).

In the case of switched Ethernet:

- The hardware topology of the network is no longer a bus, but rather a star formation (as we see with hubs);
- The software topology is no longer that of a shared bus, but again a star formation;
- The communications between two individual elements of the same level (pairs) are thus isolated (unlike what happens with hubs and coaxial Ethernet buses), which increases the network's overall transmission capacity;
- Each host 1/host 2 pair communicates through a sort of "virtual point-to-point link" formed by the switch;
- Communications can take place in full-duplex (simultaneous transmission and reception), and there is no longer any risk of collision. To achieve the same result with CSMA/CD technology, the transmitter listens to the network as it is transmitting, and if another node (a receiver) transmits at the same time, a collision takes place and it is deactivated (which is incompatible with the full-duplex mode);
- Distances are no longer constrained by the signal propagation speed/time (there are no longer any collisions to detect, as there are in a CAN); they are constrained only by the line losses (signal attenuation in the cables).

Switches and bridges

A bridge works as an interrupter, and has only two ports. Switches have more than two ports, each port being connected to a network segment. They are hardware elements that can be used to create unique collision domains by segmenting the general network. They can operate in half-duplex or full-duplex mode. When switches are used in full-duplex mode, there is no risk of collision in any segment. Smart, fast switches can be used to create a real-time communication network, using the IEEE 802.3 standard.

Although switches are devices whose operation belongs to the data link layer (layer 2 in the OSI model), modern switches are also capable of performing the data-based switching functions found in layers 3 and 4. Layer-3 switches can operate on information supplied by IP (such as the version of IP, the source/destination address, and the service type). Level-4 switches can operate on the basis of the source/destination port, or even higher-level application information.

An example of the application of this technology in an automobile is given in Figure 5.7.

5.1.5 Variants of the physical layer of industrial Ethernet

Lower layers of the OSI model (layers 1 and 2)

Figure 5.8 illustrates the series of layers and sublayers in the OSI model.

It should be remembered that, in general, during communication between two nodes on a network, the data stream created at the transport layer (level 4 in the OSI model) (for example: TCP segments) is encapsulated multiple times. Figure 5.9 illustrates the different layers of encapsulation before a message is sent:

- Firstly, when it is sent to the network layer (level 3 in the OSI model), a message is broken up into packets by a host of communication protocols (for example: to be used by the Internet, using the specific Internet Protocol, or IP). It should be noted that IP variants are part of a long string of Internet protocol suites, and allow unique addressing for all connected terminals;

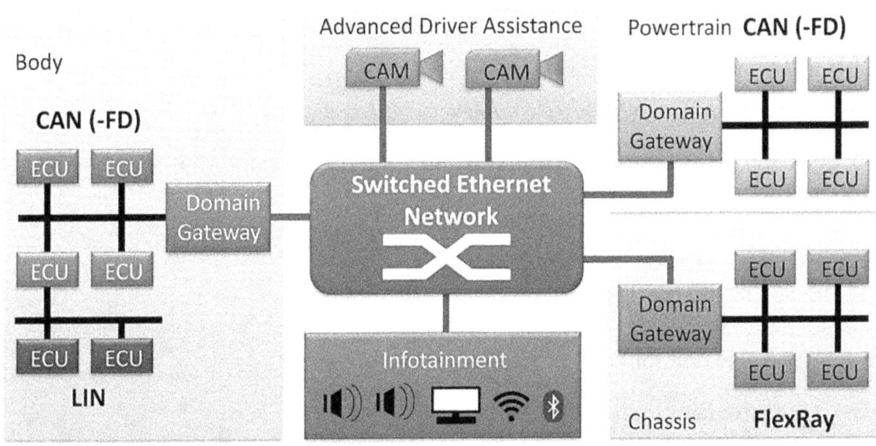

Figure 5.7 Example of application in an automobile.

	PDU	Layer	
Upper layers	Data	7	Application
		6	Presentation
		5	Session
	Segment/Datagram	4	Transport
Hardware layers	Packet	3	Network
	Frame	2	Data link
	Bit	1	Physical

Figure 5.8 Layers and sublayers in the OSI model.

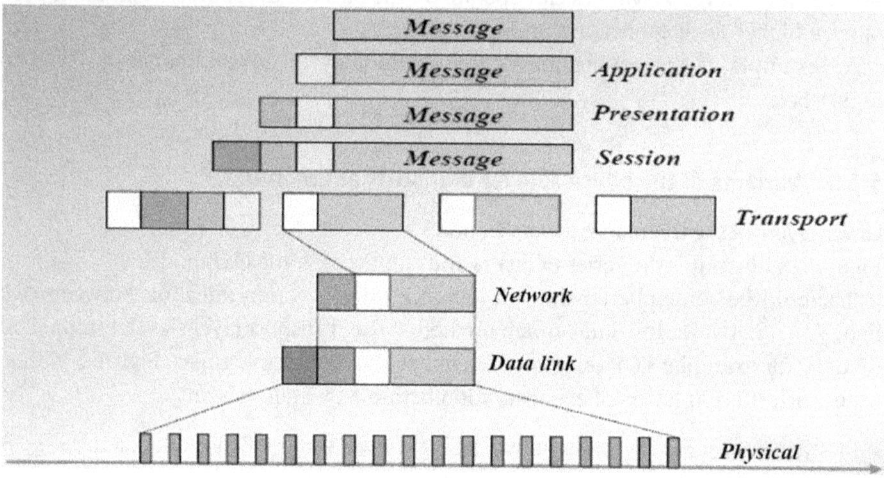

Figure 5.9 Successive layers of encapsulation before a message is sent.

- These packets are then passed to the data link layer (level 2) and are once again packaged up into frames (Ethernet frames, for example), and finally transmitted over a physical, hardware layer, typically referred to as the "PHY" layer (level 1 in the OSI model).

Layer 2 is well known, with its two sublayers LLC and MAC.

On the other hand, layer 1 of Ethernet may require a somewhat lengthy further explanation.

Ethernet layer 1, physical layer ("PHY")

In IEEE 802.3, Ethernet layer 1, known as the "Physical layer," or more simply, "PHY," is divided into several sublayers, as shown in Figure 5.10. The names are as follows, from top level to bottom:

- **AUI** (attachment unit interface);
- **Reconciliation**, which maps the physical statuses (carrier loss, collision, etc.) in the MAC layer;

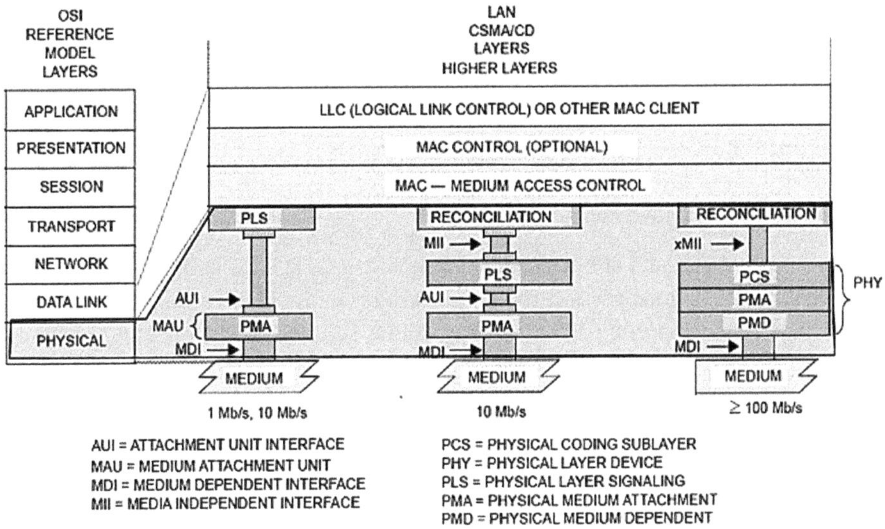

Figure 5.10 The multiple sublayers of the PHY layer of Ethernet.

- **(G)MII** (media-independent interface) (G in the case of Gigabit/s communication), which provides an *n*-bit transmission/reception interface for the MAC layer and recognizes digital datarates of 10, 100, or 1000 Mbit/s;
- The **PHY** (physical layer) entity – the part of the phsical layer that sits between the MDI (*medium-dependent interface*) and the MII (*media-independent interface*), GMII (*Gigabit media-independent interface*) or XGMII (*10 Gigabit media-independent interface*) – is made up of the sublayers PCS (*physical coding sublayer*), PMA (*physical medium attachment*), and, sometimes, the WIS (*WAN interface sublayer*) and PMD (*physical medium-dependent*). The PHY layer encapsulates the functions that transmit, receive, and manage encoded signals, which are placed on and recovered from the physical medium:
  - The PCS encodes, multiplexes, and synchronizes outgoing character streams (e.g. 4B/5B coding, etc.);
  - This sublayer is used in certain types of port to link the MII, GMII or XGMII, and the PMA. PCSs contain functions to encode data bits to pass or transmit via the PMA and decode the conditioned signal received from the PMA. There are multiple PCS structures.
  - PMA: This part of the physical layer contains signal transmission and receiving functions, and, depending on the PHY, collision detection and skew alignment (serialization of character streams, clock resetting, etc.).
  - Auto-negotiation: in Ethernet, auto-negotiation is an automated mechanism with no prior configuration between two connected elements (a station and a switch). By means of this mechanism, firstly, the connected elements declare and exchange the technical capabilities of their modes of communication, possibly measuring the quality of the connection between them, and then negotiate and select the best mode and transmission parameters that they share, such as the highest supported speed, half- or full-duplex mode, with full-duplex being

preferred for the same speed, and whether or not to use flow control. Auto-negotiation is required for Gigabit Ethernet in 1000BASE-T on copper twisted pairs;
  - PMD (physical medium-dependent sublayer).
- MDI (medium-dependent interface): RJ45, etc.;
  - PLS (physical layer signaling);
  - Medium: UTP, fiber, etc.

Before examining the use of Ethernet in automobiles (see Section 5.3), we must begin by looking at the numerous possible industrial variants of the PHY layer for Ethernet networks capable of supporting digital datarates of 10, 100, and 1000 Mbit/s and 10 Gbit/s!

Ethernet – 10 Mbit/s

10 Mbit/s Ethernet was Chapter 1 of the story. Also see Chapter 3 for information on CAN XL.

10BASE-T

10BASE-T works with a minimum of four wires (two twisted pairs, conventionally on a CAT-3 or CAT-5 cable [see Section 5.3], with an RJ45 connector). A hub or switch is at the center of the network, with a port for each node. This configuration is also used for 100BASE-T and Gigabit Ethernet (using a CAT-6 cable). Though the inclusion of a central hub might suggest a star topology, it is actually a bus logical topology (all transmitted signals are received by all connected machines). A star logical topology only occurs when a switch is used. We shall see the new potential applications for this technology in automobiles in Section 5.3.

Fast Ethernet – 100 Mbit/s

The wide range of Ethernet networks capable of a datarate of 100 Mbit/s is referred to as "Fast Ethernet." There are many variants.

100BASE-T

The term 100BASE-T refers to any variant of the Fast Ethernet standard, carrying 100 Mbit/s of data on a **twisted pair** (hence the T). There are multiple variants, including 100BASE-TX, 100BASE-T4, 100BASE-T2, and 100BASE-T1.

100BASE-TX

100BASE-TX is the predominant industrial version of Fast Ethernet on twisted pairs. This version uses **two twisted pairs** (so four wires) to communicate (because it requires one pair for transmission and another for reception in order to obtain the same bitrate in both directions), whether the link is half or full duplex with a CAT-5 or 5e cable (Figure 5.11).

The logical topology is a bus if it uses a hub or a star if it uses a switch, as is the case with 10BASE-T.

100BASE-TX – RJ45: this is the same solution, but the type of connectors used is specified: RJ45 cables.

100BASE-T4

This variant is capable of a datarate of 100 Mbit/s in semi-duplex only, on CAT-3 cables. It uses **four twisted pairs** (thus, eight wires) and is now obsolete.

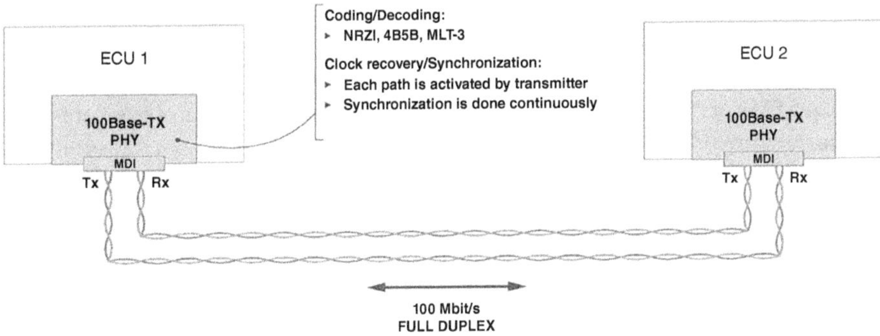

Figure 5.11 Conventional configuration of 10 or 100BASE-TX PHY.

100BASE-T1

This variant corresponds to IEEE 802.3 specification (physical layer) for full-duplex 100 Mbit/s Ethernet on a **single balanced twisted pair** (differential), so has only two wires. We shall discuss this solution in great detail in Section 5.3, on Ethernet in automobiles.

100BASE-FX

To recap, the 100BASE-FX specification describes the use of 100 Mbit/s Ethernet **on fiber-optics** (hence the F). When using twisted pairs, communication can be achieved over a distance of some 100 m. By contrast, when using fiber-optic technology, it is possible to achieve datarates of 100 Mbit/s over a distance of 2 km in full-duplex or around 400 m in half-duplex with multimode fiber optics. Currently, the use of fiber optics for this datarate appears to be out of the question for automobile applications, mainly for reasons of cost.

Gigabit Ethernet – 1 Gbit/s = 1000 Mbit/s

The broad family of Ethernet networks supporting a 1000 Mbit/s (1 Giga bit/s) datarate is called "Gigabit Ethernet." Gigabit Ethernet was originally defined in two versions: half-duplex and duplex. Note that the IEEE 802.3ab interface uses a particular coding scheme that keeps the modulation speed as low as possible and facilitates twisted pair transmission.

Half-duplex

In the half-duplex version, the medium is shared, and support access is managed by the CSMA/CD protocol. To ensure compatibility with the higher layers, the minimum frame size is kept at 64 bytes, in keeping with those layers. However, in order to preserve a reasonably high maximum length for the shared support, frames that are smaller than 512 bytes are padded with empty bytes, to make up 512 bytes. After the sending of this first frame, which allows a device to occupy the shared support, the same device is able to send other frames, this time smaller than 512 bytes, for a duration corresponding to 8000 bytes, without issue. This process of temporarily reserving the medium for a frame larger than or equal to 512 bytes ensures a high useable datarate, even if the transmission is dominated by small frames (< 512 bytes). The maximum frame size is the same as for Ethernet (1514 bytes).

Duplex

The duplex version of Gigabit Ethernet can be used to interconnect elements through a bidirectional medium (theoretical datarate of 2 × 1 Gbit/s), used by two elements at once, making CSMA/CD obsolete. Not only does this help to increase the useable datarate, but also to circumvent timeslot constraints and thus increase the maximum size of the network to the limits imposed by the transmission technology (3000 m for 1000BASE-LX on a single-mode optical fiber).

For transmissions on optical fibers (LX, SX) or on a balanced shielded twisted pair (150 m, CX), encoding on 8 bits/10 bits (2 synchronization bits for 8 useable bits) (see Section 5.3) synchronizes the transmission and reception clocks. The presence of this encoding means that the support must function at 1250 Mbit/s to achieve a useable datarate of 1000 Mbit/s.

1000BASE-X

1000BASE-X is the generic name of the modular interfaces (*transceivers*, also known as GBICs or SFPs depending on the technology they use) adapted to the medium in use (single-mode or multi-mode optical fiber, copper cables).

1000BASE-T – IEEE 802.3ab

In 1000BASE T, the "T" refers to a connection using an unshielded twisted pair.

1000BASE-T IEEE 802.3ab is a Gigabit Ethernet standard using a pair of cables. It can deliver a bitrate of up to 1 Gbit/s and operates on **four twisted pairs** (4 × UTP, so 8 wires) of category 5 (class D) or above (in accordance with EN 50173-2002) (see Figure 5.12). Each network segment can have a maximum length of 100 m.

With the 1000BASE-T standard, it is possible to exploit the existing cabling in most cases (4 × UTP, 100 m, category 5). Multiple techniques are combined to deliver a theoretical datarate of 1 Gbit/s on a physical support of that type measuring 100 m:

- Four twisted pairs (see Figure 5.13);
- Full-duplex operation of each pair at a modulation rate of 125 Mbauds, transmitting 2 bits per modulation period (per clock signal), so at a bitrate of 250 Mbit/s, giving a total of 1 byte across all four pairs, in each direction;
- PAM 5 coding (5 levels and 4 levels for the two information bits per modulation period, with the fifth used by the error correction code), supplemented by FEC (*forward error correction*);

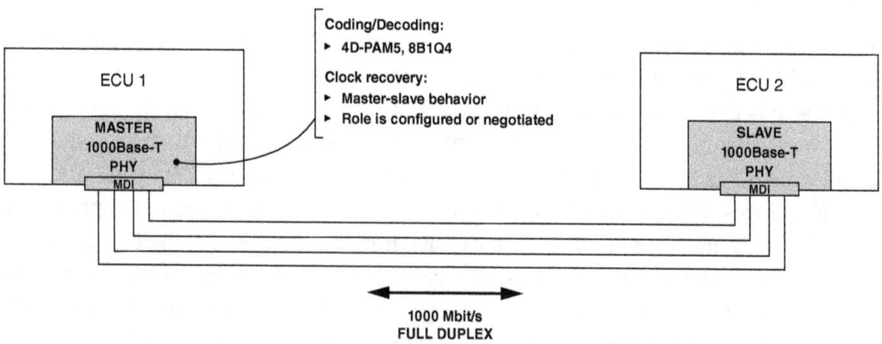

Figure 5.12 Conventional configuration of 1000BASE-T PHY.

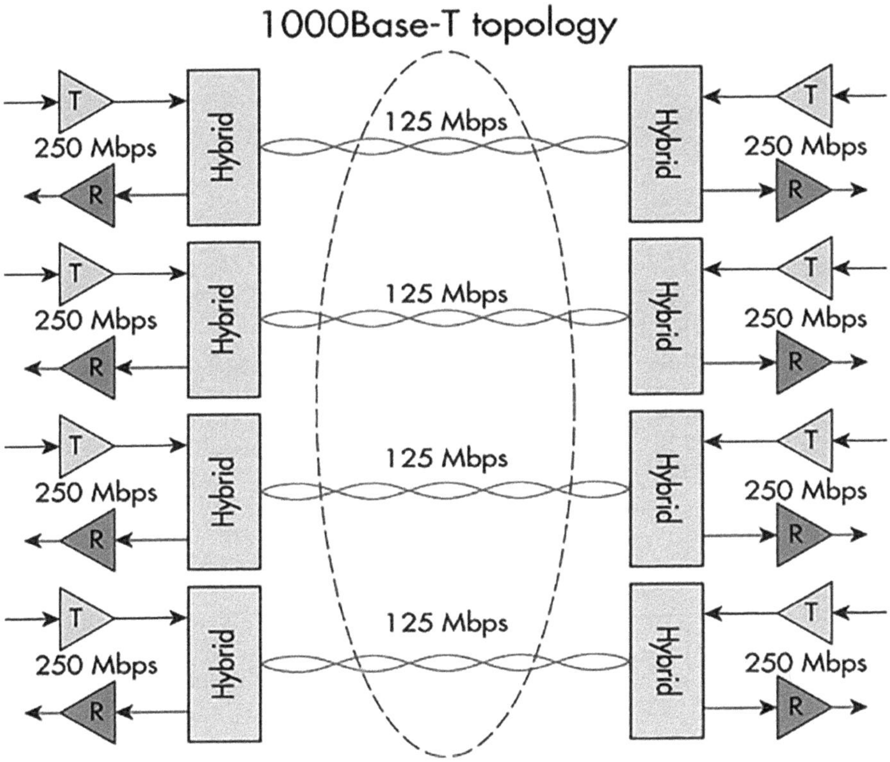

Figure 5.13 Conventional topology of 1000BASE-T PHY.

- Each twisted pair is used simultaneously for transmission and reception, so full duplex function is possible;
- The signal spectrum is adapted to the cable characteristics (with *pulse shaping*) and the received signal is balanced by nonlinear filters;
- 1000BASE-T is designed to work with Gigabit Ethernet hubs in semi-duplex mode or with switches in full-duplex mode;
- The topology is always in the form of a star, because there are no 1000 Mbit/s hubs, so switches have to be used. This solution is specified by IEEE 802.3ab;
- Auto-negotiation between 100 and 1000 Mbit/s is a requirement to use 1000BASE-T. The clock source must be negotiated, because one element must be the master and the other the slave;
- 1000BASE-T IEEE 802.3ab is compatible with 100BASE-TX, with automatic detection of Tx and Rx.

At the beginning, 1000BASE-T used four channels (four pairs of cables) for simultaneous transmission in the two directions and pulse-amplitude modulation at five levels (PAM-5). The symbol rate is identical to that of 100BASE-TX (125 Mbauds) and the 5-level immunity to signal interference is also identical to that of the 3-level signaling in 100BASE-TX in MLT-3. 1000BASE-T uses four-dimensional trellis-coded modulation (TCM) to obtain a 6 dB coding gain across the four pairs. The 8-bit data are transmitted on four pairs of copper cables and the 8 bits are expanded into four symbols of

Symbol	000	001	010	011	100	101	110	111
Line signal level	0	+1	+2	−1	0	+1	−2	−1

Figure 5.14 Example of 3-bit symbol mapping.

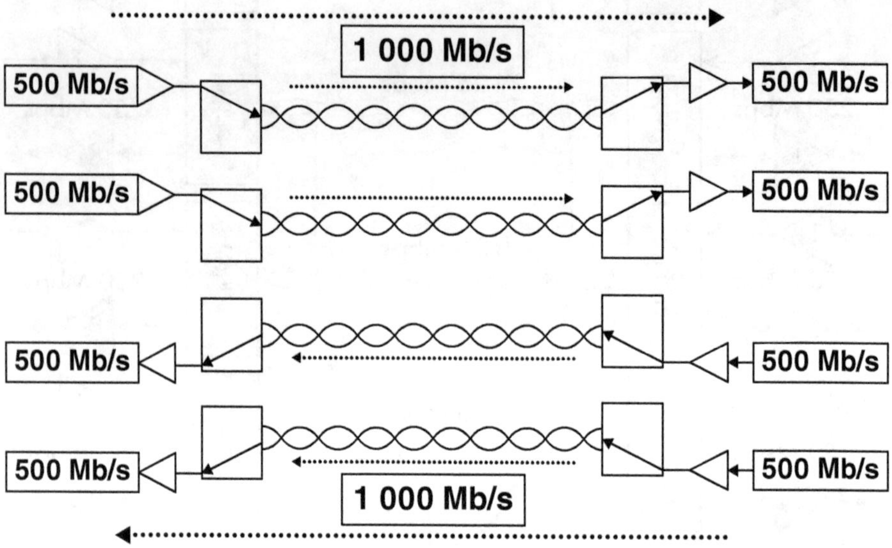

Figure 5.15 Conventional configuration of 1000BASE-TX PHY.

3 bits each, by padding based on a linear-feedback shift register. The 3-bit symbols are then mapped at line signal levels. Figure 5.14 gives an example of such mapping.

1000BASE-TX

1000BASE-TX is a physical layer similar to 1000BASE-T, created and managed by the TIA (*Telecommunications Industry Association*). 1000BASE-TX is also known as TIA/EIA854 (see Figure 5.15):

- It is designed to operate in full-duplex mode on two of the four UTPs of category-6 and 7 cables;
- It transmits data at 500 Mbits/s on two pairs and receives data at the same datarate on the other two pairs;
- Delay skew (the time difference between the slowest and fastest pairs in the same cable) becomes increasingly significant as the datarates increase.

1000BASE-TX can be implemented with simpler electronics, because it does not require hybrid circuits, or echo cancellation. Consequently, its design means that less expensive electronics can be used, but it is more rarely used, because of the high cost of category-6 and category-7 cables, and the dropping cost of 1000BASE-T products that work with category-5 cables.

1000BASE-T1 – IEEE 802.3bp

1000BASE-T1 technology is a **single-pair** version of Gigabit Ethernet, which allows for the aggregation of multiple 100 Mbit/s channels – for example: for applications with data-rates of over 100 Mbits/s, such as those produced by the cameras of an ADAS. We shall revisit this solution in greater detail in Section 5.3, on automotive Ethernet.

1 Gbit/s solutions in summary

A summary of 1 Gbit/s solutions is given in Figure 5.16.

It should also be noted that automobile networks (and industrial ones) are increasingly turning to IEEE 802.3bv, Gigabit Ethernet, using plastic optical fibers (POFs), which defines the standards and parameters of management of the physical layer, offering a reliable transmission alternative for Ethernet networks in automobiles. There are also many more variants, which are beyond the remit of this book.

Multi G – Ethernet 10 Gbit/s

IEEE 802.3ae is the version of the standard for 10 Gbit/s Ethernet. The 10 Gbit/s Ethernet standard covers seven different media types for different types of local-area networks (LAN), metropolitan-area networks (MAN), and wide-area networks (WAN). Let us examine what is relevant to our study here.

10GBASE-CX4

This standard (802.3ak) uses an InfiniBand 4 × copper cable over a maximum length of 15 m.

10GBASE-T

In this standard (802.3an), the transmission takes place in full-duplex on **4 twisted pairs** on cables of categories 6, 6 A, or 7, with a number of encoding moments as a function of the cable category chosen and the desired interference resistance, over a maximum length of 100 m. 10GBASE-T should be compatible with 1000BASE-T, 100BASE-TX, and 10BASE-T.

Overview of industrial Ethernet variants

Here are the main Ethernet variants in the industrial world, in terms of supports, physical layers, etc. (see Figure 5.17).

This completes the inventory of widely used industrial Ethernet networks.

Gigabit Ethernet	Medium	Distance
1000BASE-T	4 twisted pairs (Cat-5, Cat-5e, Cat-6, Cat-7)	100 meters
1000BASE-TX	2 twisted pairs (Cat-6, Cat-7)	100 meters
1000BASE-T1	Single balanced twisted pair	15 meters

Figure 5.16 Summary of 1 Gbit/s solutions.

Type Date	IEEE standard	Name	Datarate	Support	Distance	Cables		Comments
			Mbit/s		m	Category		
1976				Bob Metcalfe and David Boggs published *Ethernet: Distributed Packet Switching for Local Computer Networks*				
1980	DIX standard	10BASE5 or *Thick Ethernet*	10	Coaxial cable	500			50 ohms N BNC
Ethernet 1983	802.3.			Starting point for standardization				
1985	802.3a	10BASE2 *Thin Ethernet*		Coaxial cable	185			50 ohms N BNC
1987	802.3d.			*Fiber-optic inter repeater link*	1000			
	802.3 i			Star topology				
		10BASE-T		Minimum 4 wires (2 twisted pairs)	100	CAT-3 CAT-5		Full-duplex after negotiation
	802.3j	10BASE-Fxxx		Multi-mode optical fibers (62.5/125 µm)	2000			
Fast Ethernet	802.3u	100BASE-TX	100	Two pairs UTP5 STP	100	CAT-5	Semi-duplex	
		100BASE-T2		Two pairs	100	CAT-3 4,5	Semi-duplex	
		100BASE-T4		Four pairs	100	CAT-3 4,5		
		100BASE-FX and SX		850 nm optical fibers	550			
Gigabit Ethernet	802.3ab	1000BASE-T -	1000	Twisted pairs	100	CAT-5		Full-duplex

Figure 5.17 Main Ethernet variants used in the industrial world.

Type Date	IEEE standard	Name	Datarate	Support	Distance	Cables	Comments
			Mbit/s		m	Category	
		1000BASE-LH		Single-mode optical fiber	Long distance		
		1000BASE-CX		Special copper cable	25		
10 Gigabit Ethernet 1998	802.3z 802.3ae 802.3ak	10GBASE-CX4	10 000	InfiniBand copper cable 4×	15		
	802.3an	10GBASE-T –		Cable		CAT-6, 6A, 7	
		10GBASE-SR		850 nm	300		
		10GBASE-LX4		Multi-mode fiber	240 and 300		
		10GBASE-LR		1310 nm SM optical fiber	10 and 40 km		
		10GBASE-SW		850 nm optical fiber	300		

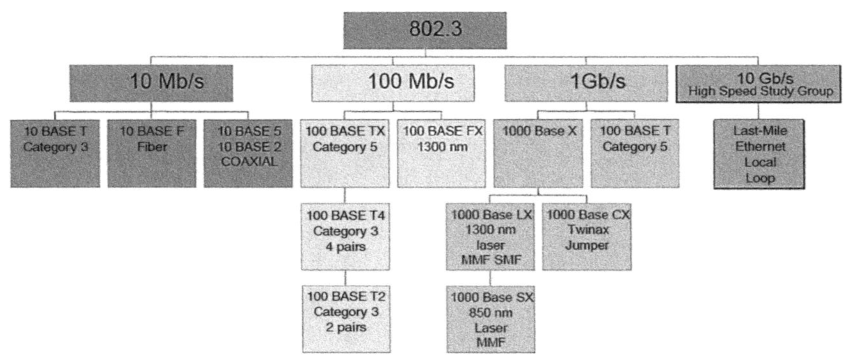

Figure 5.17 (*Continued*)

5.2 Paradigms of Ethernet physical layers and automobiles

5.2.1 Introduction

For many years, the communication networks discussed above (Section 5.1) have been developed primarily for the purpose of motor control and related tasks. Today, automakers are making a concerted effort to support all possibilities. They are adopting an increasing number of high-performing systems for gear transmission, safety, communication, and in-vehicle infotainment. The three main goals with this technology are:

- Ever-greater bandwidth;
- Connectivity;
- Economies of scale, achieved by adopting a standardized backbone for communication networks.

Automobile makers are turning to Ethernet as the basis of their networks in next-generation vehicles, to serve their current purposes and support future developments. According to Frost & Sullivan, in 2020, the number of Ethernet ports should be somewhere between 10 (entry-level vehicles) and 100 (luxury vehicles).

Because of the specific requirements of IVNs, as listed in Section 4.5, the IEEE 802.3 and IEEE 802.1 Ethernet standards, described in Section 5.1, must include extensions describing solutions that offer specific features and functions required for an automobile network. Such a network must be flexible, robust, reliable, and economical. There are multiple specifications that shape the specific evolution of the physical layer in automobile networks:

- Bandwidth;
- Determinism;
- Reliability;
- Safety;
- Cost.

Ideally, Ethernet technology for automobiles should be tailored to these conditions.

Bandwidth

The bandwidths required are increasing exponentially with the volumes of real-time data, transmission control signals, and safety signals generated by modern vehicles. Most industry experts agree that, by 2025, each subsystem, hinging on motion in a car, will require connectivity at speeds of between 20 and 2500 Mbit/s.

Reliability and determinism

Time-critical tasks, real-time systems, ADASs, systems or brake transmission control require IVNs to guarantee delivery of data packets and control signals at a specific instant, or in a known time. As we shall see in Section 5.4.2, IEEE 802.1 protocol, TSN (time-sensitive networking) will make such determinism for Ethernet a possibility. The Ethernet standards, such as IEEE 802.1AS, IEEE 802.1Q, and IEEE 802.1Qca, governing transport, switching, and management of latency-sensitive connections, will also benefit from real-time high-datarate control applications in vehicles. In addition, all these possibilities are based on the well-known IEEE 1588 PTP (precision time protocol), which provides the necessary synchronization information to each network node. In Section 5.4.2, we shall discuss TSN as a whole, which will be the final hurdle before Ethernet can truly become an automobile technology.

Safety and security

In time, vehicle networks will come to have multiple terminals that can also connect to services and resources in the Internet of Things (IoT). Thus, security against malicious cyberattacks is becoming a major concern. Fortunately, the automotive industry can draw upon the security solutions available for Ethernet in the world of commerce, protecting its networks (and buses). IEEE 802.1AE MACsec and IEEE 802.1x KeySec define measures aimed at authentication, authorization, and accounting (AAA) to secure a data system and devise management and control plans:

- Authenticate each node/device connecting to the system;
- Authenticate access rights for all transactions;
- Authenticate each packet's entry onto the network.

5.2.2 A bottleneck for automobile Ethernet = cost + EMC

Bearing the above general remarks in mind, let us now come back to the subject of the physical layer and its paradigms for use in automobiles. As we have seen, Ethernet was developed for industrial actors in the early 1970s, and we have discussed its numerous industrial variants. Today, the Ethernet layers in charge of MAC have been addressed and accepted. Now, it is a question of the physical layer.

In summary
With respect to the market applications for ACVs, the true bottleneck preventing Ethernet from being widely adopted is the difficulty in striking the right compromise between the type of physical layer and the distance, the medium technology, datarate, regulatory compliance, and so forth, all for a cost that is reasonable in the situation.

In short, it is crucial to focus, in detail, on techniques and technologies relating to coaxial cables, wiring, differential/twisted pairs (shielded or unshielded), wire gage, characteristic impedance, losses, interference, RF pollution, EMC, wire weight, price, etc., and the impact on other factors, such as fuel consumption, pollution, environmental performance, etc. Therefore, one cannot help but attach a great deal of importance to methods and techniques for encoding and processing the electrical signals used in the PHY layer. Consequently, we shall examine these factors in great detail, after a detour into the highly technical matters that are essential for understanding the subject at hand.

5.2.3 Choice of physical layer in automobiles

Regulations/datarate/bit coding/spectrum/radiation Various geographic regulations

Regulatory requirements pertaining to the spectral masks of RF transmissions, and measurements relating thereto (e.g. CISPR 25), which apply to vehicle design, are extremely strict. Moreover, they vary considerably depending on the geographic region or country where the vehicle will be driven. Yet the very point of a vehicle is to be able to travel! Thus, there is a very wide range of spectral mask/transmission requirements to be satisfied. In practice, though, these requirements may lead, either to overly restrictive system designs, applied to multiple geographic markets, or to multiple specialized designs that are individually targeted to certain geographic markets. Finally, it is often necessary for the chosen system to be easily adaptable and configurable to fit different requirements. This may see multiple spectral masks developed by the multiple automakers. Alternatively, it may lead to multiple design requirements and spectral mask scenarios for a specific vehicle from a particular automaker (see Figure 5.18).

Radiation and network topology

The process of design of a system conforming to a spectral mask presents numerous challenges and compromises. More specifically, a device in the PHY layer must be designed for use in automobile applications and may need to be applied to different types of installations, represented by different makers.

Consider the example of different automakers using different types of cabling (e.g. twisted pairs, optical fibers, etc.) to interconnect their electronic systems. As one might fear, control of RF emissions may be considerably more difficult when an automaker decides to use unshielded copper cables.

Beyond the geographic aspects and types of cables used, there are many other factors that impact on RF emissions/pollution. For example: the design of the vehicle bodywork (using metal, resin, plastic, etc.), the internal layout, the placement of the cabling within the vehicle, the presence of other electronic equipment in the same vehicle, and other factors, may influence the desirable emission thresholds. Consequently, there is little chance of applying a single spectral mask requirement.

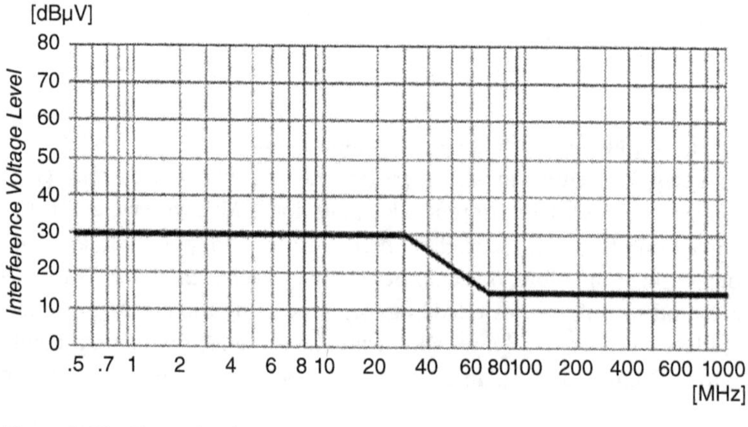

Figure 5.18 Example of an RF spectral mask.

Sensor types	Camera	Radar	Lidar	Sonar	GPS	Audio video
Raw bitrates required	~ 20–40 Mbit/s	~ 10–100 kbit/s	~ 10–70 Mbit/s	~ 10–100 kbit/s	~ 50 kbit/s	~ 200 Mbit/s

Figure 5.19 Overview of bitrates needed in automobile applications.

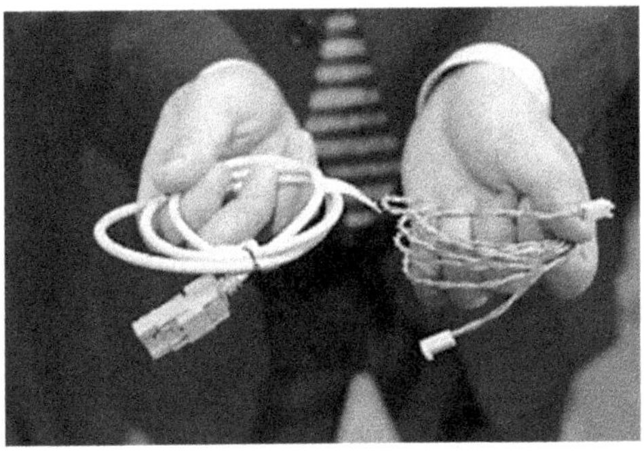

Figure 5.20 Comparison between an LVDS cable (left) and an unshielded twisted pair (right).

Required bitrates

Let us recap (see Chapter 2) the values of the bitrates typically required by audio and video applications (cameras, radars, lidars, TV, DVD, etc.) and telematic applications designed for automobiles (see Figure 5.19).

Physical medium

The physical medium of the network is one of the key elements, because its cost, its weight, etc. are among the most important parameters (Figure 5.20). To briefly sum up, "inexpensive" is best! Obviously, this immediately counts out fiber-optic technologies, coaxial cables, standard sheathed or shielded cables, etc., leaving only ordinary wire – or at least, these solutions are *almost* discounted, because despite the cost consideration, the technology must be able to satisfy certain qualities and specifications:

- The maximum/minimum lengths of communication networks;
- The material used for the wire, its diameter or gage, its resistivity, its resistance, its weight;
- The type of insulation used, and thus its thickness, and hence its capacity and, ergo, the bandwidth of the line;
- The number of dB of line losses;
- The intended maximum datarate (in baud), so the speed and acceptable propagation delay;
- Line ending wave reflection (in VSWR), so the line's impedance, compliance with the specification, or adaptation, its effect on the eye pattern and its bit error rate (BER);
- Pollution and radiation, which are the reason for differential two-cable lines and attacks on the symmetric key;
- Sensitivity (susceptibility) to parasitic signals, which, again, are the reason for differential two-cable lines and attacks on the symmetric key;

- This, in principle, leads to unshielded cables;
- *Urbi et orbi* crosstalk (i.e. crosstalk between two wires in the same pair and crosstalk between the wires of two different pairs). This gives rise to twisted lines, and thus the step of the spires of the twisted cable;
- The minimum possible radius of curvature of the cables so they can fit into the conduits; etc.

All of these factors must be taken into account just for a piece of cable! However, those in the know in the field would explain that everything has a cost, and everything has weight, so it consumes fuel or electricity; therefore, it pollutes the atmosphere, and in this day and age, these things must be taken into account.

Unshielded twisted pair An *unshielded twisted pair* (UTP) is a transmission line formed of two conductive wires, twisted around one another in a helical structure (see Figure 5.21).

There are multiple reasons to use this particular configuration:

- Keeping distance between the wires, meaning that a characteristic impedance for the pair can be defined. Thus, it can be adapted, and at the end of the line, it can help eliminate undesirable reflections of the signals. Geometric constraints (insulator thickness/wire gage) generally mean the impedance stays at around 90–100 ohms, depending on the system;
- Reduction of crosstalk, because the higher the number of twists per meter, the lower the chance of crosstalk. The average number of twists per meter (thus, the step of the twist) is part of the specification of the cable, but when there are multiple pairs present within the same cable, each pair is twisted slightly differently so as to avoid crosstalk;
- Reduction of interference, by rejection of a common mode by the use of symmetric differential signals.

Categories of cables

Ethernet cables are standardized into various categories on the basis of signal integrity, bandwidth, etc. We shall now briefly examine the main categories:

- Category 3: defined in the ANSI/TIA/EIA-568B standard and used for Fast Ethernet at 100 Mbit/s. This type of cabling is gradually being abandoned in favor of category-5 or higher cables;
- Category 5: in UTP, Category 5 cables offer a datarate of up to 100 Mbit/s over a distance of 100 m, with a 100 ohm charge. This category can be used with 100BASE-TX and 1000BASE-T systems. Category 5 is made up of three types of cables with four twisted pairs (see Figure 5.22):

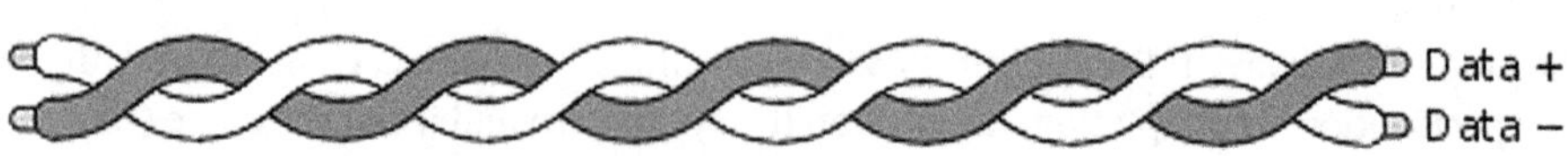

Figure 5.21 Diagram of a twisted pair.

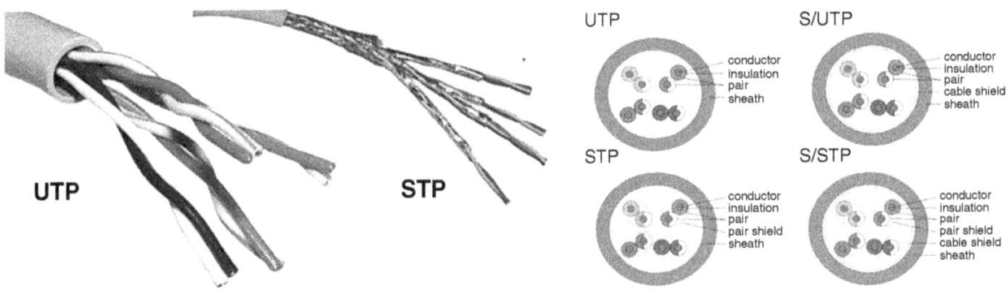

Figure 5.22 Category 5 with three types of cables, having four twisted pairs.

Property	Unit	Nominal Value	Tolerance
Characteristic impedance @ 100 MHz	Ω	100	± 15
Nominal characteristic impedance @ 100 MHz	Ω	100	± 5
DC-Loop resistance	Ω/m	≤ 0.188	
Propagation speed	c	0.64	
Propagation delay	ns/m	4.80–5.30	
Delay f < 100 MHz	ns/m	< 0.20	
Capacitance at 800 Hz	pF/m	52	
Inductance	nH/m	525	
Corner frequency	kHz	≤ 57	

Figure 5.23 Technical features in UTP.

- FTP (foiled twisted pair), sheathed, and with an impedance of 100 ohms: an FTP cable is simple aluminum foil wrapped around the four twisted pairs, which are protected by an external sheath;
- STP (shielded twisted pair): each twisted pair is individually shielded;
- UTP (unshielded twisted pair), with an impedance of 100 ohms (the most widely used and least expensive cabling). It is not shielded. Its technical features for UTP are as follows (see Figure 5.23):
  - The conductor is soft tin-plated copper wire;
  - The conductor diameter is 0.5 mm;
  - The external diameter of the insulated conductor is 0.9 mm;
  - The conductors are in spiral-wound pairs;
  - The maximum twist step is 150 mm.

- Category 5e (*enhanced*)/class D: this cabling is capable of speeds of up to 1 Gbit/s. It is a type of cabling that allows a bandwidth of 100 MHz. The standard is an adaptation of category 5 (resistance < 9.38 Ω/100 m, mutual capacity < 5.6 nF/100 m, mass capacity < 330 pF at a frequency between 1 kHz and 1 GHz, impedance of the differential pair adapted to 100 ohms ± 15% for a frequency between 1 MHz and 1 GHz, and propagation rate < 5.7 ns/m at 10 MHz. The type of shielding and length pairing are not specified);
- Category 6/class E: this is a type of cabling that offers a bandwidth of 250 MHz or more (ANSI/TIA/EIA-568-B.2-1 and ISO/IEC 11801 ed.2);

- Category 6a/class Ea: this is an extension to Category 6, with bandwidth of 500 MHz (ANSI/TIA/EIA-568-B.2-10). It can be used for 10GBASE-T over a distance of 90 m;
- Category 7a/class Fa: these cables have a bandwidth of 1 GHz and allow a datarate of up to 10 Gbit/s.

In short, these are the reasons (primarily economical) for the medium in the form of unshielded twisted pairs and their features. Now, we only need to send signals on those cables that satisfy the prescribed radiation/RF pollution patterns.

Terminology

Before engaging in a serious discussion, so that everyone is on the same page in relation to terminology, and to avoid the confusion that can so easily arise in this field, let us take the time to examine the specific definitions of a number of terms of digital transmission, supported by the documentary example of Figure 5.24.

Transmission rate – baud

Depending on the data transmission techniques, a symbol may have more than two states, so it can represent more than a binary number (a binary number always refers to one or other of only two states). For this reason, the value of the “transmission rate” in bauds is often lower than the raw bitrate in bits/s.

Example of “transmission rate” Let us use a metaphor to illustrate the difference between the signaling datarate (in baud) and the bitrate (in bit/s). Consider a man with a semaphore flag, who only moves his right arm once every second. Thus, the signaling datarate (baud) is “one symbol per second” – i.e. “one baud”. However, the flag may be held in any of eight distinct positions – straight up, at 45°, 90° and 135° to the left, straight down (this is the resting state, when no signal is being sent), and at 135°, 90°, and 45° to the right. Each signal position (or symbol) transmitted carries three bits of information, because three binary numbers are needed to encode these eight states. Thus, the bitrate is “three bits per second”. In addition, if the signalman were to use, for example, different patterned flags and both arms at once, then by using combinations of all these factors, more symbols could be sent per second, each conveying multiple bits, and thus producing a much higher datarate.

Consider, for example, an initial digital signal created by a simple binary series:

1 0 0 0 1 0 1 0 1 0

- Which is clocked at a bitrate = 1000 Hz;
- Whose period is $\Delta = 1$ ms;
- Thus, which has a modulation rate of $R = 1/\Delta = 1000$ bauds.

We wish to transmit these binary signals over a transmission line. For *x* reasons (not detailed here) of the choice of medium type (coaxial cable, paired cables, etc.), price, transmission line quality, etc., for the transmission of these signals on the medium, the decision is made to code this series of bits (using a specific line code that we shall not delve into in this example) using a coding technique that may have four distinct electrical levels during the elementary time Δ of the duration of bit 1 or 0 (see Figure 5.25), and thus has a valence of $V = 4$.

	Symb.	Units	Formulae	Definitions	Example
				Channel	**At 100 Mbit/s**
Bandwidth	W	Hz		Frequency bandwidth (measured at –3 dB)	
Signal-to-Noise Ratio	S/N	dB	S/N dB = 10 $\log_{10}$ S/N	N = *white noise*	
Transmission capacity	C	Bit/s	C = W.$\log_2$ [1+S/N]	"C" is the quantity of information that can be transported over a transmission channel per unit time (N.B. in this formula, S/N is not expressed in decibels)	
Initial bit					
Bit duration	Δb	Second		Time between two bits	10 ns
Bit rate	fb	Hz	fb = 1/Δb		100 MHz
Coding to transmission channel					**MII output**
Symbol	A symbol represents a unit of data (a symbol may be made up of one or more binary numbers or a whole number of "bits"). A theoretical definition of a symbol is a "waveform, state, or important condition of the communication channel that persists for a specific period of time." A symbol (in any form or physical representation) is a unit of data, where the data may be represented by the transitions between the symbols, or even by a sequence of numerous symbols. A transmitter sends symbols over the channel and the receiver must detect the sequence of symbols in order to reconstruct the transmitted data.				4 bits
Symbol duration time, or Unit Interval	Δs	Second		Δs is the duration of a symbol, or duration of multiple symbols. For example: in the case of line coding corresponding to 1000 pulses per second, the "symbol duration time" is 1/1000 second = 1 millisecond.	40 ns

Figure 5.24 Detailed documentary example of definitions of terms.

	Symb.	Units	Formulae	Definitions	Example
Symbol rate	fs	Symb/s	$fs = 1/\Delta s$	fs is the *symbol rate*.	25 Msymb/s
Modulation rate or Baud rate	R	Baud or Symb/s	$R = 1/\Delta s$	Rm (in baud) is the modulation rate (quantity of information transmitted per elementary moments), i.e. the number of transitions or changes of state (waveforms or signaling events) of a signal or symbol injected into the transmission medium in a second using a line code or a digitally modulated signal. The baud is the unit of the number of symbols transmissible per second. For example: a *symbol rate* of 1000 symbols per second is a "baud rate" of 1 kBd = 1000 Bd. Another example: if N bits are transported per symbol and R is the raw datarate, the symbol rate can be calculated by: $f\_symb = R/N$ In the case of a line code, the symbol rate is the pulse rate in pulses/second. Each symbol can represent or transport one or more bits of data. N.B. The symbol rate is linked to the raw bit rate, expressed in bits/second, but they are not the same thing.	25 Msymb/s or 25 Mbaud
Coding to transmission channel					**MLT 3**
Valence of a signal	V			Signal valence is the number of distinct significant states used in a transmission per elementary time interval (moment) Δb.	3

Figure 5.24 (*Continued*)

	Symb.	Units	Formulae	Definitions	Example
Quantity of information	Q	Bit/symb	Q $= \log_2 V$	The notation used is as follows: – M, the probability of appearance of content of a message on the basis of a certain number of possible/available messages (the quantity 1/M generally represents the valence V, V = 1/M); – p, the logarithmic base used (typically 2, because information content is measured in bits); – "proprietary information" of M = I(M), which is a function of its probability; $I(M) = -\log_p M = \log_p (1/M)$ Shannon's theorem on the probability of appearance of the content of a message/symbol shows that the quantity of information Q contained in the elementary moment, injected by a signal with valence V is equal to: $Q = \log_2 V$ valence of the signal V For example: if the signal is binary (bit) and is subject to ternary modulation, the probability of appearance of a certain symbol will be M = 1/3 and the quantity of information (measured in bits) of any of the ternary symbols will be equal to $\log_2(3)$, meaning that a ternary symbol will contain Q = 1.58 bits. It should be noted that $\log_2(M)$ is the value in bits for an M-ary symbol (bin-ary, tern-ary, etc.).	1.58

Figure 5.24 *(Continued)*

<table>
<tr><th></th><th>Symb.</th><th>Units</th><th>Formulae</th><th>Definitions</th><th>Example</th></tr>
<tr><td>Bitrates of a transmission channel or transfer rate</td><td>D</td><td>Bits/s</td><td>$D = R \times Q$
$D = R \times \log_2 (V)$</td><td>D is the datarate at which data are transferred over the transmission channel. It is the maximum number of bits transmitted per second.
In each elementary moment Δ, we measure the number of bits transmitted, or else determine, on the row, the number of different values or states that exist.
Measuring raw bitrate D:
– D is the product of the baudrate per information capacity per symbol (in bits).
– D (in bit/s) = Rm (in baud) × $\log_2 (V)$ (in bits)
Note: only in the event that only 1 bit per elementary moment is transported (V = 2) do we obtain: R = D, because $\log_2(2) = 1$.
For example: a modem at a modulation rate Rm of 1200 baud, depending on the modulation used, may have a maximum datarate D of:
<table>
<tr><th>Modulation type</th><th>Bits per symbol</th><th>V</th><th>$\log_2 (V)$</th><th>D bit/s</th></tr>
<tr><td>Pure binary</td><td>1</td><td>2</td><td>1</td><td>1200</td></tr>
<tr><td>4-QAM</td><td>2</td><td>4</td><td>2</td><td>2400</td></tr>
<tr><td>8-QAM</td><td>3</td><td>8</td><td>3</td><td>3600</td></tr>
<tr><td>16-QAM</td><td>4</td><td>16</td><td>4</td><td>4800</td></tr>
</table></td><td>39.5 Mbit/s</td></tr>
</table>

Figure 5.24 (*Continued*)

	Symb.	Units	Formulae	Definitions	Example
Line code efficiency		%		The line code efficiency is the value of the fraction: $$\text{Efficiency} = \frac{\text{Log}_2(\text{number of possible sequences})}{\text{Log}_2(\text{number of usable sequences})}$$ For example: MLT-3: – pure binary line code (1 bit and possible cases = 2); – MLT-3 (3 separate levels and 3 usable cases) This gives an efficiency = $\log_2 2/\log_2 3$ $= 1/1.58$ $= 0.631$ For example: 3B-2T: – binary line code "3B" (3 bits and possible cases = 8) – 2T (3 ternary symbols taken 2 by 2; usable cases 9) This gives an efficiency = $\log_2 8/\log_2 9$ $= \log_2 2^3/\log_2 3^2$ $= (3 \times 1)/(2 \times 1.58)$ $= 0.949$	0.631 or 63.1%

Figure 5.24 *(Continued)*

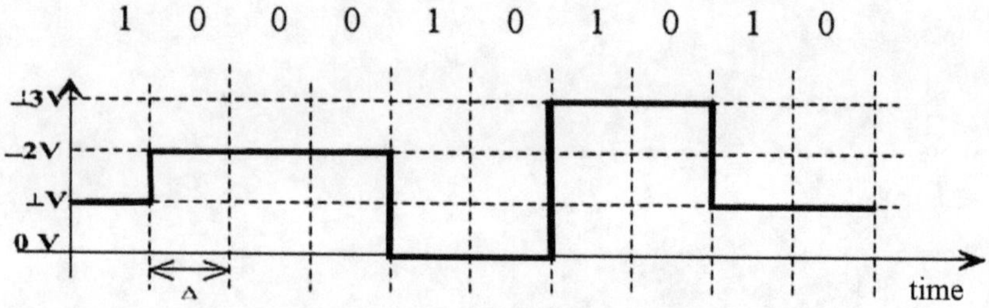

Figure 5.25 Coding technique with four distinct electrical levels.

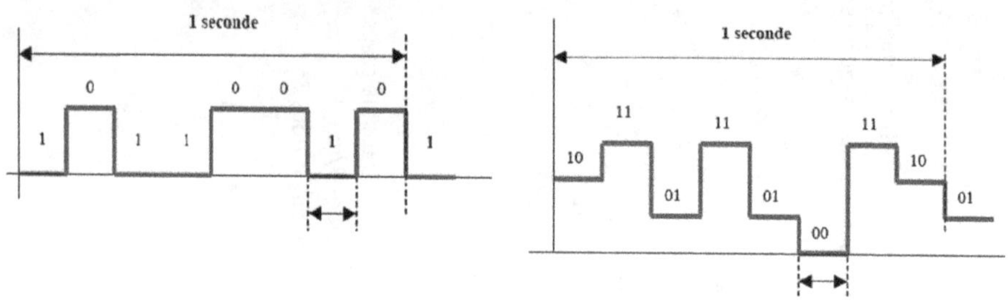

Figure 5.26 Examples of coding (example 1 on the left and example 3 on the right; see Figure 5.27 for details).

Transmission coding		Units	Example 1	Example 2 not illustrated	Example 3
Signal valence	V		2	3	4
Quantity of information	Q	Bit/s	1	1.58	2
Modulation rate symbol rate	R	Baud	1000	1000	1000
Bitrate of channel	D	Bit/s	$1000 \log_2(2)$ = 1000 × 1 = 1000	$1000 \log_2(3)$ = 1000 × 1.58 = 1580	$1000 \log_2(4)$ = 1000 × 2 = 2000

Figure 5.27 Examples of coding – details.

During the time Δ (duration of the bit indicated in the figure), the probability of occurrence of a certain electrical level is ¼ and the quantity of information contained in the elementary moment Δ due to the coding valence $V = 4$ will therefore be $\log_2 4 = 2$, measured in bits. This means that, in this case, with a (quaternary) symbol of duration $\Delta = 1$ ms (i.e. a modulation rate of 1000 symbols/s = 1000 baud) and containing 2 bits, the bitrate on the line can be $D = R \log_2 V = 2000$ bits/s.

Also see the examples in Figures 5.26 and 5.27.

Bit coding Generally, the form and bit coding are chosen on the basis of numerous parameters: efficiency, noise resistance, collision management facility, etc., and, in the case of interest to us here, especially the shape of its baseband spectrum, given its intended propagation over an inexpensive medium (unshielded wire, which inevitably emits radiation), and the type of formatting and modulation we wish to apply later on. Therefore, in terms of radiation, a pure sinusoidal signal (a single spectral line!) would be best. This being the case, the simplest bit code that can be used is conventional NRZ!

Often, for transmission-related reasons (line quality and/or length, datarates, etc.), to the initial bitstream produced by the purely applicative part of the system, we must apply specific treatments to make it compatible with the hardware that will transport it. These strategies are known as "line coding," "block line coding" and "signal modulation." All of this happens by progressing gradually from dealing with bits to data blocks to symbols and finally to a datarate, expressed in baud, and coding efficiency.

Line code

In a communication system, *line code* (also known as digital baseband modulation or digital baseband transmission) is used for transmission. This technique is often used for transporting digital data. Note that a line-coded signal and a signal produced on a terminal may differ.

Purpose of line code Depending on the types of media, datarates, distances, etc. intended, there are different goals that may be served by line code.

Reliable clock recovery and receiver synchronization Generally, a maximum length is imposed on a generated sequence, and/or a constraint as to the maximum/reasonable number of consecutive ones or zeros. When the signal reaches the receiver, the timing information is recovered by observing the transitions in the received sequence. Thus, by ensuring the sequence does not exceed a given length, clock recovery is made easier. It should also be recognized, however, that too short a sequence may compromise detection and synchronization.

Error checking It is best to choose a line code whose internal structure allows for good error checking because, if synchronization is less than perfect, the signal to be decoded will, amongst other things, display suboptimal or non-existent differences in amplitude between the different digits or symbols used in the line code, which increases the likelihood of errors in the received data.

DC component Most communication channels either cannot, or struggle to, transport or maintain the value of the DC component over long distances. The DC component is also known as DC bias, DC polarity, or DC coefficient (DC being "direct current"). The simplest line code, called unipolar encoding, has a DC component, and produces an excessive error rate in these systems. The majority of line codes eliminate the DC component and are therefore said to be DC balanced, zero-DC, DC-free, zero-polarization, DC-equalized, etc. There are many different ways to eliminate the DC component, by using constant-weight or balanced coding – put differently, each nibble of the transmitted code is corrected so that its transmission produces an equal number of positive and negative levels, so the average level is zero. For example, the so-called Manchester code and the interleaved two-of-five (ITF) code use a paired disparity code (the AMI, 8B10B, 4B3T, etc. codes [see below]), or use a scrambler – for example, the scrambler specified in RFC 2615 for 64B/66B encoding.

Signal spectrum – transmission, EM pollution, and EMC

So that the signal spectrum and the associated radiation conform to the regulations in force, depending on the transmission line (cable, coaxial cable, differential pairs, etc.), it is very common – indeed, absolutely necessary – to modify the physical representation of the initial information in order to transport it. For the numerous reasons discussed above, before being sent over transmission lines, the source bit, typically encoded in NRZ, whether or not it has undergone line coding, is generally modified once more (for example, modulated). In our case, it undergoes baseband modification (so without frequency transposition), by amplitude modulation (AM). However, there are numerous treatments that can be applied to signals. Depending on the intended media types, the operation is often divided into two parts:

- Enhancement of the code for transmission;
- Modulation of signals on the line.

These line coding operations on the signal generally reflect the requests and technical requirements specific to the transmission lines used, such as twisted pairs (shielded or unshielded) or fiber-optic technology, and the consequences of those technologies in terms of RF pollution. These requirements are unique for each medium, because each of them behaves differently in terms of interference, distortion, bandwidth, and amplitude loss.

Line coding

The digital signal must then be transported over a medium by an amplitude signal and time-discrete values, whose pattern/waveform of voltage or current used must be suitable and optimal in relation to the specific properties of the physical transmission link/channel. Line coding in the true sense is now applied, to represent the ones and zeros of the digital data on the medium. The most common types of line coding are unipolar, polar, and bipolar encoding and Manchester code.

Line coding and block line coding or "kB/nB"

An often-used type of line coding is "kB/nB," which is a form of *block coding* of data. Its purpose is to transpose initial groups of "*k* bits" into new binary groups (or blocks) of "*n* bits." There are two possible solutions.

n > k For example: in 100BASE-TX Fast Ethernet, "4B/5B" coding, from the ANSI X3T9.5 standard for FDDI (fiber distributed data interface), would divide the incoming bit stream – e.g. 1 0 0 0 0 1 0 1 1 1 1, etc. – into 4-bit blocks, and the transcoding table (a dictionary) is used to assign each 4-bit block a new 5-bit block, before sending it over the network (see Figure 5.28).

There are multiple reasons for such (trans)coding.

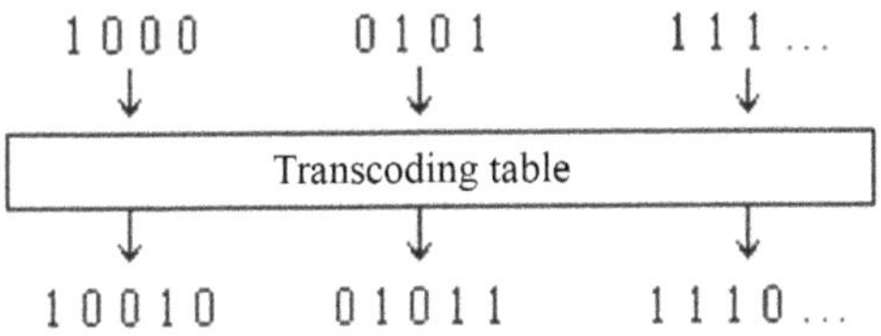

4-bit block	4B5B symbol
0000	11110
0001	01001
0010	10100
0011	10101
0100	01010
0101	01011
0110	01110
0111	01111
1000	10010
1001	10011
1010	10110
1011	10111
1100	11010
1101	11011
1110	11100
1111	11101

Figure 5.28 Transcoding table.

Synchronization For each block of four consecutive bits, an equivalent 5-bit nibble taken from this transcoding table is chosen, so as to ensure that there are at least two transitions per 5-bit block and the stream to be transmitted contains no more than two consecutive and/or successive zeros. This equivalent nibble always has a minimum output density of 1 bit. This bit redundancy and the absence of long series of zeros safeguards the possible quality of the transport and ensures that there will be periodic transitions in the signal to allow the receiver to maintain synchronization with the incoming data stream.

In addition, 4B/5B encoding leaves a number of extra symbols (5-bit blocks) unused. These can be used, amongst other things, to define the transmission control information or other functions, such as indicating the start or end of the packet. Even when blocks that could cause transmission problems – such as 00000 – are eliminated, there are still a number of words left over. Special characters, in addition to usable data, may be included in the transcoding table without requiring a special state of the signal, as happens with Manchester code. The information will be easier to transmit once it has been encoded in NRZI or MLT3.

DC component Coding ensures that the DC component is smoothed out.

Spectrum The overhead associated with the presence of the extra bit in 4B/5B transcoding increases the frequency and datarate of the output signal. If, for example, the incoming bits arrive at the transcoder input at a datarate of 100 Mbits/s (so the 4-bit blocks would have a duration of 4 × 10 ns), so as not to lose the thread of the information, the outgoing 5-bit streams must also be transmitted in a 40 ns period. Thus, the new bitrate for the output will be f_bit_output = f_bit_incident × 5/4 = 125 Mbits/s. If the medium used (e.g. a Category-5 cable) supports only 100 Mbits/s, this datarate will have to be reduced by another strategy in order to use the medium.

"4B/5B" associated with NRZI line coding Pressing on with our example, in the case of 100BASE-FX Fast Ethernet, now at 125 Mbits/s in the wake of 4B/5B transcoding, to use category-5 cables that support a maximum bitrate of 100 MHz (twisted pair, which can transport a signal whose frequency is higher than 100 m), it is necessary to use binary line coding of at least 2 Hz/bit to halve the datarate – for example, NRZI, which reduces the datarate from 125 to 62.5 Mbauds.

"4B/5B" associated with MLT-3 line coding Still in the case of 100BASE-TX Fast Ethernet at 125 Mbits/s, before transmission on the line, we can add a final coding of the data flow with an additional sublayer called the "medium-dependent sublayer." An example is MLT-3 line coding, which divides the datarate by 4, resulting in a datarate of 31.25 MBauds (see Figure 5.29).

n < k One of the possible solutions for automobile Ethernet uses 4B/3B coding, which transposes initial symbols of 4 bits into new symbols arranged in blocks of 3 bits. The transposition from 4-bit to 3-bit symbols is done on the fly, so does not require a specific dictionary. It has no impact whatsoever on the bitrate, but raises the symbol rate by a ratio of 4:3.

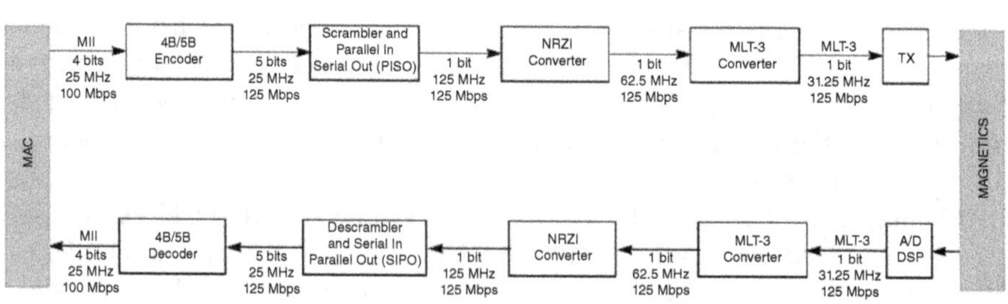

Figure 5.29 Use of MLT-3 line coding (source: Microchip).

Modulation/post-modulation of bits and "kB/nT"

After the preliminary "kB/nB" line coding operation, the signal is generally prepared before being submitted to a physical transmission channel. Sometimes, even if the appearances and features of the channels differ widely, these preparations may be sufficiently similar to allow the same line code to be used.

The most common preparations of physical channels are:

- The encoded signal can be directly sent over a transmission line, in the form of variations in the voltage or current (often using differential signaling);
- The encoded signal (baseband signal) is once again formatted (to reduce its frequency band) and is then modulated (change of frequency) to create an "RF signal" that can be sent through free space.

The bits output by the line coding, notably for reasons of EMC/RF pollution, must be symmetrical in mass, so that, at all times, the two wires in a pair radiate equal and opposite electromagnetic fields. To achieve this, a range of techniques can be used.

"*NRZI*" – non-return to zero inverted – coding The signal is inverted from +V to −V for "1" in the stream, whereas there is no inversion if the value is "0" (see Figure 5.30).

With NRZI coding, note that the transmission of long series of zeros gives rise to a lengthy period without a transition.

Consider the example of the following bitstream, which gives the fastest period of output coding:

a) 1 0 1 0 1 0 1 0 … If each bit has a duration of 10 ns, the bitrate is 100 Mbits/s;
b) + − + − + − + − … The fastest period of NRZI coding is therefore half that of the clock, which is a frequency of 50 MHz. Data encoded in this way can be transmitted at a maximum of 2 bits per Hz. The bitrate is therefore twice the maximum frequency of the coded signal:
   - Example: Fast Ethernet Application (100BASE-FX), FDDI.

"kB/nT" coding

In a block code system, the incoming bitstream is divided into binary blocks and then each of these is converted into a data symbol block. The ternary pseudocode known as "kB/nT" remedies these problems, where "k" is the number of bits and "n" is the number of ternary symbols per block.

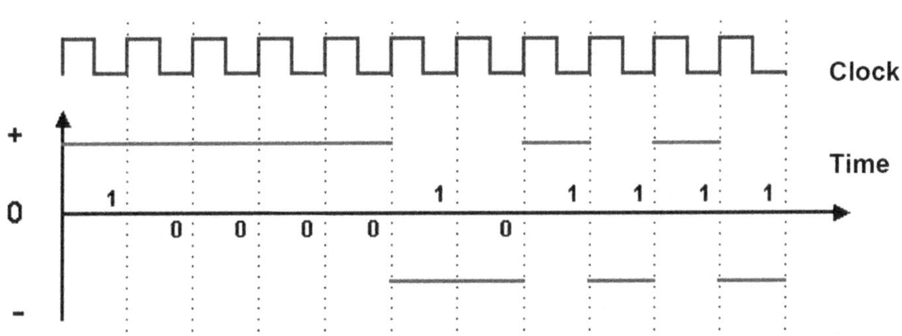

Figure 5.30 NRZI coding.

	"3B/2T" coding			
	3B	**2T**		
0	000	0 0	A	1
001	0 +	B	2	010
0 –	C	3	011	+ 0
D	4	100	+ –	E
5	101	+ +	F	6
110	– 0	G	7	111
– +	H			– –
F	8 digits	8 binary values	9 ternary values	A 9-letter alphabet

Figure 5.31 3B to 2T look-up table.

"3B/2T" "3B/2T" coding transposes a 3-bit block of binary information ("binary triplets") into two ternary elements. Of course, the three available ternary elements are the conventional "+V, 0, and –V" and, taking these three basic ternary elements two by two, it is possible to form strings of two ternary elements that can represent nine possible values of an alphabet, whereas the original binary triplets represent only eight possible values. Therefore, we must establish a "look-up table" – a map, coded and/or aligned – to convert three binary bits into two ternary elements (see Figure 5.31).

In this regard, the ninth ternary value is redundant information in the representation of the binary triplets and can be used, for example, as a marker for the start and/or end of the stream, to flag a transmission error, or to resolve the ordering of ternary symbol streams on a single twisted pair connection. "Start-of-stream delimiters" (SSDs) and "end-of-stream delimiters" (ESDs) may be inserted into ternary streams. Inactive signals may also be inserted before the SSD and after the ESD of the next data frame on the MII.

Another example: line coding can serve to slow down the datarate of the line to facilitate communication over a longer distance. In this regard, the redundancy of line codes can be minimized to extend operating ranges. For example: using PAM-3 for transmission on a twisted pair may allow for an appropriate level of redundance and signal-to-noise ratio (SNR).

"4B/3T" "4B/3T" coding, which is a reasonable compromise between all these competing objectives, has been widely used in numerous applications. One complication in the "kB/nT" line code is the need for the decoder to know the boundaries of the "*n*" ternary symbol blocks (see Figure 5.32).

PAM (pulse amplitude modulation)

Multi-level signaling plays an important role in line coding techniques, as it enables us to increase the bitrate offered by existing infrastructure cables and provides fast symbol transfer rates in comparison to those observed on lines.

The principle behind multi-level signaling is to use a more extensive alphabet of "m" symbols to represent data, so that each symbol represents more than one bit of data.

	Example of "4B/3T" coding			
	Input Binary block	Output		Digital sum
		Ternary block		
		Mode A	Mode B	
	4B	3T	3T	
0	0000	+ 0 –	+ 0 –	0
1	0001	– + 0	– + 0	0
2	0010	0 – +	0 – +	0
3	0011	+ – 0	+ – 0	0
4	0100	+ + 0	– – 0	±2
5	0101	0 + +	0 – –	±2
6	0110	+ 0 +	– 0 –	±2
7	0111	+ + +	– + +	±3
8	1000	+ + –	– – +	±1
9	1001	– + +	+ – –	±1
10	1010	+ – +	– + –	±1
11	1011	+ 0 0	– 0 0	±1
12	1100	0 + 0	0 – 0	±1
13	1101	0 0 +	0 0 –	±1
14	1110	0 + –	0 + –	0
15	1111	– 0 +	– 0 +	0

Figure 5.32 4B to 3T look-up table.

Thus, the number of symbols that need to be transmitted is less than the number of bits and therefore the bandwidth is reduced. Pulse amplitude modulation (PAM), which is widely used in baseband digital data transmission, is a type of signal modulation where, whatever its form, the signal of the incoming message is sampled at regular intervals, and each sample obtained is made proportional to the amplitude of the signal, modulating at the moment of sampling. These sampled pulses can then be transmitted directly or modulated by a carrier wave before transmission.

For example: a PAM-4 modulator takes an incident information symbol formed (mapped) of two bits at once, and transposes it into a signal whose amplitude is modulated into four associated possible voltage levels (for example: −3 V, −1 V, +1 V, and +3 V), during a specified wave period Tp. The signal is demodulated by detecting the amplitude of the carrier at each period Tp of the symbol (see Figure 5.33).

There are two main types of pulse amplitude modulation:

- Single polarity PAM: in this case, fixed and appropriate DC polarization is added to the signal to ensure all pulses are positive;
- Double polarity PAM: the pulses are positive and negative.

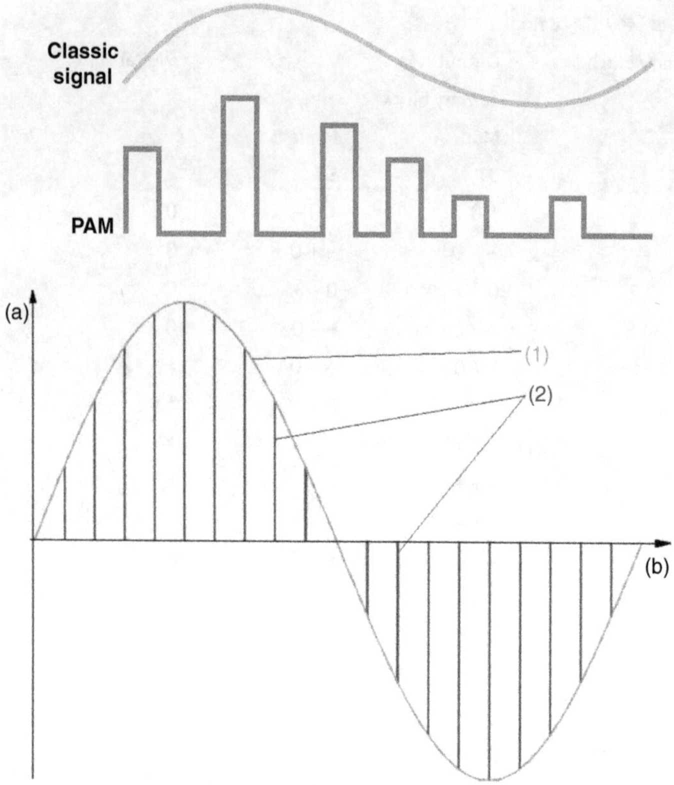

Figure 5.33 Principle of PAM.

Note
Typically, a flat-top PAM generator is used, because, during transmission, the noise that occurs at the top of the transmission pulse can easily be filtered out if the PAM pulse is flat topped.

Out of the numerous forms of modulation generally referred to as "PAM", there are a host of amplitude modulation variants, with bit coding on multiple analog levels (3, 4, 5, etc.), such as differential biphase encoding, bipolar encoding, and second-order interleaved bipolar encoding (see Figure 5.34). Most Ethernet variants use PAM, constellations.

One of the simplest, of course, is three-level PAM. PAM-3 signaling involves one of three states or conditions at all times: power level, phase position, and pulse duration or frequency. Thus, it is "ternary." "PAM-3" is similar to second-order interleaved bipolar encoding, giving signals whose amplitude alternates between +V, 0, and −V, and whose shape, viewed from (very) far, may be roughly sinusoidal. The Ethernet modes using PAM-3 are 100BASE-T4, and particularly 100BASE-T1 and 1000BASE-T1, which operate with differential pairs.

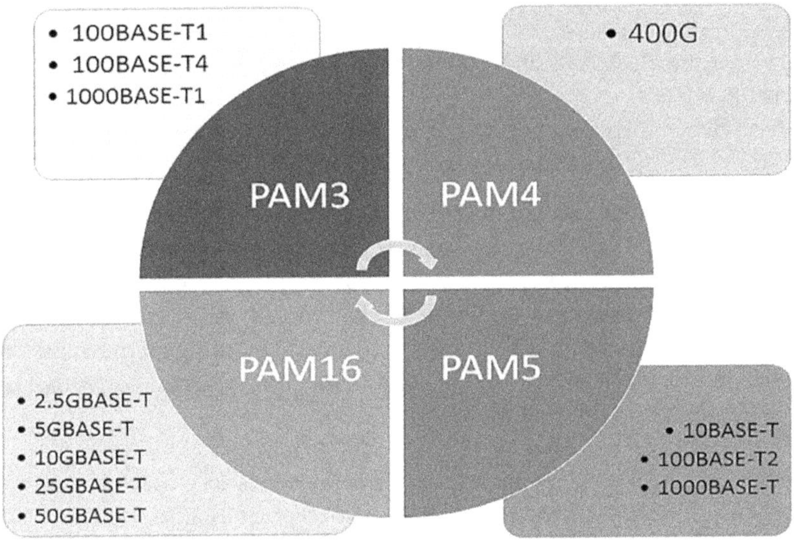

Figure 5.34 Variants of PAM encoding in Ethernet technology.

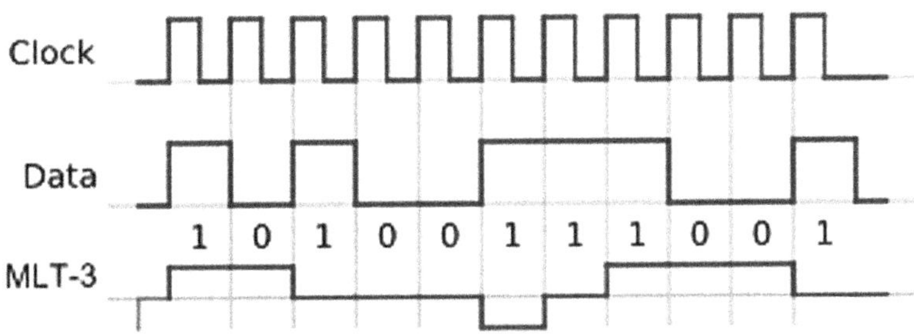

Figure 5.35 Principle of MLT-3 coding.

In the OSI model, PAM introduces a final additional sublayer of coding, called the *medium-dependent sublayer*, to the data stream before it is transmitted over the physical line.

Moving on, let us briefly examine MLT-3 coding, which is also well known and widely used.

MLT-3 – multi-level transmit – coding MLT-3 is a line code that uses three voltage levels: +V, 0, and –V. As Figure 5.35 shows, the principle behind this type of coding is extremely simple: the transmitted level is shifted to the next state when the bit value is 1, and remains unchanged when the bit value is 0. It should be noted that encoding long sequences of zeros may cause clock loss or dephasing on the receiver. An example of MLT-3 coding of the binary series is 1010 0111 001 etc.

MLT-3 is similar to NRZ, and its coding efficiency is 1 bit/baud. However, the MLT-3-encoded signal also requires four transitions (bauds) to complete a full cycle (from "0" to "+V", from "+V" to "0", from "0" to "–V", and from "–V" to "0"). Its maximum datarate occurs during a series of successive 1s.

Example:

- 1111111111: if the clock runs at 100 MHz, each bit lasts for 10 ns, so the bit frequency is 100 Mbits/s.
- +0–0+0–0, etc.: in this case, the fastest period of the MLT-3-encoded signal is ¼ of that of the bit clock, so has a frequency of 25 MHz.

The bitrate is then four times that of the maximum frequency of the encoded signal.

Thus, the maximum fundamental frequency of an MLT-3-encoded signal is reduced to a quarter of the bitrate. In view of this peculiarity, it is possible to greatly decrease the necessary frequency (and thus the bandwidth) of the transmitted signal for a given bitrate, by using the three states, rendering that signal transmission more suitable for the use of inexpensive copper wires, rather than the costly coaxial cables.

A few examples Certain versions of Ethernet communications are examples of the use of PAM technologies. For instance: 100BASE-T2 Fast Ethernet uses five levels of PAM (PAM-5), and 100BASE TX Ethernet shares the same principle of block coding of the signal, "4B/5B," then applies NRZI coding, and finally chooses the MLT-3 sublayer to encode data streams before transmission over the physical support (see the examples in Figures 5.36 and 5.37).

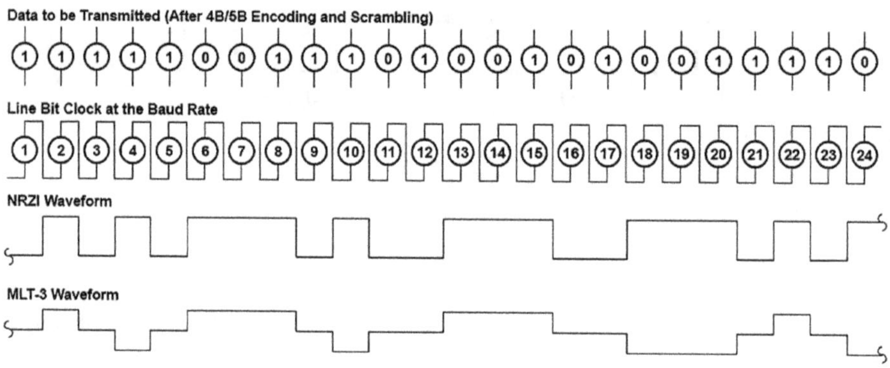

Figure 5.36 Example of MLT-3 encoding.

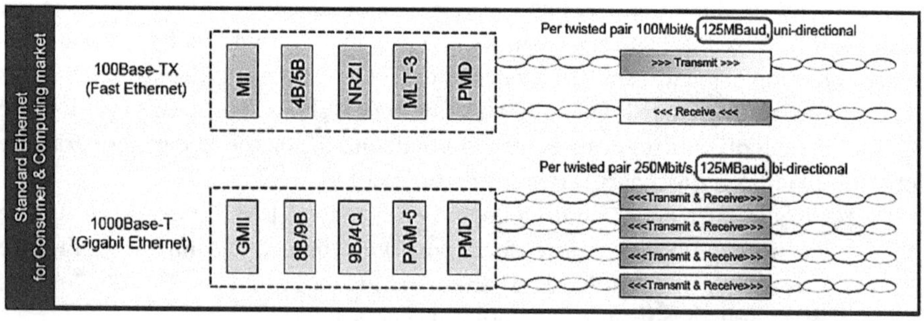

Figure 5.37 Examples of NRZI/MLT3 and 8B/9N/PAM-5 encoding.

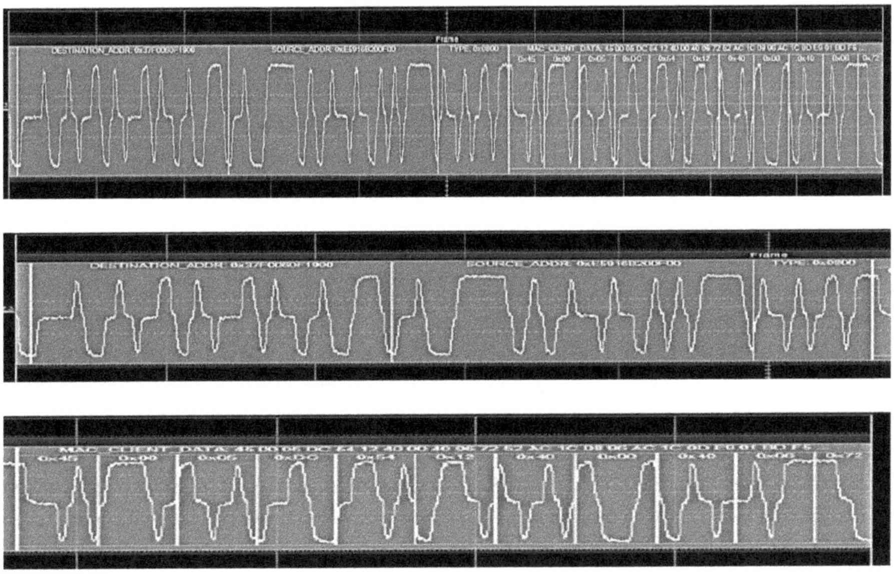

Figure 5.38 Examples of signals in 100BASE-TX Ethernet (readings taken on a LeCroy oscilloscope).

Examples of transmitted signals Figure 5.38 presents a few concrete examples of the shapes of transmitted signals.

Baseband spectrum

In the frequency domain, obviously, the spectrum in MLT-3 requires a lesser bandwidth than most other binary or ternary interfaces, is more compact, and can more comfortably fit into the spectral masks/patterns mentioned above. At the same transmission rate, MLT-3 offers a better compromise between bandwidth and quantification, and emits less electromagnetic interference. Thus, it allows us to transmit data at a high datarate on lines with a smaller bandwidth – on inexpensive copper wires, which are acceptable for use in the automobile market!

Effect of transmission line

The line – whether or not the wires are twisted or adapted – always produces the effect of a low-pass filter set at around 6 dB/octave in its amplitude/frequency ratio. The removal of all spectral redundancy renders the signal rather vulnerable to linear distortions. This distortion is a well-known phenomenon: intersymbol interference.

Filtering – radiated spectrum

Obviously, ultimately, there is no secret: the radiated spectrum must satisfy the local regulations and the intended application. Plainly, it is often necessary to review or correct one's copy and revisit the idea hundreds of times to find the best compromise, because there is never only one solution to a problem – that would be too easy!

Indeed, up until now, our explanations have been only theoretical, but every day, in real cases of automobile applications, the environmental media surrounding transmission lines change and evolve. One day bodywork is made of metal and the next resin;

the placement of wires and harnesses are different from one model to the next and often even different in two vehicles of the same model! Thus, despite the effects of the lines and such different metal screens (chassis, engines, etc.), if we wish to avoid degrading the digital information transmitted, it is necessary to have a filter whose rolloff coefficient at the cutoff frequency at least satisfies "Nyquist's first law." Therefore, we need to include low-pass filters or band-pass filters to make the signal spectrum fit the pattern, without compromising quality. On this note, we must mention the eye pattern and the bit error rate.

Eye pattern

The eye pattern and the size of its aperture are well known to digital transmission aficionados,[1] but in the case of automobiles, an in-depth study of the impact of the environmental conditions (mainly temperature) on signal integrity is necessary.

For example: Figure 5.39 compares the eye pattern for a 20 m cable (unshielded twisted pairs) at ambient temperature with the same cable at a higher temperature – 105°C or 125°C – frequently encountered under the hood of a car.

As we can see, at high temperatures, the smaller aperture of the eye indicates an increase in intersymbol interference (ISI), due to limitations of bandwidth of the cable

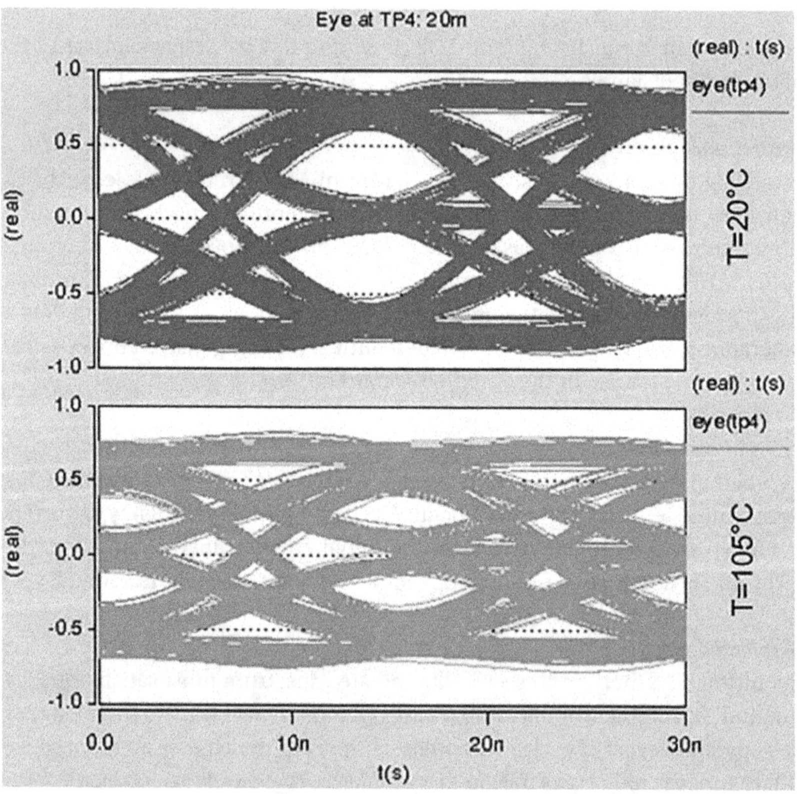

Figure 5.39 Eye pattern for a 20 m unshielded twisted pair at ambient temperature of 20°C and at a temperature of 105°C under a car hood.

that has evolved. Provided the cable length remains less than 20 m, the reduction of the SNR can generally be compensated by adaptive equalization on reception, without compromising the bit error rate.

Bit sampling and bit sample point

Let us briefly return to the general virtues of the aperture of the eye pattern. It can be examined from two different perspectives.

Vertical The vertical dimension of the eye aperture at the time of sampling of the signal received by the network receiver node gives a good indication of the level of noise beyond which there is a danger of bit errors occurring. When the sampling point does not coincide with the maximum vertical aperture, the tolerable level of noise is, of course, less. In sensitive cases such as those of applications of unshielded twisted pairs, this means that, at the receiver, an electronic architecture must be carefully tailored to optimize the instant of sampling of the incident signal.

Horizontal The wider the lateral aperture of the eye pattern, the less susceptible the system is to sampling clock offset (poor quality of synchronization) and to all sorts of jitters (e.g. crosstalk between cables in a strand/bundle). In addition, that lateral aperture is strongly linked to the time responses and stiffness of the filters (pulsed response, whether or not the signals rebound, good line adaptations, etc.) and is highly sensitive to temperature variations. Ultimately, on the basis of all of the above, with some careful calculations, we can draw conclusions as to the quality and total BER of the digital transmission.[2]

Bit error rate (BER) To clarify our ideas and elucidate the orders of magnitude at issue here, using PAM-3 and a signal-to-noise ratio of 16 dB, generally a bit error rate of between 10^{-7} and 10^{-10} can be obtained, even without direct error correction. With an additional margin of 6 dB in the SNR, that value goes to 10^{-25}. In closing, to compensate for the loss of 6 dB of the signal-to-noise ratio in PAM-5 in comparison to MLT-3, a costly error correction code (trellis modulation) must be applied when the signal is received.

This concludes the important discussions, drawing together all the parameters in digital communications as a function of the choice of medium. In the wake of those discussions, we can finally assess whether the global concept we have chosen is "roadworthy" (which is normal for an automobile) or not for the intended application!

The transceiver

In transmission A transmitter for the Ethernet PHY layer, which satisfies the various emission patterns, and limitations on EM pollution and noise, will include the following:

- A carrier signal modulator, calibrated to incoming data;
- A wave-shaping module;
- A digital-to-analog converter (DAC);
- A phase-locked loop (PLL);

- A line driver;
- A PHY control module.

Let us briefly examine the main building blocks.

Modulator The modulator performs pulse amplitude modulation. It can operate at multiple symbol rates, multiple power transmission levels, and multiple modulation levels.

Control module and its registers A special module runs the wave-shaping system. By means of its configuration registers, it can tailor the wave-shaping to the specific spectral mask requirements in place.

Wave-shaping The modulated signal generated by the modulator is then sent to a wave-shaping module. To achieve maximum immunity to noise whilst reducing emissions at the top of the authorized frequency band, the wave-shaping module applies a filter to the signal, adapting the shape of the transmission wave to suit customized needs, so that the spectrum associated with the signal can fit into the required transmission mask/pattern. As the filter is programmable, the module can apply other wave-shaping functions on a case-by-case basis, in relation to local RF regulations. Note that, for lack of good spectral shaping, the transmitted signal level (power) needs to be reduced to fit into a particular transmission mask. This signal reduction has immediate and important impacts, firstly, on the level of noise resistance and, secondly, on the communication range/distance.

Digital-to-analog converter The transmission signal thus generated and tailored by the wave-shaping module is passed to the digital-to-analog converter (DAC), which then passes the analog transmission signal to the line driver for transmission to the receiver. The reception part of this component includes a fixed or adaptive inverse filter corresponding to the wave-shaping function to optimize the decision as to the best SNR for the receiver and to improve the BER. The normalized power spectral density (PSD) can be shaped using analog filters, either an internal part or an external ancillary to the integrated circuit. For example, on output from the line driver, if necessary, passive filters can be applied for a particular datarate, to further reduce the side lobes or the edges of the main lobe of the PSD.

Line driver for unshielded twisted pairs The physical layer technology offered by BroadR-Reach (see Section 5.3.1 below) is appropriate for use in automobiles. It conforms to the main requirements of automobiles in terms of EMC, robustness, and the strict environmental conditions on automobile applications, allowing for the use of unshielded twisted pairs and of Ethernet at a competitive cost. Note that in relation to the Fast Ethernet standard, the nomenclature is considerably reduced. A diagram of the system is presented in Figure 5.40 (1) and (2). Figure 5.40 (3) goes a step further in reducing the number and cost of components, by using capacitive coupling in the form of coupling capacitors instead of a conventional transformer (blocking the DC component of the spectrum). The physical hardware layer (common-mode choke [CMC], connector, and UTP cable) is very similar to a FlexRay or a CAN link.

In receipt

Adaptive equalizer By structure, a link cable acts as a low-pass filter and attenuates the high frequencies more than low frequencies, which distorts the received signal. The adaptive equalizer is a digital filter that continually minimizes the mean square error (MSE) of the signal coming out of the slicer (division of the signal into slices on the basis of their voltage level). The constant adaptation of the equalizer coefficients means that the optimal set of coefficients is always produced for a given length or quality of cable.

Echo and crosstalk suppressors

Crosstalk occurs when there is coupling (magnetic or capacitive) between the wire carrying the signal and neighboring wires. Interference between the emitted signals and those received on the same wire results in the phenomenon of echo. Suppressors cancel out echoes and crosstalk produced by simultaneous transmission and receptions.

Radiated spectra

The spectrum of the transmitted signal in conventional Fast Ethernet is much higher than the authorized level in the FM radio band. It is for this reason that the solution BroadR-Reach functions with PAM 3 at 66.66 MBauds, which allows for a very low spectrum in the radio band, at 88–108 MHz.

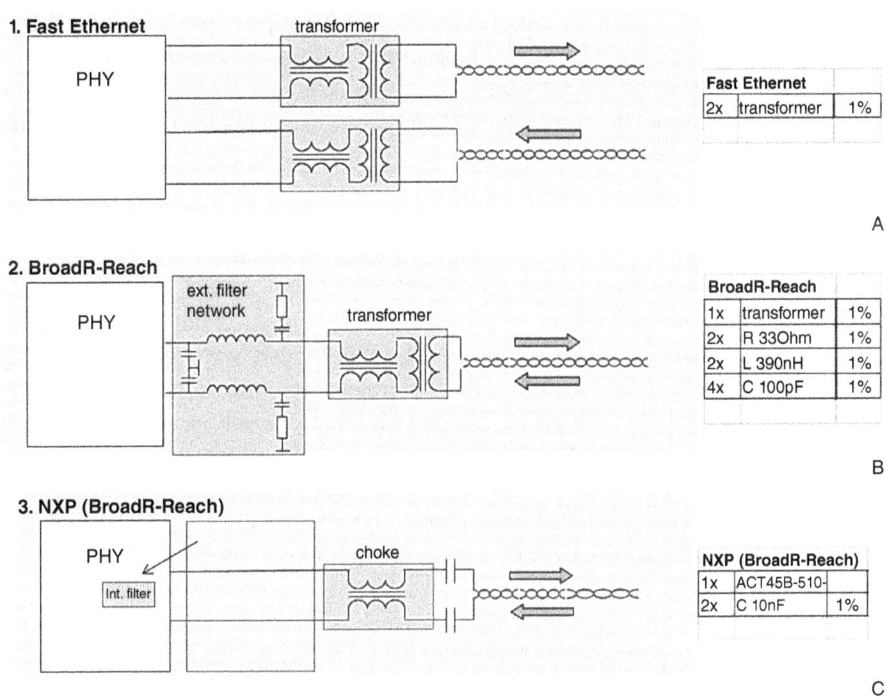

Figure 5.40 Examples of physical layers for 100 Mbit/s applications.

In addition, in order to realize the full potential of the technology, often referred to as "OABR Ethernet" (OPEN Alliance BroadR-Reach), and to allow Ethernet to be used widely in automobiles, the use of common-mode choke for the PHY layer is encouraged.

5.3 The variants of Ethernet PHY used in automobiles

Before going any further, it is important to provide readers with a solid grasp of the industrial variants of Ethernet that have been in place for a long time, in order to understand the origins of the solutions adopted and used for applications in automobiles, which we discuss in this section.

5.3.1 100 Mbit/s Ethernet in automobiles

"Non-choice" of 100BASE-TX Ethernet in IEEE 802.3u

Automotive applications impose more stringent requirements on electronic systems and their components, in comparison to those in place in the general public and industrial domains. In particular, the requirements are stricter in terms of EMC, radioelectric pollution (ISO 11452), and environmental conditions. As previously mentioned, in relation to the performances of the physical layers, and in spite of numerous factors in the automotive world (two pairs of wires, high datarate, radiation, price, etc.), 100BASE-TX-IEEE 802.3u Ethernet (100 for 100 Mbit/s, T for twisted pair) is not a satisfactory solution. It is incapable of delivering a bitrate of 100 Mbit/s on an inexpensive, "single unshielded twisted pair" medium for automobile applications, and conform to the numerous local regulations and standards on EM radiation. However, this version is the closest to the technology envisaged for numerous applications in automobiles, though improvements have had to be made in order for it to be adopted.

The proposed solution "BroadR-Reach"

As we have seen, there are numerous industrial Ethernet variants (with differing datarates, physical layers, frames, etc. – see Section 5.1), and myriad requirements for applications in automobiles expressed for the future, highly technical flights of fancy and numerous conventional treatises in *Signaling Theory and Signal Processing*. Historically, in view of this situation, the integrated circuit manufacturer Broadcom Corporation (notably specializing in Ethernet) developed a technical solution known as "BroadR-Reach Ethernet." It is designed for use in automobiles, offering "Low cost, Cheap connectors, Low weight, Longer cables" to extend the range of an Ethernet connection up to 500 m at 100 Mbit/s (as opposed to the usual limit of 100 m), and supports the transmission of Ethernet frames in full-duplex mode, with unshielded twisted pairs.

The operational principle behind BroadR-Reach is described, in part, in the patent *System and method for transmit signal pulse shaping in automotive applications* and, in part, in the Broadcom document – *BroadR-Reach Physical Layer – Transceiver*

Specification for Automotive Applications, dated May 2014. This technology has since garnered the support of the OPEN Alliance (*one-pair Ethernet*) (see Section 5.3.7) and has been approved for use in automobiles.

A little technical detail

In brief, the physical layer of the OPEN Alliance BroadR-Reach (OABR) used at 100 Mbit/s is full-duplex (simultaneous transmission and reception), on a *single unshielded twisted pair* (S-UTP). The technologies are similar to those used every day in CAN and FlexRay. Though the PHY layers in consumer technology generally offer a line length of less than 10 m, automobile technology is optimized to support over 100 m of cable. The challenge, then, is to find wave-shaping solutions and receiver equalizers that are optimized for such a length of cable, and can comply with the strict constraints on emissions and immunity in the automobile market.

To use BroadR-Reach, the layers above the MAC interface in the OSI model (physical media attachment – PMA) are unchanged, and only the implementation of the PHY layer needs to be replaced, in relation to those of the OSI model. Otherwise, the global topology is identical to that of IEEE 802.3 Ethernet (see Figure 5.41).

At both ends of the line are specific BroadR-Reach PHY integrated circuits, which send and receive data in both directions at once.

To bring the RF radiation and electromagnetic compatibility (EMC) to levels compatible with CISPR 25 without compromising on bitrate, BroadR-Reach technology is designed and based on a hybrid of the 100Base-Tx and 1000Base-T variants, adapted to automotive applications. Thus, the physical layer of BroadR-Reach "BR-PHY," due to all the subtle factors explained at great length in the foregoing chapters, uses specific solutions for line coding and modulation. At 100 Mbit/s, we can summarize:

- Firstly, the databits are generated and ordered with a clock at 100 Mbit/s;
- Data packets from the MII (medium-independent interface) are in the form of initial 4-bit symbols (nibbles), ordered with the clock at 25 MHz;
- A 4B/3B binary-to-binary data conversion is then applied. The original nibbles are shaped into a new data stream made up of 3-bit (3B) packets. These 3-bit symbols

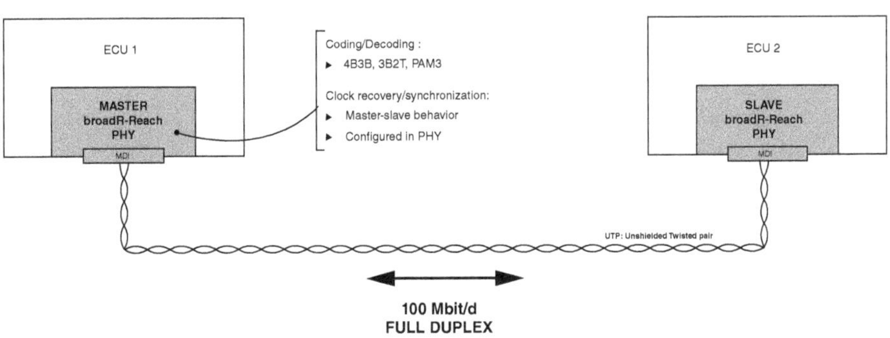

Figure 5.41 Physical layer in BroadR-Reach.

are transmitted over the course of a clock period equal to 25 × 4/3 = 33.3 MHz, which represents a new symbol rate of 33.33 Msymb/s;
- The bits of these symbols are then scrambled, and subjected to BR-PCS coding, being turned into a new stream of 3-bit symbols, from pairs of ternary symbols into two ternary elements with binary-to-ternary line coding "3B/2T";
- These ternary pairs are then multiplexed into a serialized stream of symbols at 66 Mbauds;
- Finally, PAM-3 is used to transmit the signal over an unshielded twisted pair. The symbol encoding method used in BR-PHY is 1D-PAM3. Blocks of one-dimensional (1D) ternary codes in "BR-PCS Transmit" are transmitted at three voltage levels (PAM-3) (+1, 0, and −1). A symbol is transmitted at each symbol period.

In summary

Figure 5.42 offers an overview of the successive stages of encoding.

The shape of the electrical signal sent over the Ethernet network is similar to that shown in Figure 5.43.

BroadR-Reach automotive Ethernet can therefore be summarized as follows (see Figure 5.44):

• MII – frequency bits	f bit	= 100 Mbits/s
• 4-bit symbol	F symb 4	= 25 Msymb/s
• 4B/3B transposition	F symb 3	= 25 × 4/3 = 33.33 Msymb/s
• 3B/2T transposition		= 33.33 × 2 = 66.66 Msymb/s
• PAM 3		= 66.66 Mbaud

Figure 5.45 shows an example of a path compatible with *BroadR-Reach* and Figure 5.46 shows the detail of how it works.

100 Mbit/s BroadR-Reach can be viewed as a light version of the 100BASE-T solution, adopting two-way communication. The reduction of the datarate to 66.6 MBauds means that unshielded twisted pairs can be used. Figure 5.47 offers an overview of this situation.

Main applications of 100 Mbits BroadR-Reach

The main applications of BroadR-Reach are based around advanced driver assistance systems (ADASs). Remember (as seen in Section 3.3), ADASs can be passive or active:

- **A passive ADAS** simply provides drivers with alerts or warnings. Examples include:
  - Lane departure warning;
  - Reversing camera applications.
- **An active ADAS** goes further than warnings, and actually takes control to apply corrective action (e.g. by recentering the vehicle in its lane or emergency braking): For example: an assisted parking ADAS.

At a datarate of 100 Mb/s, BroadR-Reach is most suitable for:

- Passive ADAS applications, because active ADASs require lossless video, in order for the image recognition algorithms to function properly; if the vehicle has to recognize potential dangers in the fairly near distance, the algorithms will need greater precision, with a complete lossless incoming video;

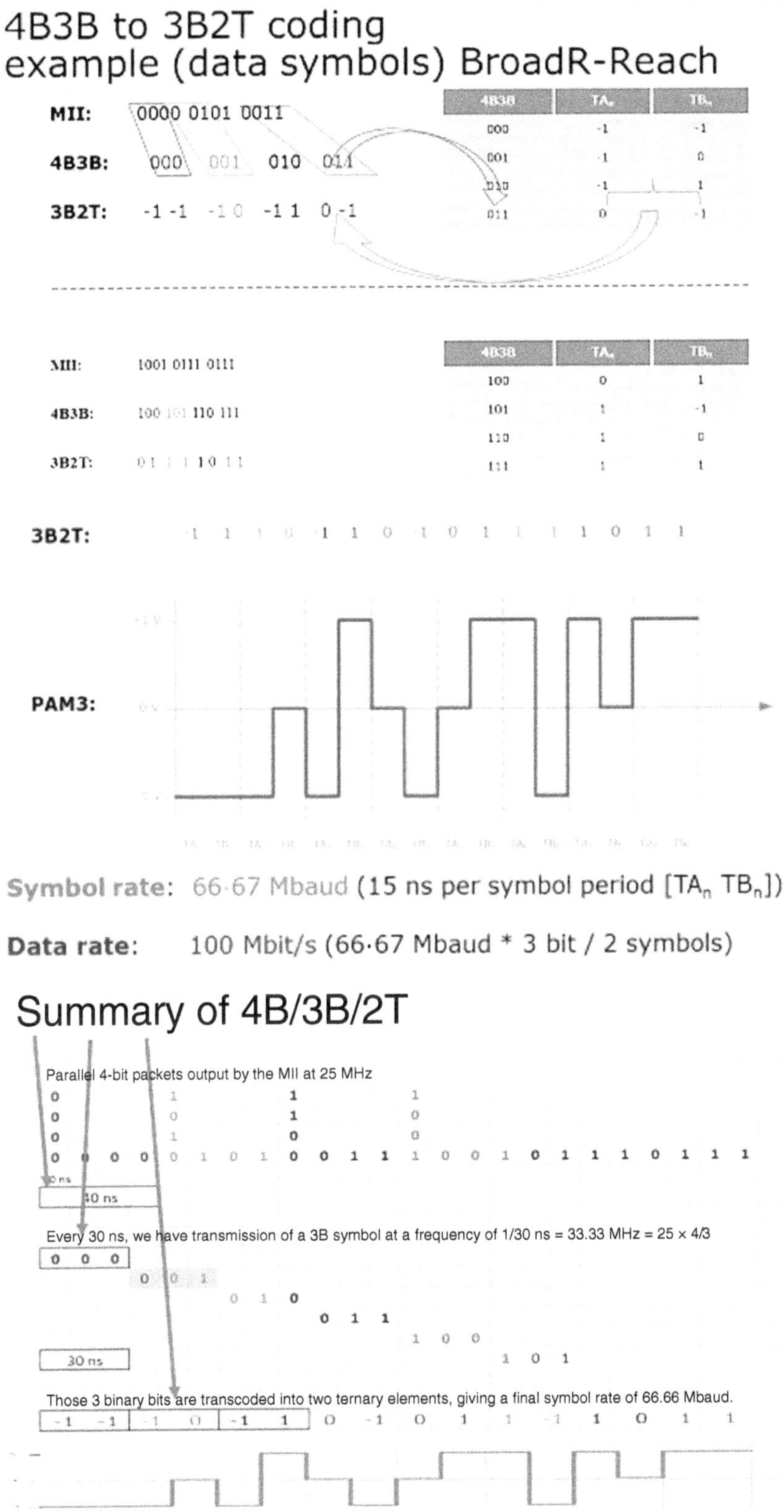

Figure 5.42 Overview of encoding in BroadR-Reach.

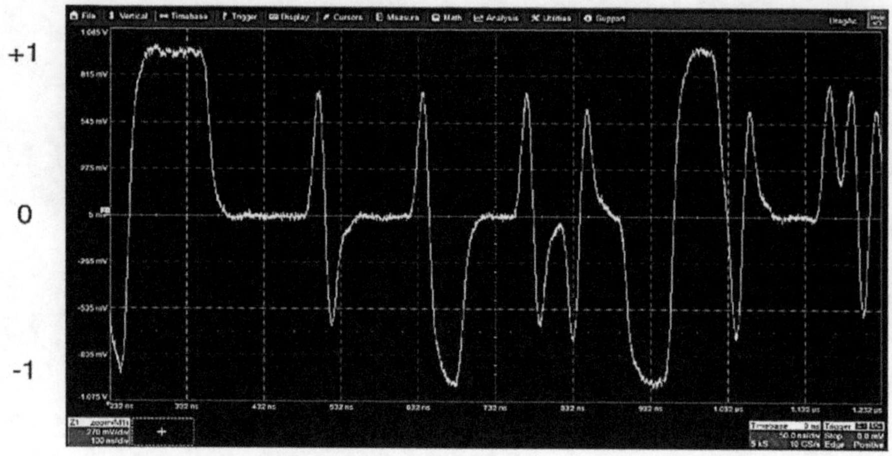

Figure 5.43 General shape of BroadR-Reach signal.

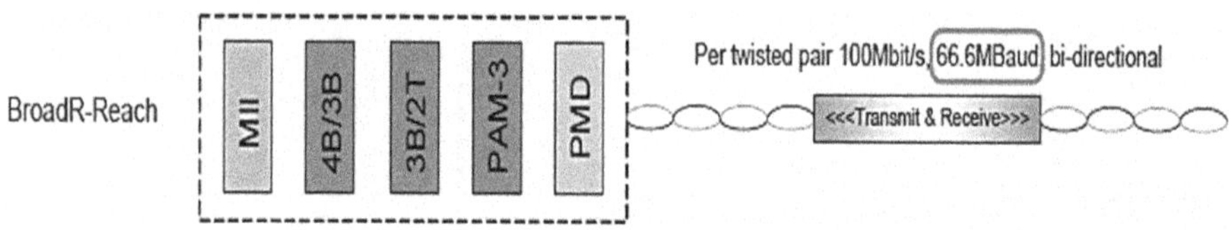

Figure 5.44 Overview of the BroadR-Reach solution.

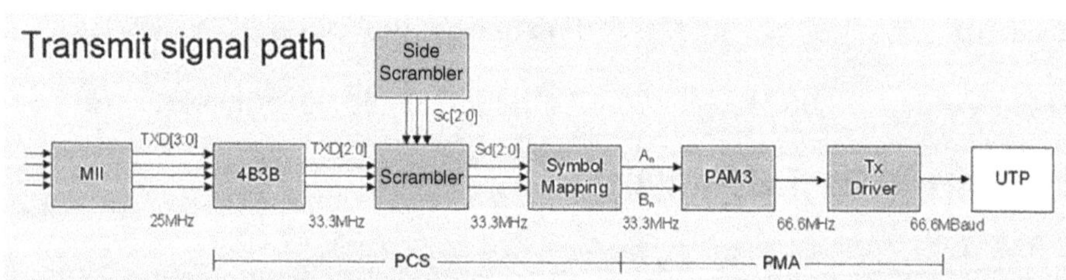

Figure 5.45 Example of a complete BroadR-Reach channel.

- Infotainment applications, because it is more profitable than media-oriented system transport (MOST) or low-voltage differential signaling (LVDS). The greatest savings come in the cabling. Indeed, LVDS coaxial cables often serving for analog camera connections are fairly thick, heavier, and costlier. With the switch to digital cameras, BroadR-Reach offers the benefits of twisted pair cables and a higher bandwidth.

Upstream	**The bit signal is generated upstream by a clock at 100 MHz**	**The bit duration is 10 ns and the bit frequency is 100 MHz**	**Thus, the bitrate is 100 Mbit/s**
Medium-Independent Interface MII	The MAC layer of Ethernet is linked to the PHY layer by the synchronous parallel MII, which transmits 4-bit nibbles at the rate of 25 MHz to achieve a datarate of 100 Mbit/s	The MII determines the maximum theoretic bitrate for all versions of 100 Mbit/s Fast Ethernet	At output from the MII, all nibbles of 4 bits (symbols) are transmitted every 40 ns, so at a rate of 25 MHz
4B/3B	This stream of 4-bit packets is converted into a series of new 3-bit packets	The new 3-bit packets, with a duration of 30 ns, are therefore shorter	These 3-bit symbols are output at the rate of 25 × 4/3 = 33.33 MHz
Scrambler	The 3-bit packets are then mixed to randomize the codes		The DC component is rebalanced, and long sequences of zeros are eliminated to ensure precise timings without line coding redundancy
3B/2T **symbol mapping**	The mixed 3-bit packets, still with a period of 30 ns, are converted into 2T codes and sent to the modulator for PAM-3	The 2T signal, made up of two parts, has a duration of 30 ns, so each part has a duration of 15 ns	The signal is then output at a frequency of 33.33 × 2 = 66.66 MHz
PAM 3	The modulator creates the signal to be transmitted containing the three electrical levels corresponding to the 2T codes		

Figure 5.46 Details of how the BroadR-Reach path works.

	Mbit/s	**PAM level**	**MSymb/s**	**MHz**	**Duplex**	**Cables**	**Distance**
100Base-T	100	3	125	65–80	Half	2	
BroadR-Reach	100	3	66.67	33.33	Full	*Single unshielded twisted pair* (UTP)	*Up to* 15 m
						Single shielded twisted pair (STP)	*Up to* 40 m

Figure 5.47 Differences between 100Base-T and BroadR-Reach.

100BASE-T1 – IEEE 802.3bw

The counter-proposal to IEEE, a variant (or amendment) of Ethernet, known as 100BASE-T1 – IEEE 802.3bw at 100 Mb/s emerged a short time later (in 2015) with a goal largely similar to BroadR-Reach from BroadCom.

We may therefore legitimately wonder why this specification was published. The reasons were that demand for Ethernet on twisted pairs was arising in other industrial areas, such as industrial automation and avionics, and the proprietary technology BroadR-Reach originally developed by Broadcom was only deliberately designed for automotive applications. On the other hand, the benefits offered by BroadR-Reach, supported by the OPEN Alliance and certified for use in automobiles, were sufficiently attractive for the IEEE to create its own version of the specification, drawing a great deal of inspiration from the original. Thus, the specifications BroadR-Reach Ethernet and 100Base-T1 share a common environment, and many of the same ecosystems.

Note: 100Base-T and 100Base-T1 serve the same specifications. The suffix "-T1" is an important distinction; it refers to the automobile Ethernet standard, which must not be confused with 100Base-T.

The 100BASE-T1 standard describes the specifications of the PHY layer and the management parameters at certain operational rates. 100BASE-T1 uses a common medium access control (MAC) system and management information base (MIB). This specification:

- Is devoted to the transport of Ethernet frames at 100 Mbit/s;
- Operates in full-duplex mode;
- Operates, point to point, on a single balanced twisted pair;
- Specifies the baseband used for the medium;
- Describes 3-bits-per-symbol encoding (PAM-3);
- Indicates that the twisted pair must support 66 MHz;
- Allows for a maximum segment length of 15 m;
- Specifies no particular connector, which accounts for the applications of RJ45 connectors, for example;
- That the data encoding technique 4B/3B is used by 100BASE-T1 in converting the 4-bit (4B) nibbles from the MII, with a clock at 25 MHz, to 3-bit (3B) nibbles, with a clock at 33.333 MHz; etc.

What are the differences between BroadR-Reach and 100BASE-T1?

BroadR-Reach, supported by the OPEN Alliance for use in vehicles, and IEEE 100BASE-T1, are both dominant in the world of automotive Ethernet. This may lead to confusion, because they are almost identical, and there is a great deal of overlap. As these two standards are intended to be interoperable, they are often interchangeable in applications. However, there are two small but significant differences that must be noted, which are in the PHY layers, because at all higher levels, the Ethernet stack is the same:

- The series of tests of the physical medium attachment (PMA) in 100BASE-T1 includes a test for the *transmit peak differential output*. This parameter is not explicitly defined in the BroadR-Reach specification;
- 100BASE-T1 exhibits a number of differences in the timing protocol for wake commands, to make these periods shorter.

In addition, the IEEE benefitted from the work of the OPEN Alliance (specifically, the PHY technologies in 100BASE-T1 and 1000BASE-T1 mean that Ethernet can easily replace earlier systems – Figure 5.48).

Let us now look at solutions operating at 1 Gbit/s.

5.3.2 Automotive Ethernet at 1 Gbit/s

History seems to have repeated itself somewhat, with the 1 Gbit/s solutions of BroadR-Reach and the IEEE.

1 Gbit/s BroadR-Reach

In line with the 100 Mbit/s solution, the 1 Gbit/s BroadR-Reach proposition, building on the 1000BASE-T specification, was as follows:

- The PCS uses a transmission code to improve the transmission characteristics of the data to be transferred on the link and to support the transmission of control characters and data;
- The encoding defined by the transmission code ensures that sufficient information is carried in the PHY bitstream to enable clock recovery at the receiver;
- RS (Reed-Solomon) coding also safeguards the probability of detecting frame errors that can occur during the transmission and reception of the information.

During transmission:

- PCSs map the 80-bit signals from the GMII into blocks of 81 bits inserted into the RS frame (81B-RS coding and vice versa for reception);
- The 80B/81B (or similar) encoders need an overhead of 1 extra bit in addition to the useable data:
  - A single "0" at the start of the packet labels it as a data packet;
  - A single "1" at the start of the packet labels it as a control packet;
- This coding, with only one extra bit to distinguish between data and control packets, generally requires an additional error correction code for transmission;
- The longer the data blocks, the less necessary is the overhead coding, and the latency of the PHY layer is increased;
- The acceptable compromise of 80B/81B coding has been adopted.

Furthermore:

- There must be an effective way of encoding the bitstream into PAM-3 symbols;
- The simplest way is to encode the 3 binary data bits (8 possible combinations) into 2 ternary symbols (9 combinations, with the "00" not being used);

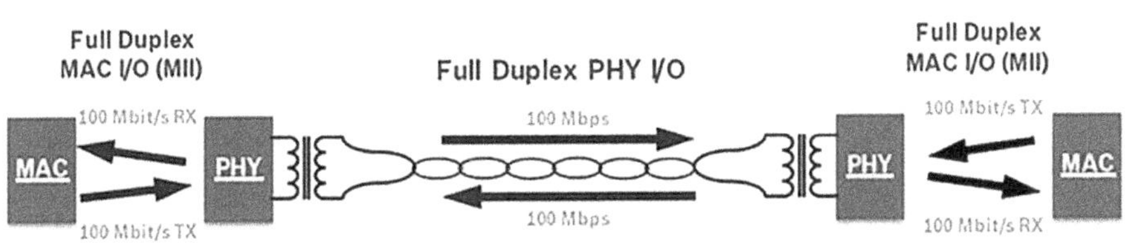

Figure 5.48 Physical layer 100BASE-T1 (source: OPEN Alliance).

- However, in principle, this encoding does not produce a DC-free BLW-free line. Thus, this encoding requires a sort of preprocessing (for example: scrambling the bitstream).

In summary, the final choice is as follows (Figure 5.49):

- Bit coding 80B/81B;
- Line coding 3B/2T;
- Modulation PAM-3.

1000BASE-T1 – IEEE 802.3bp

The IEEE has also issued its own industrial specification for two-way transmission at a datarate of 1 Gbit/s, over simple twisted pairs of copper wires for automotive and industrial applications, as was the case with 100 Mbits/s. The specification 1000BASE-T1 – IEEE 802.3bp was published in 2016.

The PHY layer of 1000BASE-T1 operates at a datarate of 1 Gbit/s in full-duplex (with echo cancellation) over a single pair of balanced copper wires, satisfying the requirements (in terms of EMC, temperature, etc.) of industrial and automotive environments. It also supports two types of network segments:

- Type A: an automobile segment (line) using copper cabling over a balanced single unshielded twisted pair, over a distance of 15 m;
- Type B: an additional line using balanced copper cabling of less than 40 m in length (up to four pairs) to support applications that require a further physical range, such as industrial control units, automation, and transport (airplanes, trains, buses, and trucks).

The physical layer in 1000BASE-T1 also uses PAM-3, and transmits data at 750 MBauds. A 15-bit scrambler is used to improve the performances in terms of EMC. The signals TX_D, TX_EN, and TX_ER (10 data cycles) from the GMII are encoded together, using 81B encoding to reduce the overload. To preserve a BER less than or equal to 10^{-10}, 1000BASE-T1 PHY adds 396 bits of Reed-Solomon forward error correction (RS-FEC) to each group of 45 blocks of 81B (containing 450 bytes of GMII data). PAM-3 mapping, the scrambler, RS-FEC, and the 81B encoder/decoder are all part of the PCS.

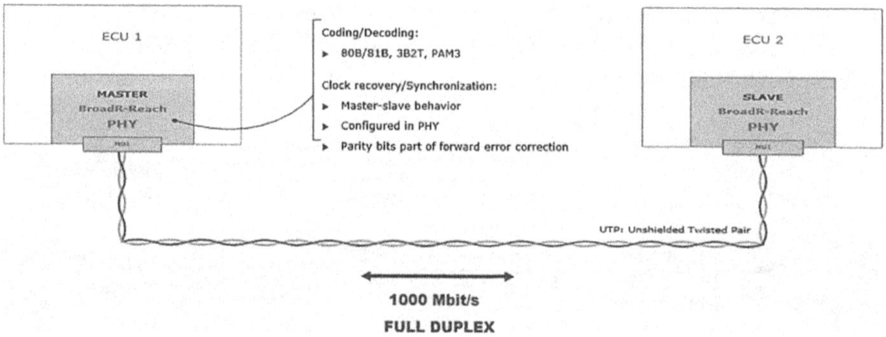

Figure 5.49 Physical layer of 1 Gbit/s Ethernet.

Applications of 1 Gbit/s Ethernet

The applications of 1 Gbit/s Ethernet in the field of automobiles are as follows:

- Main communication network;
- Smart antenna;
- Driver assistance systems;
- Infotainment;
- Optimized battery charging (OBC); etc.

For example: it is logical to use a 1000BASE-T1 solution in a vehicle when we wish to connect multiple types of cameras to the same network (vision applications, reversing camera, video and entertainment, with or without image compression), so with different definitions and therefore completely different datarates. Generally, automakers accept compressed video for video entertainment applications, but not for the capture and analysis of computerized images. This is why LVDS has, for some time, been the preferred connectivity method for uncompressed video transfer, but is tending to be replaced by very high datarate Ethernet.

An HD-quality uncompressed image requires a datarate of over 100 Mbit/s. Thus, in order to use 100BASE-T1 at 100 Mbit/s, video compression must inevitably be used. Such compression causes, firstly, the degradation of image quality and, secondly, latency caused by the compression and decompression algorithm, and also cost. On the other hand, unlike with LVDS (i.e. a point-to-point connection), 1000BASE-T1 allows the following simultaneously (see Figure 5.50):

- Transport of uncompressed digital video – for example, a 720p30 image (720 horizontal pixels, progressive scan, and 30 frames per second);
- Simultaneous handling of multiple (compressed) HD video feeds, with 4K resolution (with "little or no" compression);
- Transport of higher-quality images for vision, the console cluster, and HD instruments;
- Management of multiple video feeds, with an increasing number of cameras inside the vehicle.

Summary of 1 Gbit/s

Figures 5.51 and 5.52 offer an overview of what has been said in the foregoing discussion.

In closing, Figure 5.53 shows the main differences between 100 Mbit/s and 1000 Mbit/s Ethernet.

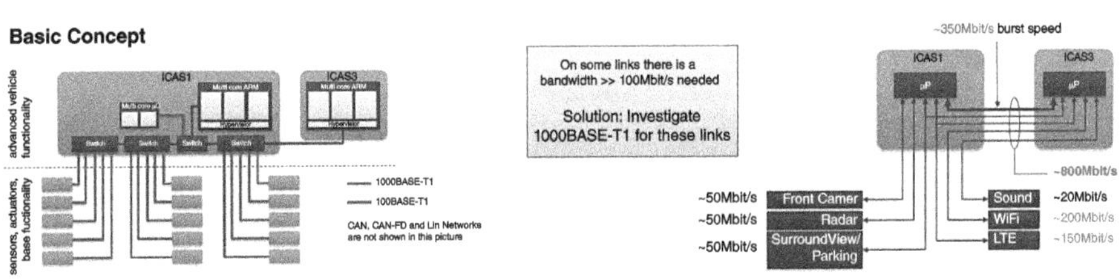

Figure 5.50 Examples of applications implemented by VW and Marvell.

	Mbit/s	*PAMlevel*	MSymb/s	MHz	Duplex	Cables	Distance
1000Base-T	1000	5	125	65–80	*Full*	Four	100 m
1000BASE-T1	1000					*Single, balanced twisted pair*	15 m
BroadR-Reach	1000	3			*Full*	*Single, balanced twisted pair*	15 m

Figure 5.51 Summary of 1 Gbit/s Ethernet solutions for automobiles.

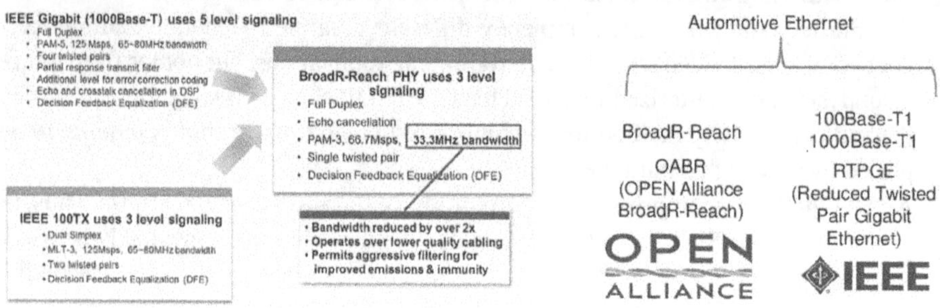

Figure 5.52 IEEE to BroadR-Reach migration.

5.3.3 Multi-Giga Ethernet in automobiles

Multi-Giga – IEEE 802.3ch

The Multi-Gig Automotive Ethernet PHY group is in charge of the transport of Ethernet frames at datarates over 1 Gbit/s in a vehicle. IEEE 802.3ch (*Standard for Ethernet Physical Layer Specifications and Management Parameters for Greater Than 1 Gb/s Automotive Ethernet*) is formulated to cater to the ever-increasing bandwidth needs, in particular for applications in connected vehicles, advanced driver assistance systems (ADASs), and onboard infotainment equipment. This standard uses the specification of medium access control (MAC) and the management information base (MIB). The carrier sense multiple access with collision detection (CSMA/CD) protocol in MAC specifies half- or full-duplex operation. The medium-independent interfaces (MIIs) offer an interface for architectural and facultative implementation of selected entities in the PHY layer. The physical layer encodes the frames for transmission and decodes received frames with the specified modulation for the operation rate, transmission medium, and segment length in question. Other specifications include protocols for control and management, and supply of energy over the twisted pair for certain types of PHY.

Multi-Giga applications

Driven by the need for increased capacity and speed to serve the needs of increasingly autonomous, heavily connected vehicles, a revolution in automobile networking is already under way. The high-speed automobile Ethernet technologies 1000BASE-T1 and Multi-Giga facilitate the rapid data transfer necessary for autonomous driving, and offer automakers next-generation data transfer solutions. In time, the capacity of

		100 Mbits/s			1000 Mbits/s	
		BroadR-Reach	**100BASE T1**	**100BASE TX**	**1000BASE T1**	**1000BASE T**
IEEE		n/a	802.3bw		802.3bp	
		Full-duplex	Full-duplex	*Dual simplex*		Full-duplex
Speed	Mbits/s	100	100	100	1000	1000
Cable		Cat 5	Cat 5		Cat 5e	
Signaling	*levels*	3	3	3	3	5
	coding	PAM3	PAM3	MLT 3	PAM3	PAM 5
	@Msps	@33.3	@66.66	@125	@750	@125
PHY		USTP	USBTP	2 TP	USBTP	4 TP
Distances	m		15			

Figure 5.53 Main differences between the 100 Mbit/s and 1000 Mbit/s Ethernet variants.

Automotive Ethernet will exceed 10 Gbit/s. The next generation of vehicles will carry eight or more cameras, lidar, radar and V2X (vehicle-to-everything), generating enormous data feeds and, in so doing, posing major challenges for the vehicles' internal networks. These requirements, associated with the need to transfer multiple types of information in the context of economic models linked to connected vehicles, will see terabit-level data processing become commonplace. In addition, it has been projected that the cameras and screens will increase the number of high-speed connections in vehicles from 150 million in 2020 to 1.1 billion by 2030, which has already begun with 1000BASE-T1 Ethernet. The expected synergies will focus on PHY designs at datarates between 1.25 and 28 Gbit/s.

5.3.4 Automotive Ethernet at 10 Mbit/s

There is no mistake – we are talking about a 10 Mbit/s version of Ethernet! Why have we included this section at the end of the chapter?

The trend in automobiles is towards "Full Ethernet," so that is all that was missing – the unexpected solution to fill the gaps that CAN, CAN FD, CAN XL, and FlexRay cannot resolve in 10 Mbit/s operation. Thus, IEEE 802.3 has produced a 10 Mbit/s Ethernet technology called 10SPE (10 Mbits/s single pair Ethernet).

10BASE-T1S – IEEE 802.3cg

Part of the 10BASE-T1 family of standards, IEEE 802.3cg – *Physical Layer Specifications and Management Parameters for 10 Mb/s Operation and Associated Power Delivery over a Single Balanced Pair of Conductors* – is divided into two:

- 10BASE-T1L: 10 Mbit/s transmission on a *single balanced pair*, up to a distance of 1 km;
- 10BASE-T1S: short-range 10 Mbit/s transmission on a *single balanced pair*, conforming to the requirements of automobile systems.

These technologies are designed to offer a collision-free system, deterministic transmission on a multi-point network, which allows for the use of a single pair to connect the sensors and ECUs (Figure 5.54). A number of automakers (VW, for example) see this as a way to make the vehicle Full Ethernet, by replacing today's CAN, CAN-FD, CAN XL, and FlexRay networks with 10BASE T1S or 100BASE-T1.

The primary characteristics are:

- 10 Mbit/s, 12.5 MBauds, 4B/5B, DME signal encoding, PAM 2;
- Limited to 8 nodes over 15–25 m;
- Stub length limited to 10 cm;
- Collision avoidance layer (ACV), facilitating the use of the full bandwidth of 10 Mbit/s;
- Arbitration by means of a round-robin procedure (functioning like a turnstile), ensuring each node has medium access in deterministic time.

5.3.5 Power over Ethernet – PoE IEEE 802.3bu

The standardization body, the IEEE, has also published 802.3bu-2016, which defines the specifications and parameters necessary for the addition of a power signal over an Ethernet link (*Power over Ethernet* [PoE]) formed of a single twisted pair of copper wires that also carry data. This standard defines a power transmission protocol that supports various levels of voltage and different power classes for each supply voltage with detection functions to identify the components to be powered, and fault protection. There are two subvariants:

- IEEE P802.3bu (1 pair power over data lines [PoDL], which uses the wires defined in IEEE 802.3bw);
- IEEE P802.3bp, which offers maximum efficiency for the cable bundle, because it can carry power and data on the same twisted pair.

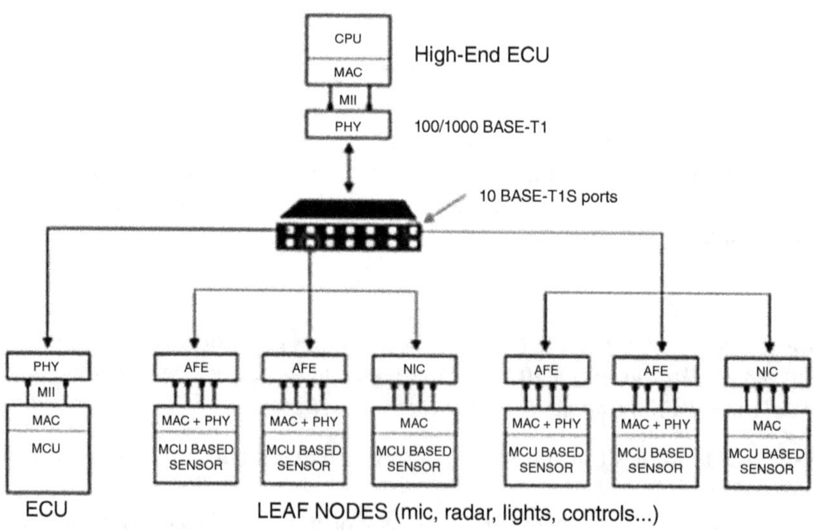

Figure 5.54 Field of use of 10 Mbit/s Ethernet.

5.3.6 General overview of Ethernet in automobiles

Figure 5.55 summarizes the Ethernet standards used in automobiles.

5.3.7 OPEN Alliance (One-Pair EtherNet)

The OPEN Alliance *(One Pair EtherNet Alliance)*, founded by the industrial Special Interest Group (SIG), (Broadcom, NXP, Freescale, Harman International, Hyundai Motor Company, and BMW Jaguar Land Rover), was formed to optimize the PHY layer for automobiles, with a datarate of 100 Mbit/s, and encourage the establishment, development, and widespread adoption of 100 Mbits/s Ethernet. The Alliance now has over 100 members and its influence has grown to include other automakers and manufacturers.

In order to establish an international standard for the PHY layer of BroadR-Reach, the OPEN Alliance is working on standardizing the components and on compliance testing for this technology. The SIG caters for the needs of industry to improve vehicle safety, comfort, and infotainment, whilst considerably reducing network complexity, cabling, and costs. Another aim of the SIG is to pull together the requisites for the technologies of the future, such as *reduced pair Gigabit*. Finally, AUTOSAR addresses the Ethernet automobile in the stack of software layers.

OPEN Alliance compliance tests

The OPEN Alliance's compliance test structure is open to any company that wishes to test its components' compatibility with BroadR-Reach technology. The independent UNH-IOL (University of New Hampshire InterOperability Laboratory) is at the disposal of OEMs and automakers whenever they choose to adopt this technology.

Why is there such enthusiasm for "BroadR-Reach Ethernet?"

As we have shown in great detail, the choice of line coding solutions such as 4B/3B or 80B/81B and PAM-3 or PAM 5 in the medium-dependent sublayer, and a host of small additional points chosen for BroadR-Reach Ethernet, have helped reduce modulation rates (in bauds) and radiation, to function with an inexpensive physical layer (UTP) whilst preserving a high bitrate. Thus, Ethernet:

- Has sparked a new interest in autonomous and/or connected vehicles;
- Contributes to a high reuse factor for components, software, and tools;
- Should be used as widely as possible in consumer products and other fields besides automobiles.

Vehicles evolve in electronic environments that are increasingly sophisticated and densely connected. In this environment, there is a burgeoning demand for a networking technology managing high-datarate applications without being excessively expensive. Ethernet succeeds on a mass market, because it has been possible to impose a consistent standard across the automotive industry, whilst reducing connection costs (up to 80%) and cabling costs (up to 30%) with respect to LVDS.

IEEE 802.3			
	Symbol	**Symbol rate SR**	**Code group**
	The smallest unit of data transmission on the medium. Symbols are unique to the coding system employed.	The total number of symbols per second transferred to or from the Medium Dependent Interface (MDI) on a single wire pair.	A set of encoded symbols representing encoded data or control information.
10BASE-T1S		12.5 MBaud	DME, PAM 2
100BASE-T4,	Ternary symbols	25 MBaud	A set of six ternary symbols that, when representing data, conveys an octet.
100BASE-X	Binary symbols or code-bits	125 MBaud	A set of five code-bits that, when representing data, conveys a nibble.
100BASE-T2	Quinary symbols	25 MBaud	A pair of PAM5 × 5 symbols that, when representing data, conveys a nibble.
100BASE-T1	Ternary symbols	66.666 MBaud	A set of ternary symbols that, when representing data, conveys three bits.
1000BASE-T	Quinary symbols	125 MBaud	A vector of four 81BQ4 coded quinary symbols that, when representing data, conveys an octet.
1000BASE-X			A set of ten bits that, when representing data, conveys an octet.
1000 BASE-T1			
Multi Giga			

Figure 5.55 Summary of Ethernet standards used in automobiles.

5.3.8 Electronic components for automobile Ethernet

Of the various components currently on the market, below we present a few examples of switches and line drivers for 100 and 1000 Mbit/s Ethernet.

BroadCom

BroadCom offers a range of components for BroadR-Reach Ethernet. These include standalone PHY circuits and switch circuits with integrated PHY. Each device complies with the EMC requirements and the typical temperature ranges for automobiles, and complies with the TS16949 and AEC-Q100 standards currently in force.

Transceiver

The BCM89810 is a full-duplex integrated circuit for 100 Mbit/s Ethernet, including packet encoding and functioning as both transmitter and receiver in the BroadR-Reach PHY layer, so the system can work with a UTP. This circuit combines: digital circuits; adaptive equalizers; analog-to-digital converters; elimination of phase-lock loops; encoders and decoders; echo-, noise- and jitter suppressors; line drivers; and all the other necessary circuits (see the overview in Figure 5.56).

Its main features are as follows:

- Simple port transceiver for 10/100BASE-T;
- Operates at 100 Mbits/s on UTPs;
- Integrated end-device resistors;
- Remote detection and diagnostics of split pairs or short circuits;
- Low emissions of EMI and high EMC – good immunity;
- Compatible with the temperatures found in automobiles;
- Complies with the standards ISO 11452-5 stripline, ISO 11452-4, IEC 61000-4-2, and AEC-Q100;
- Complies with CISPR 25 (radio disturbances);
- Automatic detection and correction of signal polarity;
- Auto-negotiation and link establishment with other BroadR-Reach-compliant partners.

BCM8988x (1000BASE-T1 PHY) and BCM8955x (secure switch) are designed to operate with a UTP.

Switches

The switches used are:

- BCM89200 – 4-Port switch with two integrated BroadR-Reach PHY;
- BCM89500 – 7-Port switch with four integrated BroadR-Reach PHY;

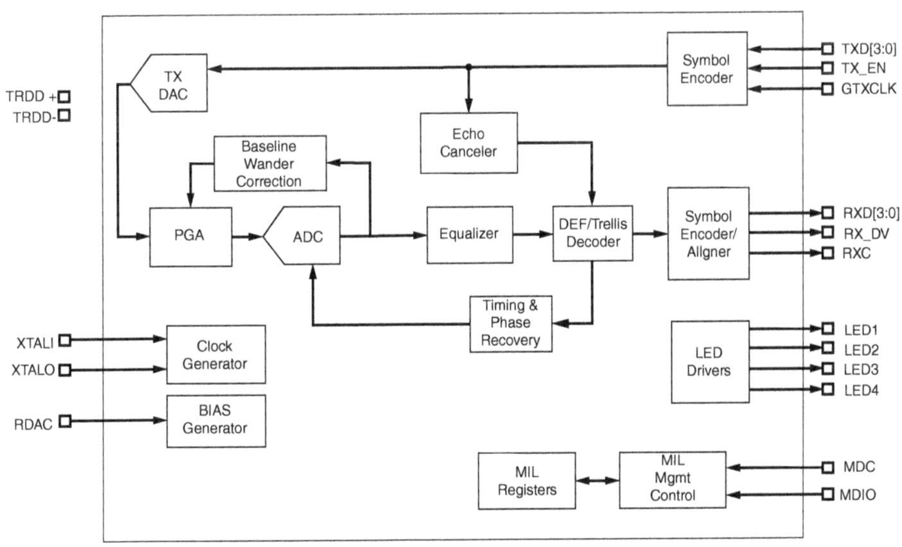

Figure 5.56 Overview of the BCM89810 transceiver.

- BCM89501 – 7-Port switch with five integrated BroadR-Reach PHY;
- BCM89610 – single-Port 10/100/1000BASE-T PHY.

All these circuits are designed to support IEEE 1588 and IEEE 802.1AS.

Marvell

Transceiver

The 88E1680 transceiver is a low-consumption 8-port 10, 100, or 1000 Mbps transmitter-receiver circuit, which offers *Energy-Efficient Ethernet* (EEE) compliant with IEEE 802.3az. It allows users to deploy an extensive range of cabling infrastructure. It supports a MAC QSGMII (Quad-SGMII) interface, functioning at a datarate of 5 Gbit/s.

Switch

Marvell's 1 Gbit/s Ethernet secure switch, 88Q5050 (Figure 5.57), is an 8-port Gigabit Ethernet switch compatible with IEEE 802.3 and 802.1 (AVB/TSN). It supports 100BASE-T1 and 1000BASE-T1, and offers four fixed ports 100BASE-T1 from IEEE, and a configurable selection of six additional ports, including 1 100BASE-T1 port, 1 100BASE-TX port, 2 MII/RMII/RGMII ports, 1 GMII port, and 1 SGMII port.

Security system As vehicles are becoming more autonomous and more connected, and the demand for mobility is growing, they are increasingly vulnerable to security breaches and to new types of cyberattacks. Marvell's circuit is equipped with security mechanisms, designed to:

- Guard against malicious attacks such as hacking of data held in a connected vehicle (or received from the outside environment), transferred over hardwired or wireless networks, and prevent any compromise of the data being transported;
- Protect connected vehicles applications from any danger of attack;
- Guarantee optimal security by means of a whitelist/blacklist system, with barriers to prevent cyberattacks such as DoS (Denial of Service) on the Ethernet ports.

It carries a Deep Packet Inspection (DPI) engine that analyzes the content to detect any attacks and intrusions, and a Trusted Boot mechanism to ensure a high level of security in the hardware architecture.

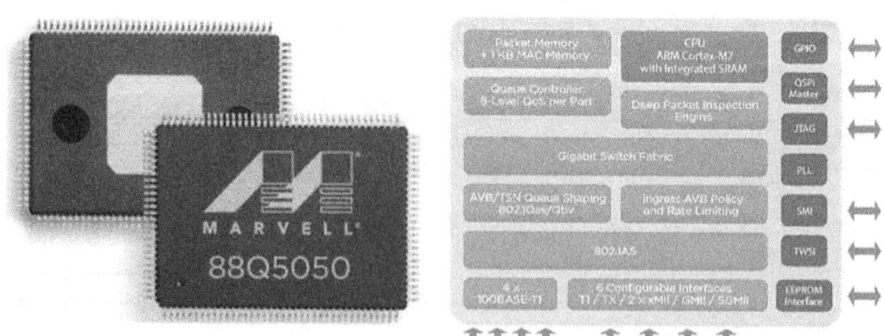

Figure 5.57 1 Gbit/s Ethernet secure switch – Marvell, 88Q5050.

NXP

Transceiver

NXP produces TJA110x – a family of Ethernet IEEE 100BASE-T1 transceivers for automobile applications, which, through each port, can transmit and receive up to 15 m on unshielded twisted pairs. The signal integrity and noise resistance are obtained by proprietary wave-shaping of the PAM-3 signals, optimized for PSD (Power Spectral Density, discussed above), and thus more efficiently reducing RF emissions. Filtration systems also improve the rejection of the Nyquist frequency (at 33 MHz) and finally optimize the EMC of the output stages MII/RMII. These circuits are suited to applications such as ADASs, infotainment, and communications.

- Automotive Vision Systems;
  - Front View Camera – Smart Rear-View Camera;
  - Surround View and Sense Park Assist System;
- Automotive Radar Systems;
- V2X and AVB;
- Infotainment/Sound Systems.

The main performances of this circuit are:

- HVQFN 36 package (6 × 6 mm) (see Figure 5.58);
- Operating temperature between –40 and +125°C;
- Automotive Qualified – OPEN Alliance TC-10 compliant sleep/wake up, ISO 26262 ASIL-A, ISO 7636 Transient Pulses, IEC 61000-4-2 +/− 6kV ESD;
- Low-voltage power supply (3.3 V) and low power consumption;
- Improved pulse-shaping stages;
- Output stage optimized for capacitive coupling (UTP) over 25 m;
- Remote circuit wake-up over Ethernet cables;
- Remote detection and diagnostics of split pairs or short-circuits;
- Detection of low supply voltage;
- Numerous advanced diagnostic functions.

Switches

SJA1105 is a five-port Ethernet switch compatible with IEEE 802.3. Each of the five channels can be individually configured to function in MII, RMII, and RGMII mode, up to 1000 Mbit/s. It is divided into two variants, SJA1105EL and SJA1105TEL, which offer the possibility of connecting a mixture of switches, microprocessors, and other PHY devices;

- SJA1105EL caters for Ethernet and AVB;
- SJA1105TEL includes additional functions to support time-triggered Ethernet (TTEthernet) and time-sensitive networking.

Figure 5.59 shows an example of how to connect switches, microprocessors, and PHY layers using Fast Ethernet and Gigabit Ethernet. The high-datarate line driver SJA1105s is able, with ease, to cascade multiple networks that can be used in different automotive scenarios, such as gateway applications, domain controllers, Audio Video Bridging (AVB), or interconnecting multiple computers.

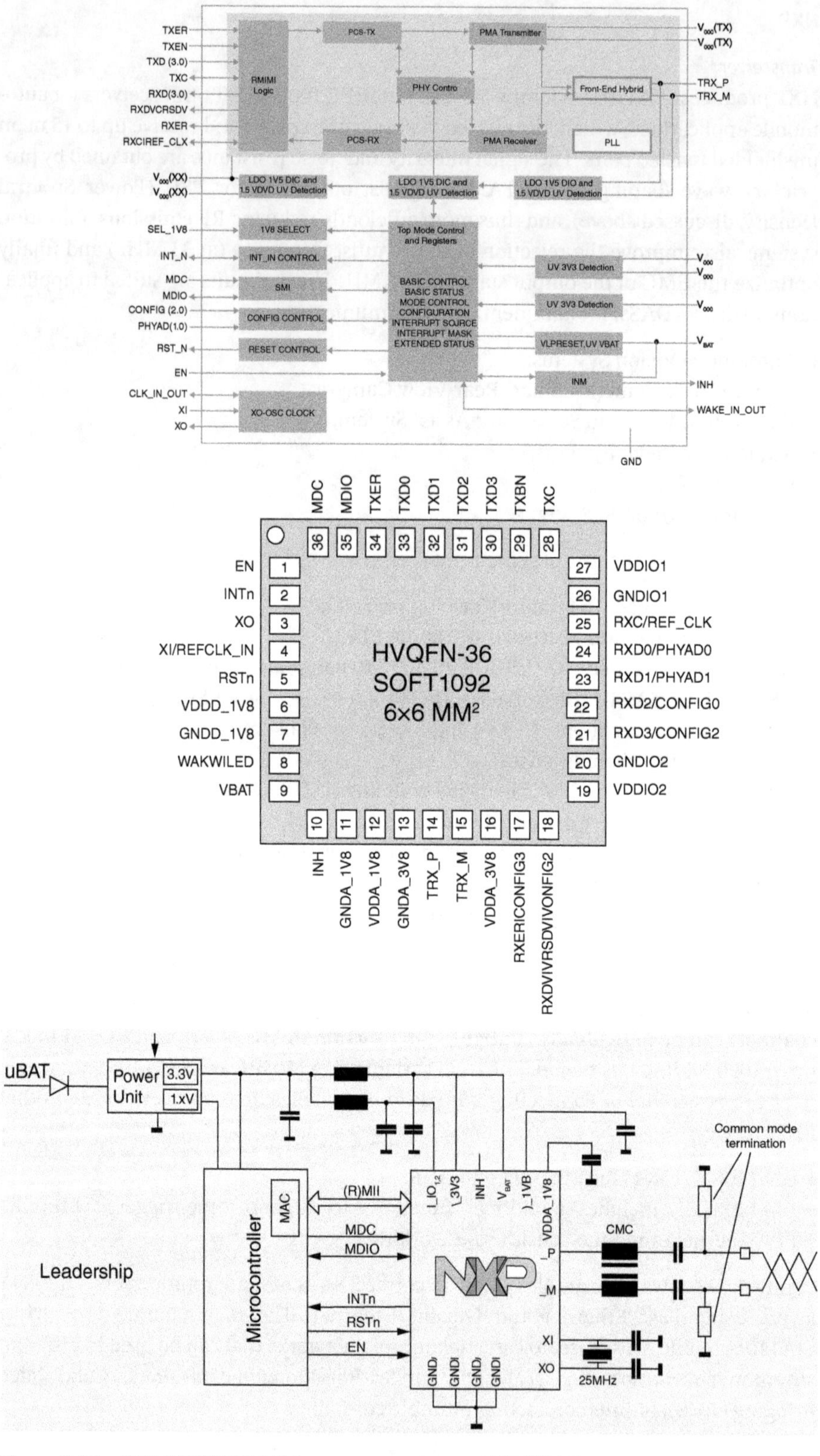

Figure 5.58 IEEE 100BASE-T1 Ethernet transceivers – NXP, TJA110x.

To offer readers a concrete idea, Figure 5.60 shows an example of an evaluation card, comprising:

- An LS 1021A ARM Cortex 7 microcontroller (with 525 ball pins), controlling the SJA 1105TEL switch, which controls the four Ethernet TJA 1101 line drivers;
- An SJA 1105TEL switch with 802.Qbv time-aware shaping;
- Four TJA 1000 100BASE-T1 physical layers;
- A further microcontroller (not shown here): LPC 17xx with 801.AS.

Finally, to avoid showing undue favor to any particular company, there are numerous other suppliers in the market (Texas, STm, etc.). For further information, readers are invited to consult their websites.

5.4 Deterministic, real-time, and automotive Ethernet

The previous section offered a detailed examination of layer 1 of the OSI model, the "PHY layer," of automotive Ethernet. There are many other peculiarities that must be taken into consideration when using Ethernet for applications in autonomous and connected vehicles. Notably, when sending frames with the standard protocol, they

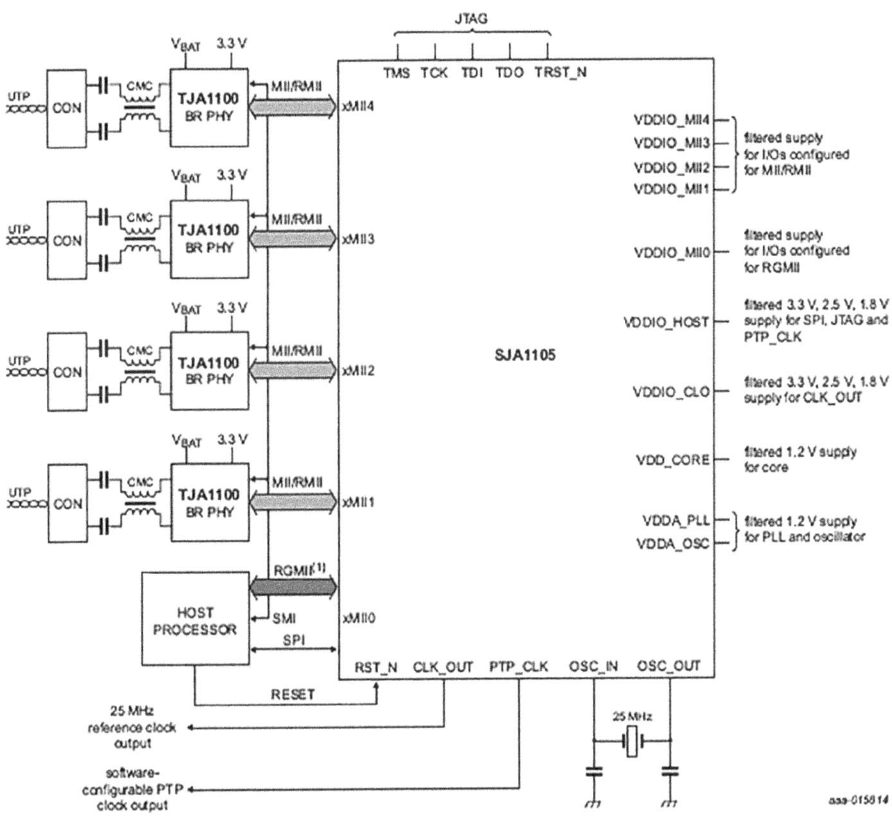

Figure 5.59 Five-port Ethernet switch compatible with IEEE 802.3 – NXP, SJA1105.

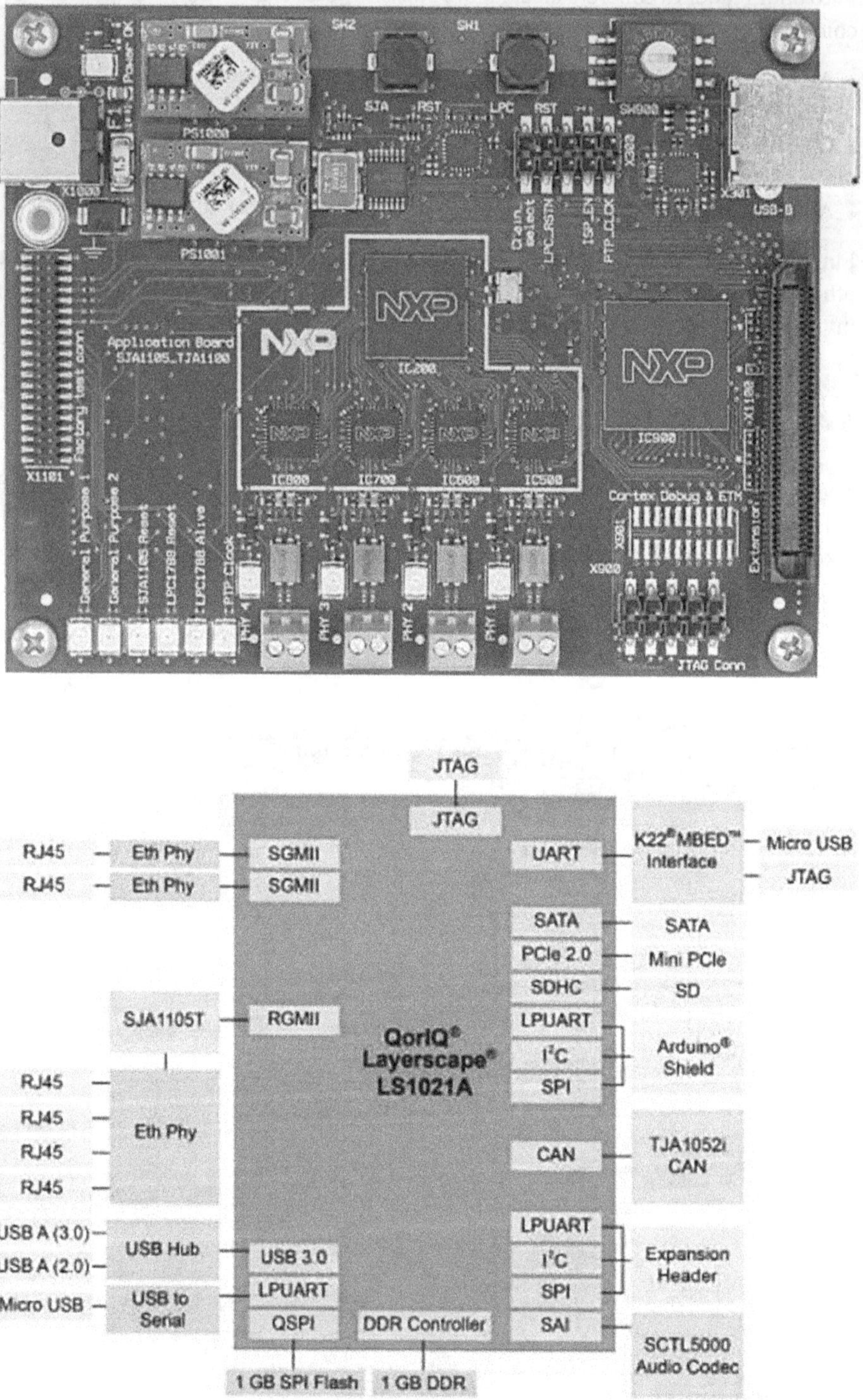

Figure 5.60 Example of applications.

may arrive at any time, and when there are switches and their buffers along the signal path, this does not help at all! In short, such a potential delay cannot be tolerated when designing autonomous vehicles with safety features. Amongst other things, automotive applications require:

- Datarate and bandwidth;
- Short, and known, latency;
- Priority management;
- Deterministic, real-time operation.

All these management operations, which take place at layer 2 or higher in the OSI model, were originally implemented by the working group in charge of audio/video management for ADASs (see Section 5.4 – AVB). They were then extended to all of the vehicle's control and safety systems. We shall now examine this situation.

5.4.1 Deterministic and real-time Ethernet, or TSN

Structurally speaking, the conventional standard Ethernet frame is not remotely deterministic. In principle, in a specific order, Ethernet frames are sent from a specific source address to a specific destination address. With a little luck, the frames will arrive, but may be delayed, depending on the path they take, and in accordance with the *best-effort* principle. The best-effort principle means that the network service will attempt to deliver messages to their destination, but offers no special quality function in terms of retransmitting the message if it is corrupted or packets go astray. In short, there is no guarantee of delivery.

In numerous applications, there are stringent requirements on delays, to ensure that real-time data transmissions can serve the purposes of the applications. This new technology, called "*time-sensitive networking*" (TSN), transforms standard Ethernet communication technology, where messages "get there when they get there" into technology that "offers synchronization guarantees for time-critical applications." Thus, a degree of determinism is introduced into the IEEE 802.1 and IEEE 802.3 Ethernet networks, with restricted latency and extremely slight jitters.

Numerous real-time communication technologies based on conventional Ethernet have been in place for a long time. They offer short cycle times and the necessary guarantees to carry the data in the desired time. However, in order to achieve real-time communication, they sadly include additional technical mechanisms, such as protocol enhancements (increased bandwidth, etc.), which are often incompatible with one another (for example: EtherCAT, PROFINET IRT, and SERCOS III).

TSN represents a reliable and standardized evolution of communication technology. It is only possible when the communication infrastructure is capable of providing two essential services at once on the same network:

- Strict and reliable real-time communication capable of activating demanding applications on a large scale, with flexibility distributed throughout the networks;
- A high datarate to deal with an enormous amount of data from the sensors and the background tasks.

The TSN solution that satisfies these requirements has an important place in communications infrastructures containing a large number of connected elements, in which the communication requirements become even more diverse.

TSN components

When we speak of the TSN solution, there are five main components:

- The TSN stream: this is the term used to describe time-critical communications between end devices. Each stream has strict time requirements that must be observed by the network elements and each stream is identified individually by those network elements;
- End devices: the end devices are the source and destination of the TSN streams. These devices execute an application that requires deterministic communication. They are also known as talkers and listeners;
- Bridges: also called Ethernet switches. For TSN, bridges are able to transmit TSN Ethernet frames in accordance with a set schedule of transmission and receipt of Ethernet frames;
- Central network controller (CNC): for TSN, the CNC acts as a proxy for the network (the TSN bridges and their interconnections) and the control applications requiring deterministic communication. The CNC defines the transmission schedule for all TSN frames. The CNC application is generally supplied by the vendor of the TSN bridges;
- Centralized user configuration (CUC): the CUC is an application that communicates with the CNC and the elements, and also represents the control applications and elements. The CUC makes requests of the CNC to establish deterministic communications (TSN stream), with specific requirements in place. The CUC is an application that is "vendor-specific" (proprietary). Generally, the vendor of TSN elements supplies a CUC to go with them.

5.4.2 TSN (time-sensitive networking)

The working group *Time-Sensitive Networking* (TSN) under IEEE 802.1 (formed in November 2012, by renaming the existing *audio video bridging task group*) has developed a set of standards that define the mechanisms for transmission of time-sensitive data over Ethernet networks. Unlike standard Ethernet with IEEE 802.3 and Ethernet Bridging with IEEE 802.1Q, Virtual LAN time plays a crucial role in TSN networks.

The majority of extensions to IEEE 802.1Q relate, in particular, to transmission with very low latency and high availability. The potential applications include the convergence of networks between audio/video streaming and real-time control flows that are used in automotive control systems. All of this will be discussed in Section 5.4.2.

It should be noted that work in this field is also being done by the Avnu Alliance – an industrial group set up especially to define the requirements of conformity and interoperability for TSN network elements.

Simply put

- IEEE 802.1Q operates in layer 2 of the OSI model;
- IEEE 802.1Q TSN defines technology that is capable of deterministic messaging over standard Ethernet;
- TSN is a technology belonging to layer 2 of the OSI model;
- TSN is centrally managed, and offers guarantees of delivery and reduced jitter by using a time scheduler for real-time applications that require determinism;

- TSN is an Ethernet standard:
  - The dispatch decisions made by TSN bridges use the contents of the Ethernet header, rather than the IP address;
  - TSN can be used in any environment and Ethernet frames can transport any type of message, for any industrial application, and are not limited to Internet Protocol;
- TSN is designed to allow for deterministic communication over standard Ethernet. Up until now, the market in deterministic communication used non-standard technologies or non-standard Ethernet. Before the advent of the IEEE 802.1 TSN standards, the Ethernet standard had no pure layer 2 with deterministic possibilities leading to new levels of connectivity and optimization, and savings on cost;
- TSN technology is fundamentally based on, and hinges on, time:
  - TSN offers a way to ensure that information can travel from point A to point B in a fixed, predictable time;
  - TSN provides improved efficiency, and there is the implicit condition that devices in this network using TSN (end devices and bridges) must share a common understanding of time;
- *Precision Time Protocol* (PTP) is used to maintain that common understanding of time. The PTP profiles chosen to work with TSN are IEEE 802.1AS and IEEE 802.1ASRev.

802.1 TSN standards

The various TSN standards necessary to build a complete real-time communication solution are specified by IEEE 802.1, and can be divided broadly into three categories. Each of these specifications can be used alone, and they are usually autonomous. However, when they are used together, TSN can realize its full potential as a communication system. The three fundamental components are:

- *Time synchronization*: all elements participating in real-time communication must have a shared understanding of time;
- *Scheduling and traffic shaping*: all elements participating in real-time communication must obey the same rules in scheduling, processing, and routing of the communication packets;
- *Selection of communication paths, path reservation and fault-tolerance*: all elements participating in real-time communication must obey the same rules in the selection of communication paths, reservation of bandwidth, timeslots, and the ability to use multiple paths simultaneously, with low fault-tolerance.

We shall now examine each of these three main entities in turn.

Time synchronization

All participants (end devices, controllers, Ethernet switches, etc.) in a real-time communications network need to establish end-to-end transmissions whose latencies have precise, strict, and non-negotiable limits. They also need to have a shared understanding of time and, consequently, need to synchronize their clocks. It is only by having synchronized clocks that all network elements will be able to function in unison and carry out the required operation at exactly the intended moment. In principle, in TSN networks, time synchronization can be achieved with various technologies:

- In theory, it is possible to equip each element and each switch with a GPS clock or radio clock. However, this technique is costly, first of all, and, second, there is no guarantee that access to the GPS or radio clock, or a satellite signal, will always be available. For example: if the Ethernet network is installed inside a moving vehicle, or under the Earth's surface (underground, in a road tunnel, in a parking lot, etc.);

- In view of these constraints, the "time" is generally taken from a central time source, set directly by the network itself. In most cases, this is done by using IEEE specification 1588: *precision time protocol* (PTP), which is universally applicable. PTP uses Ethernet frames to distribute time synchronization information. In addition, the TSN working group of IEEE 802.1 specified a profile for IEEE 1588, called IEEE 802.1AS. The purpose of this profile is to restrict the list of different options for IEEE 1588 to less critical, more flexible, options that can apply to automobile, domestic, or industrial networks.

Scheduling and traffic shaping

Scheduling and traffic shaping allow different classes of traffic, with different levels of priority, different requirements in terms of bandwidth, and end-to-end latency, to coexist on the same network.

Bridging, in accordance with IEEE 802.1Q, uses a strict priority system containing eight different levels of priority. In this protocol, these priorities are indicated by the values of the three bits in the PRIO field of the header "tag" of an 802.1Q VLAN Ethernet frame (see Figure 5.61).

These priority levels mark distinctions between different specificities of network traffic. Alas, even at the highest of the eight priority levels, there is no absolute guarantee of end-to-end delivery. This is due, amongst other things, to the effects of buffering in the Ethernet switches. If a switch has begun transmitting an Ethernet frame on one of its ports, then even if another frame in its buffer has a higher priority, it must wait to finish transmitting the current frame before being able to transmit another. With switched Ethernet, there is no way of avoiding this non-determinism. This represents no problem for applications that are not time-dependent on simple Ethernet frames (e.g. transfer of non-time-sensitive data, usually protected by other

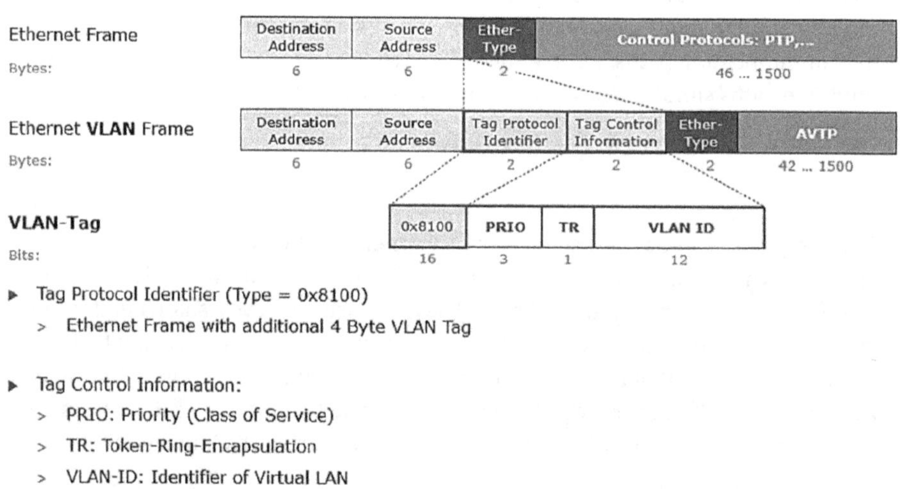

Figure 5.61 The 3-bit PRIO fields in the header tag of an 802.1Q VLAN frame.

mechanisms higher up in the protocol stack, such as the transmission control protocol – TCP). On the other hand, in environments such as autonomous vehicles, in which the control circuits operate in closed loops and safety and security applications use the Ethernet network, fast, reliable delivery of messages is of prime importance. Therefore, the strict priority system found in IEEE 802.1Q needs to be strengthened.

Different time slices for different classes of traffic

Time-aware scheduler VLAN's eight priority levels are still enforced to ensure TSN Ethernet's backwards compatibility (compatibility and interoperability with existing infrastructures and smooth migration to new technologies). However, TSN improves standard Ethernet communication by adding new mechanisms to ensure timely delivery, to satisfy the requirements of real-time operation. Indeed, with TSN, for each of the eight priorities, the user can select different treatments for the Ethernet frames, and the priorities can be individually assigned to those of the existing methods (such as IEEE 802.1Q, strict-priority scheduler) or processing methods, such as those of a traffic scheduler taking care of the timing aspects of TSN (such as IEEE 802.1Qbv).

The IEEE 802.1Qbv time-aware scheduler is designed to separate the different forms of communication on the Ethernet network into fixed-length, repeating time cycles, as is the case with FlexRay. Within these cycles, time slices can be configured and assigned to one or more of the eight Ethernet priority levels. It then becomes possible to assign exclusive use of the Ethernet network (for a limited period) to traffic classes that need guaranteed transmission, and cannot be interrupted. The basic concept is that of a conventional "time-division multiple access" (TDMA) system.

By creating virtual communication channels assigned for specific periods, time-critical (urgent) communications can be separated from the non-critical traffic. By granting exclusive medium access to devices transmitting time-critical traffic, the effects of transmission buffering by the Ethernet switches can be avoided and urgent traffic can be transmitted without non-deterministic interruptions. An example for an IEEE 802.1Qbv configuration of the scheduler is presented in Figure 5.62.

In this example, each cycle is made up of two time slices:

- Slice 1 can only be used for transmission of VLAN priority-tagged traffic;
- Slice 2 can be used to transfer the rest of the priorities needing to be sent. In that time slice, data will be transported in accordance with the best-effort principle.

The IEEE 802.1Qbv scheduler of course requires the clocks on all the network elements (Ethernet switches and end devices) to be synchronized and an identical schedule to be set up. It also requires that all devices understand which priority level can be sent over the network at a given moment in time. In addition, as packets with more than one level of priority are in slice 2, they are processed in strict order of priority, in accordance with IEEE 802.1Q. This separation into cycles and time slices for Ethernet transmissions can be taken still further, by the inclusion of other schedules/calendars or wave-shaping algorithms, such as the IEEE 802.1Qav credit-based traffic shaper (Audio and Video Bridging). It is this protocol that handles real-time flexibility in autonomous vehicles. In this particular example, IEEE 802.1Qav may be assigned to one or two of the priorities that are used in time slice 2 to further distinguish between audio/video traffic and background file transfers.

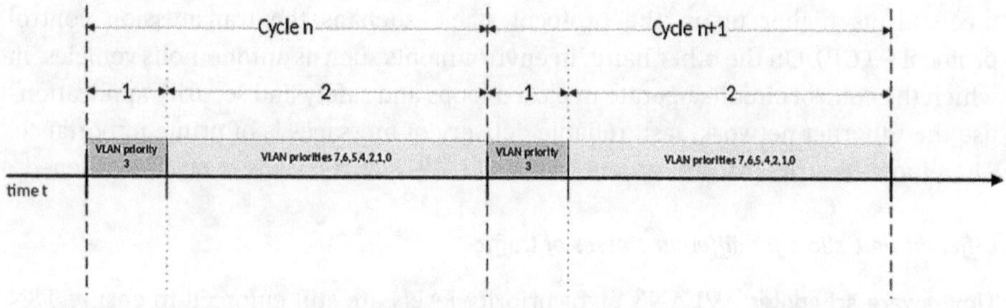

Figure 5.62 Example of a scheduler in IEEE 802.1Qbv.

The IEEE 802.1 *Time-Sensitive Networking* group specifies various schedulers and traffic shapers, which can be combined to achieve "non-reactive" coexistence of hard real-time, soft real-time, and background traffic on the same Ethernet infrastructure.

Time slices and guard bands

When an Ethernet device has begun transmitting a frame on the medium, that transmission must be fully completed before it can transmit anything else. In principle, the end of the transmitted frame includes a CRC32 checksum to ensure reliable, fault-free transmission. This intrinsic property of Ethernet networks poses a new problem that needs to be solved by the TDMA of the IEEE 802.1Qbv scheduler (see Figure 5.63):

Consider an example where, at the end of slice 2 of cycle n, a new frame begins transmitting just before the slice ends. Unfortunately, the frame in question is too big to be sent in the intended time slice. As the transmission of this frame cannot be interrupted, the frame will spill over into slice 1 of the next cycle, $n + 1$, thus partially or completely blocking the time-critical slice. This means that time-critical or real-time frames may be delayed, to the point where they can no longer satisfy the requirements of the application. This problem is highly similar to the effects of buffering that occur in non-TSN Ethernet switches. Thus, TSN must have a devoted mechanism to stop this from happening.

The IEEE 802.1Qbv scheduler must ensure that the Ethernet interface is not busy transmitting a frame when it switches from one time slice to the next. It does this by placing a guard band in advance of each time slice that transports time-critical traffic (see Figure 5.64). During that guard band, no new Ethernet frames may be transmitted, but the time may be used to complete transmissions that are already in progress. In principle, the duration of this guard band must be as long as necessary for the maximum frame size to be safely transmitted. With regard to an Ethernet frame complying with IEEE 802.3, with a single VLAN IEEE 802.1Q tag and interframe spacing, the total length is 8 bytes (preamble and SFD) + 18 bytes (Ethernet addresses, ethertype and CRC) + 4 bytes (VLAN Tag), + 1500 bytes (frame payload) + 12 bytes (interframe spacing), which gives a total of 1542 bytes.

In Fast Ethernet applications (100 Mbit/s, which is 12.5×10^6 bytes/s), the total time needed for this frame to be sent depends on the speed of the Ethernet link. Thus, the transmission time is:

$$t\_max = 1542/(12.5 \times \text{-}10^6) = 123.36\ \infty s$$

The guard band must be at least 123.36 μs long. The time that can be used for time slice 2 (the total bandwidth) is reduced by the length of the guard band.

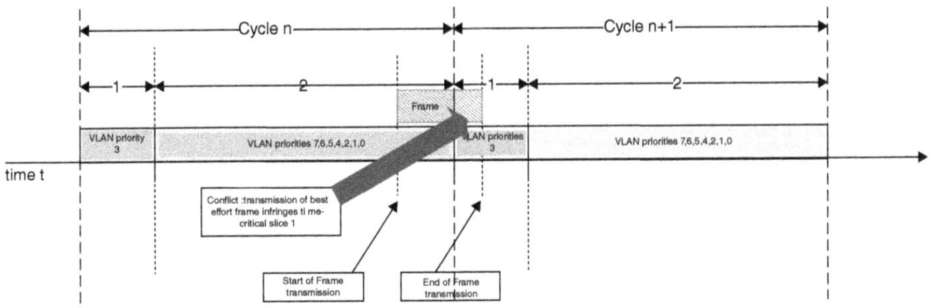

Figure 5.63 Problems due to late sending of frames.

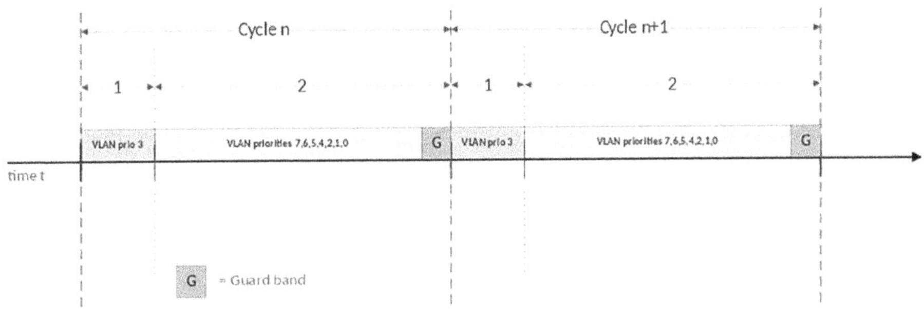

Figure 5.64 Implementation of guard bands.

Note
For ease of reference, Figure 5.64 is not to scale. In the real world, the guard band is much smaller than shown.

In this example, slice 1 always contains high-priority data, while slice 2 always contains best-effort data. Therefore, a guard band must be placed at each transition into slice 1, to protect the time slice for critical data streams. While guard bands do manage to protect high-priority time slices for time-critical traffic, they also present a number of significant disadvantages:

- The time given over to a guard band is wasted – it cannot be used for data transmission, because the Ethernet port must be silent during this time. This wasted time results directly in a loss of bandwidth for background traffic over that particular Ethernet segment;
- A single time slice can never be configured with a smaller value than the length of the guard band;
- Note: with slower Ethernet connections, and with growing guard band size, this has a negative impact on the smallest achievable time slice and, thus, on the cycle time.

Preventive length-based scheduling

To partly compensate for the loss of bandwidth due to the use of guard bands, IEEE 802.1Qbv includes a preventive frame length scheduling mechanism. This is used when store-and-forward switching is employed. Once an Ethernet frame transmitted on a port where the guard band is in place has been received in full, the scheduler checks the overall length of the frame. If it can fit completely within the guard band, without infringing on the next high-priority slice, then the scheduler can send the frame despite the guard band, thus reducing bandwidth wastage.

On the other hand, when cut-through switching is active, to minimize end-to-end latency, this prevention mechanism cannot be used, given that the total length of the Ethernet frame must be known in advance. Therefore, bandwidth is still wasted, and this does not help reduce the minimum cycle duration. Thus, such frame length aware scheduling represents an improvement, but it cannot compensate for all the drawbacks due to the introduction of a guard band.

Frame pre-emption and guard band minimization

To mitigate the negative effects of guard bands, the working groups IEEE 802.3 and 802.1 specified a frame pre-emption technique, because this technology requires modifications to the Ethernet MAC layer (under the auspices of IEEE 802.3), and changes to the management systems (under the responsibility of IEEE 802.1). Frame pre-emption is described in two different documents:

- IEEE 802.1Qbu on the bridge management component;
- IEEE 802.3br on the MAC component of Ethernet.

How frame pre-emption works Figure 5.65 illustrates how frame pre-emption works:

- When sending a best-effort Ethernet frame, the MAC layer interrupts the transmission just before the guard band starts;
- The partial frame is supplemented by a CRC[3] and is stored in the next switch to await delivery of the second part of the frame. After the high-priority traffic in the following slice 1 has been delivered, the cycle switches back to the next slice, and the remainder of the interrupted frame is transmitted.

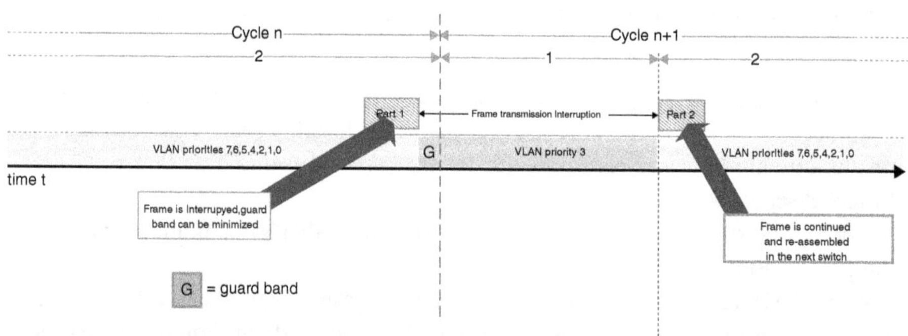

Figure 5.65 Example of frame pre-emption.

- Frame pre-emption always works on a purely link-by-link basis and only fragments from one Ethernet switch to the next, with the frame being reassembled in the next switch. Unlike with Internet Protocol (IP), end-to-end fragmentation is not supported.

In addition:

- Support for frame pre-emption must be activated individually on each element-to-element link;
- To indicate the possibility of frame pre-emption on a link, an Ethernet switch advertises it by means of the LLDP (link layer discovery protocol). When an element receives such an LLDP announcement on its port, and it is capable of frame pre-emption, it can activate the strategy. There is no direct negotiation and activation of the capability on individual elements. Any device receiving the LLDP pre-emption notice will assume that, at the other end of the line, there is an element that can understand the changes made to the frame format (in terms of the CRC32 and SFD).

Frame pre-emption significantly reduces the guard band. Guard band length now depends on the accuracy of the frame pre-emption mechanism, and on the minimum frame size that the system can still pre-empt. As the minimum size of a valid Ethernet frame is 64 bytes, IEEE 802.3br specifies that this value represents the mechanism's peak precision. This being the case, the guard band can be reduced to a total of 127 bytes: 64 bytes (minimum frame length) + 63 bytes (remaining length that cannot be pre-empted). Any larger frames can be pre-empted once again. Thus, there is no need for a guard band to protect against frames of this size.

This strategy minimizes the bandwidth lost for best-effort data transmission, and allows for shorter cycle lengths at low datarates – 100 Mbit/s or less, for example. Given that pre-emption takes place at a hardware level in the MAC, as the frame passes through, cut-through switching can also be supported, since the overall frame size is not needed, a priori. The MAC interface only checks at regular intervals of 64 bytes whether or not the frame needs to be pre-empted.

The combination of time synchronization, the IEEE 802.1Qbv scheduler and frame pre-emption is an effective set of tools that can ensure different categories of traffic can coexist on a network, and provide end-to-end latency guarantees.

Selection of communication paths, reservation, and fault-tolerance

TSN technology – notably the IEEE 802.1Qbv time-aware scheduler – was developed for use in time-critical network environments. In these networks, not only are the relevant timeslots guaranteed; fault tolerance is too. Networks in charge of applications such as elementary safety control loops or autonomous driving in vehicles must be protected against hardware faults or media network faults. For this purpose, TSN provides a fault tolerance protocol: IEEE 802.1CB. In addition to this protocol, there are high-availability protocols, such as HSR or PRP, which are specified in IEC 62439-3.

The IEEE standard 802.1Qca, on path control and reservation, can be used for manual configuration or, with supplier-specific solutions, to record the fault-tolerant communication streams across a network. The administration of large-scale TSN networks is specified in IEEE 802.1Qcc, which presents certain aspects – notably with a decentralized approach, but also a fully centralized approach that reuses the concepts of software-defined networking (SDN).

Position of TSN standards in the OSI model

Figure 5.66 shows a possible way of fulfilling layers 1 to 7 of the OSI model, with the main TSN standards listed to the right.

5.4.3 Summary of applicable standards

By way of a summary, Figures 5.67 and 5.68 list the main Ethernet standards, organized by subjects, discussed earlier.

List of standards surrounding TSN

This concludes our presentation of the different IEEE Ethernet standards that must be taken into consideration by an autonomous and connected vehicle, for its real-time, time-critical, and deterministic applications.

As a final example, since late 2019, the Japanese giant Toshiba has been supplying a range of integrated Ethernet bridge circuits for automotive applications, compatible with AEC-Q100 (grade 3), supporting up to 1 Gbit/s, IEEE specifications 802.1AS and 802.1Qav on AVB Ethernet; the TSN standards IEEE 802.1Qvb, 802.1Qbu, and 802.3br; and the interfaces PCI Express 2.0 and 1.0, I2S/TDM, RGMII, RMII, MII, and SGMII. They can be used to easily link an application processor with onboard networks, including for infotainment systems in automobiles. These circuits are encapsulated in a 9 × 9 mm casing.

The next and final part of this book gives detailed examples of software architectures to be put in place, and the development tools, tests, and simulations.

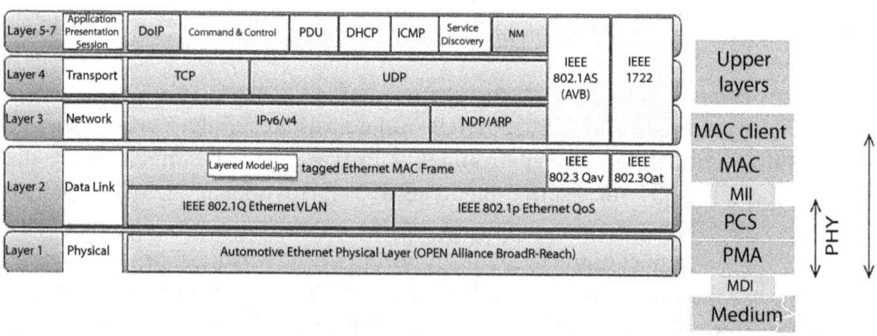

Figure 5.66 Possible fulfillment of layers 1 to 7 of the OSI model with TSN standards.

Standards relating to:	IEEE	Title	Comments
Low latency transmission	802.1BA	*Audio video bridging* (AVB) *systems*	*Intro and overview*
	802.1AS	*Timing and synchronization for time-sensitive applications* (gPTP)	*AVB key specification for global time synchronization and network setup*
	1588	*– Timing and synchronization* *– Enhancements and performance improvements*	
	802.1Qav	*Forwarding and queuing for time-sensitive streams* (FQTSS)	*Traffic shaping*
	802.1Qat	*Stream reservation protocol* (SRP)	*– Dynamic stream announcement with admission control* *– Static implementation for automotive possible use*
Time-sensitive applications	1722(a)	*Layer 2 transport protocol for time-sensitive applications* (AVTP) (a) = *automotive version*	*Covers encryption, simple A/V streams and formats, automotive message types within an A/V stream*
	1733	*Layer 3 transport protocol* (RTP)	

Then come those created by the IEEE *task group* time-sensitive networking, which is the successor of the *task group* AVB, relating to the fundamentals:

Deterministic transmissions	802.1Qbv	*Enhancements for scheduled traffic*	Forwarding and queuing
Latency reduction	802.1Qbu	*Frame pre-emption*	
	802.3br	*Specification and management parameters for interspersing express traffic*	
Safety and Security	802.1CB	*Frame replication and elimination for reliability*	*Seamless redundancy*
	802.1Qci	*– Time-based ingress policing* *– Per-stream filtering and policing*	
	802.1Qca	*Path control and reservation*	
	802.1Qcc	*– Central configuration method* *– Enhancements and performance improvements*	

Figure 5.67 List of main standards for real-time and deterministic Ethernet, divided by subject.

Standard	*Function group*	*Title*
IEEE 802.1AS-Rev	*Timing and synchronization*	*Timing and synchronization for time-sensitive applications*
IEEE 802.1Qbv	*Forwarding and queuing*	*Enhancements for scheduled traffic*
IEEE 802.1Qbu	*Forwarding and queuing*	*Frame pre-emption*
IEEE 802.1Qca	*Stream reservation* (SRP)	*Path control and reservation*
IEEE 802.1CB	*Stream reservation* (SRP)	*Seamless redundancy*
IEEE 802.1Qcc	*Stream reservation* (SRP)	*Enhancements and performance improvements*
IEEE 802.1Qci	*Forwarding and queuing*	*Per-stream filtering and policing*
IEEE 802.1Qch	*Forwarding and queuing*	*Cyclic queuing and forwarding*
IEEE 802.1CM	*Vertical*	*Time-sensitive networking for fronthaul*
IEEE 802.1Qcr	*Forwarding and queuing*	*Asynchronous traffic shaping*
IEEE 802.1CS	*Stream reservation*	*Local registration protocol*
IEEE 802.3br		*Interspersing express traffic*

Figure 5.68 List of standards.

Notes

1 Dominique Paret, *FlexRay and Its Applications: Real Time Multiplexed Network*, 2012, Wiley.

2 See Dominique Paret, *FlexRay and Its Applications: Real Time Multiplexed Network*, 2012, Wiley.

3 Each partial frame is supplemented with CRC32 to detect errors. Unlike what is normally the case with Ethernet CRC32, the last 16 bits are inverted to make the partial frame distinct from an ordinary Ethernet frame. The start-of-frame delimiter (SFD) is also modified.

6

Simulations, Applications, and Software Architectures for Automobiles

Divided into four, this final chapter describes the software environment and the hardware (tools) that need to be put in place for simulating and developing autonomous and connected vehicles, to ensure they work properly, are safe, comply with the standards, certifications, approvals, etc. A few choice examples are also presented.

The authors warmly thank VECTOR France and its CEO Jean-Philippe Dehaene, for allowing us to use the numerous documents that illustrate the points made here.

6.1 Software simulations of an autonomous vehicle and its environment

A survey run by the Society of Automotive Engineers (SAE) concluded that the main obstacles to the advent of fully autonomous vehicles on the general market are: "1, public trust in such systems and 2, uptake of the idea of autonomy." Understandably, autonomous driving systems can only be rolled out to the public once developers and automakers have demonstrated that they are capable of delivering extremely high levels of safety.

Today (in late 2021), autonomous driving systems with little or no human involvement (levels 4 and 5) are largely based on "deep learning" algorithms (see Chapter 4), which can be trained to make the right decision in almost any driving situation. However, these systems do not have satisfactory architectures and cannot yet meet all the exacting requirements that have been, and continue to be, used to validate critical safety software. Therefore, simulations and road tests tend to be based on everyday situations (that are not difficult for a human driver to handle). Of course, these scenarios are an essential part of a vehicle's development. Unfortunately, these tests cannot provide completely viable solutions for validation of autonomous driving software and systems because, according to Akio Toyoda, the President of Toyota Motors, fully autonomous vehicles must be trained over a distance of around 14.2 billion kilometers of driving to ensure they are safe. As this seems an impossible goal, we must turn to "simulation" to help train these systems.

Autonomous and Connected Vehicles: Network Architectures from Legacy Networks to Automotive Ethernet, First Edition. Dominique Paret and Hassina Rebaine.

6.1.1 Simulation

Historically, simulation has developed primarily through the use of algorithms based on mathematical equations, applied to physical phenomena (Figure 6.1).

Whether autonomous vehicles are completely autonomous (level 5) or still require some degree of human input (levels 3 and 4), they will be interacting increasingly intensely with their environment. Consequently, the modeling of onboard systems, multi-physical intersystem coupling, and models to study the behavior of vehicles (or fleets of vehicles) across large geographic regions will develop on a massive scale.

From the initial specification, through development AND fine-tuning to final validation (this sentence structure flows properly, whereas the sense is disjointed as it currently stands). In view of the ability to predict how the components and systems will behave, manufacturing processes help to efficiently produce optimization loops and determine the optimal parameters or shapes to use in relation to the desired performance criteria (crash protection, vibration, fatigue, aerodynamics, etc.). Using computer systems, digital simulation predicts the behavior of real-world phenomena on the basis of mathematical formulae. These predictions can then be applied to assist with design, fine-tuning of controlled systems, development, and validation of new modes of transport. An autonomous vehicle's simulation platform must integrate the hardware, electronics and onboard systems, and software, so as to be able to accurately simulate the complete autonomous driving systems in a fraction of the time and for a fraction of the cost for a genuine road test. Thus, it must be able to cover:

- Simulation of all the systems, all exchanges on the networks, all sensors, including the cameras, radars, lidars, sonars, etc.;
- Multi-physical simulation of the physical and electronic components;
- Analysis of the system's functional safety;
- Design and automatic generation of certified security codes for the onboard software.

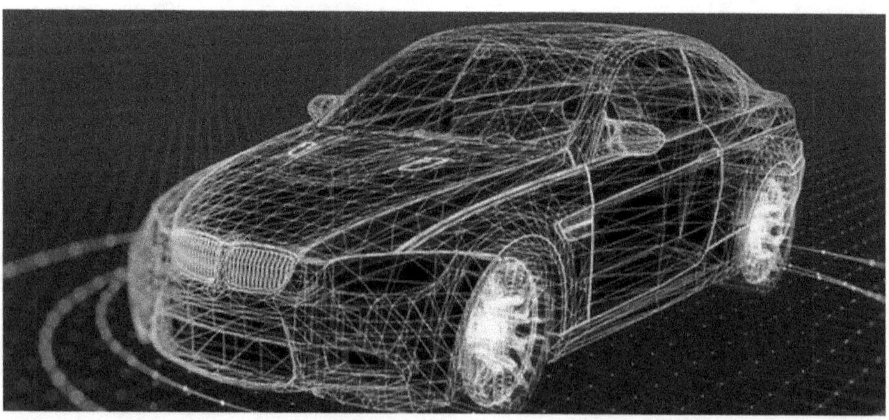

Figure 6.1 Example of a simulation image.

Such a platform serves to validate the safety of autonomous driving systems (including cybersecurity). It depends on:

- Firstly, the onboard software for location, perception, motion, planning, and execution of the control signals;
- Secondly, the sensors, whose simulation can be integrated into a closed-loop simulation environment that interacts with traffic simulators, which allows thousands of scenarios to be run virtually.

In addition, this type of simulation raises the question of how the various players involved (automakers, OEMs, etc.) will manage to:

- Feed back all the data from the vehicles' instruments and sensors (e.g. multiple cameras, radars, and lidars), which often require very high datarates, bandwidths, and acquisition rates;
- Store a huge number of stimuli (test cases).

Figure 6.2 presents the approximate digital datarates required for the signals produced by the different sources.

Let us now look at the problems involved in data acquisition.

6.1.2 Synchronous acquisitions at high datarates

In light of the above, automakers and OEMs must use hardware to perform synchronous acquisition of these different types of data, and software tools communicating with that hardware to configure the logging of huge volumes of data in different formats (physical, raw, and object-oriented data). Such data, uploaded to a server, is necessary for studying, testing, and validating a large number of functional simulation scenarios.

Source	Data type	File type	Format	Datarate (Mbit/s)
Front Camera	Video	Video	.LVDS	500
Side Radar (×4)	Data	Ethernet	.PCAP	300
Front Radar	Data	Ethernet	.PCAP	100
ECU ADAS	Data	Ethernet	.PCAP	70
Other data	CAM FR-FC	Ethernet – CAN-FD	.PCAP	20
	CAM SR (×6)	Video	.avi	320
	CAM R	Video	.avi	80
	Lidar	Ethernet	.PCAP	10
	GPS	Video	.kml	10
			total	~1500 Mbit/s
				~11 Gbyte/min

Figure 6.2 Examples of values of datarates of binary signals from the different sources that need to be logged.

The challenge, then, is to implement all these different systems in the central computer, which will run the software for data acquisition, measurement, and configuration of the log (see Figure 6.3).

For these purposes, it is essential that the computer used be high-performing, with multiple core architecture, operating at GHz rates, and have sufficient hard-drive memory. With this in mind, it is often necessary to connect a series of PCs together, and use an external unit with very high capacity, with a CPU. This external hard disk is referred to as a BrickPC.

- Example: Intel® Core™ i7-5700EQ (4 × 2.6 GHz) (see Figure 6.4).

6.1.3 Software solution for measurement and calibration

A software tool offers synchronous acquisition of various types of data (from radars, cameras, GPS, communications bus (CAN, LIN), etc.) – an example is the CANape software made by Vector). This software is equipped with an XCP driver, so can use this protocol to decode incoming acquisitions and calibrate these diverse data (see Figure 6.5).

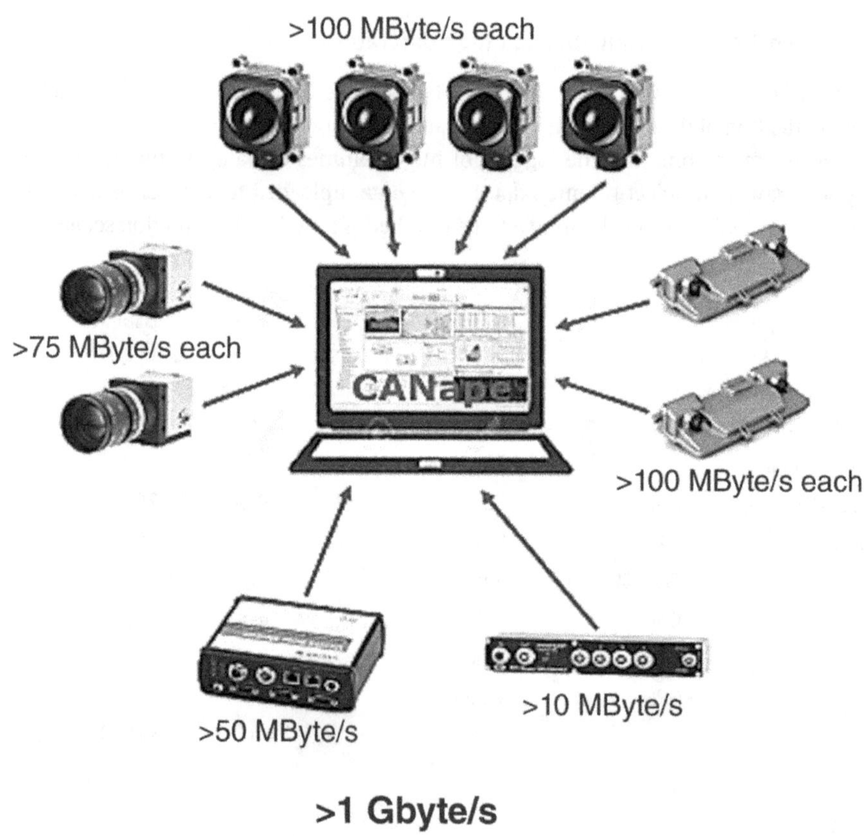

Figure 6.3 Implementation of the data acquisition structure.

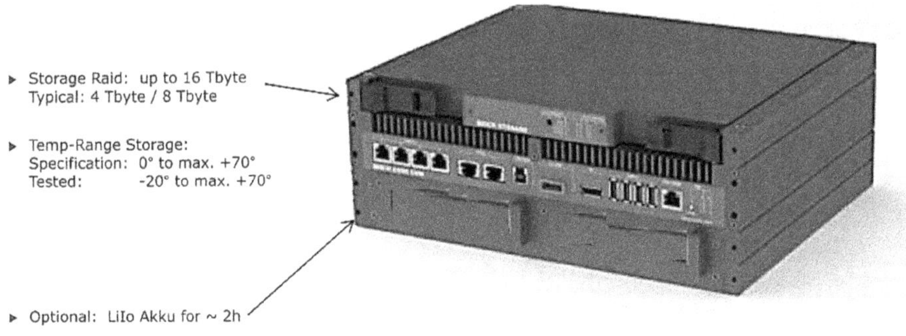

Figure 6.4 Example of a BrickPC.

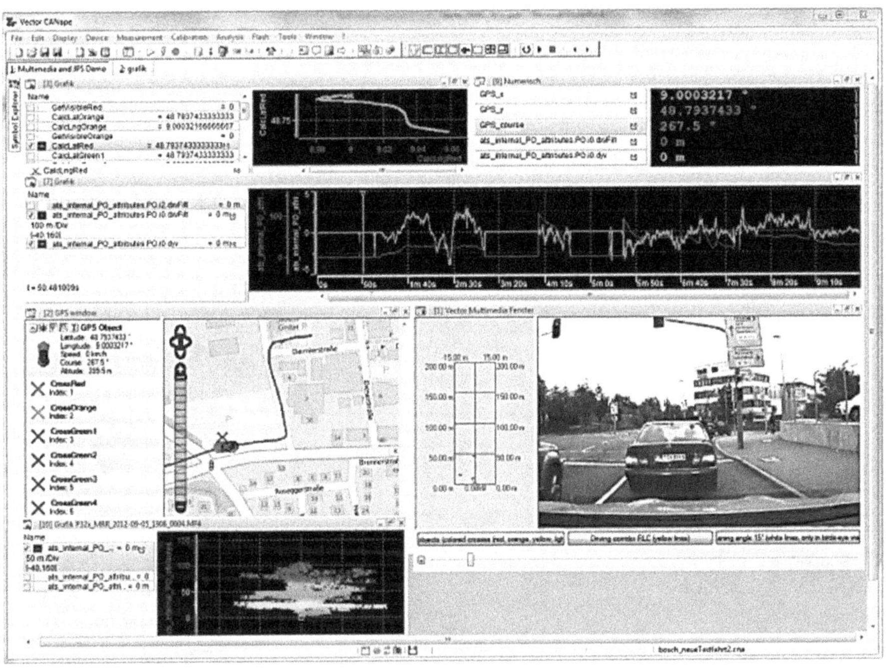

Figure 6.5 CANape – an example of a data acquisition tool.

By accessing the internal parameters of the ECU memory, the applications can be kept up to date. With this calibration function, it is possible to configure the ECU for data fusion, for use by the ADAS, while the vehicle is in motion, to simulate and target the level of performance required of the system in real-world conditions. The data from the various sensors, cameras, radars, etc., are superimposed upon the videos, and compared to the results obtained by the ADAS's ECU. Thus, users can calibrate the system to achieve perfect superposition, as shown in Figure 6.6.

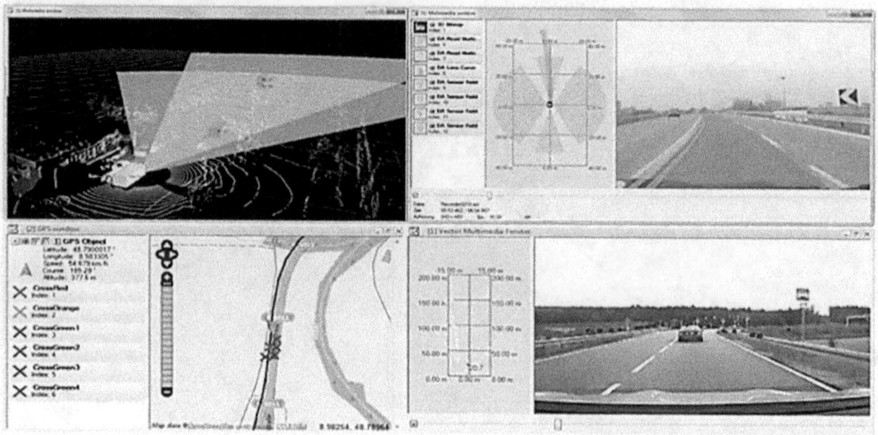

Figure 6.6 Example of calibration.

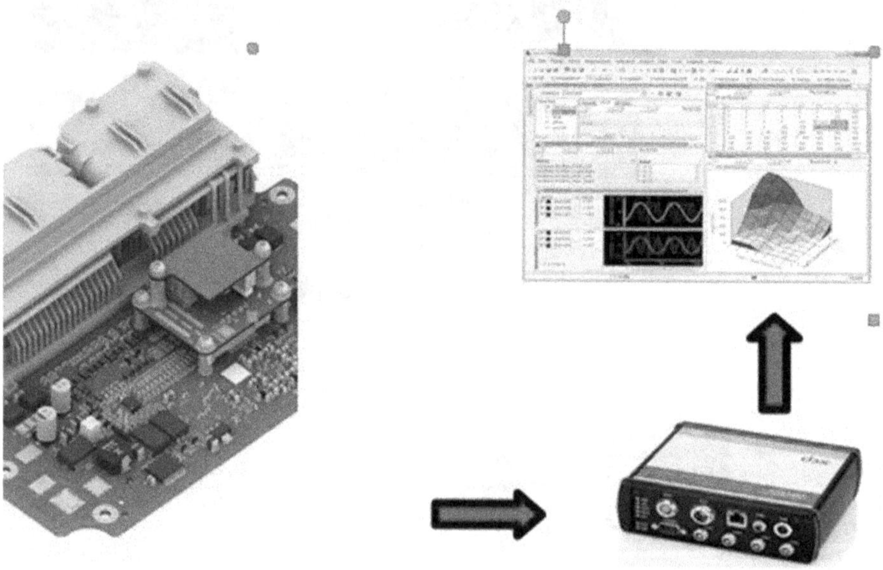

Figure 6.7 Example of an interface (VX1000).

6.1.4 Hardware solution for high-datarate acquisition

The hardware interface must be capable of acquiring high volumes of data at very high datarates (50 MByte/s). To achieve such high acquisition rates, it is necessary to collect the data through the computer's Debug interface, connected to the ECU's memory. Thus, the interface is connected, firstly, to the ECU's debug interface and, secondly, to the Ethernet port of the PC that hosts the measurement/calibration tool mentioned above. The data measured by the debug interface are then converted by that interface and transmitted using the protocol "XCP on Ethernet" to the measurement and calibration software (such as CANape software with Vector's VX1000 interface, as shown in Figure 6.7).

6.1.5 Solution for 1 Gbyte/s data logging

OEMs have the option of logging data from their sensors, radars, lidars, and cameras, using the acquisition interface and a very high-capacity external hard disk, with a CPU set for GHz operation, such as the BrickPC mentioned earlier. That BrickPC can also host measurement and calibration software – see Figure 6.8 for the acquisition of radar data at a rate of 1 Gbit/s, via a high-speed HSSL2 VX1000 interface.

This manipulation can be performed by every OEM for the sensor they wish to use, and by means of a switch and compressor, manage simultaneous contextual acquisition by multiple cameras – see Figure 6.9. Such a system could handle up to 8 Tbytes per day.

In such systems, the data gathered by a front-facing camera, for example, can be encoded in a variety of ways: as raw data, physical data, or objects, as shown in Figure 6.10. To display the stream, encoded as raw data, on various display units, a high-throughput digital video interface such as an FDP Link is needed in this case.

Generally, an automaker must acquire data from the various sensors mounted around the vehicle, such as those shown in Figure 6.11A. As these different data

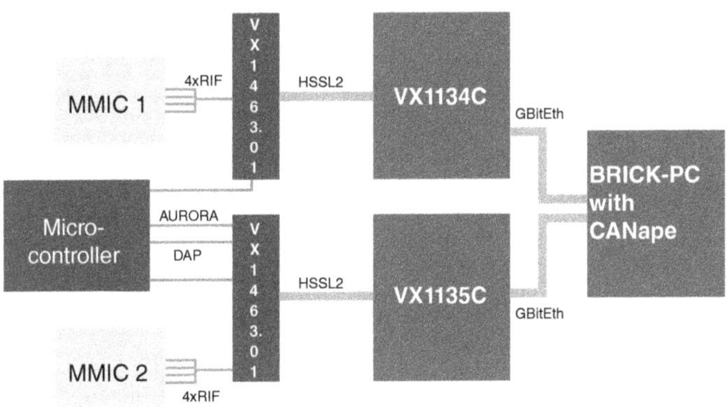

Figure 6.8 Data-logging hardware.

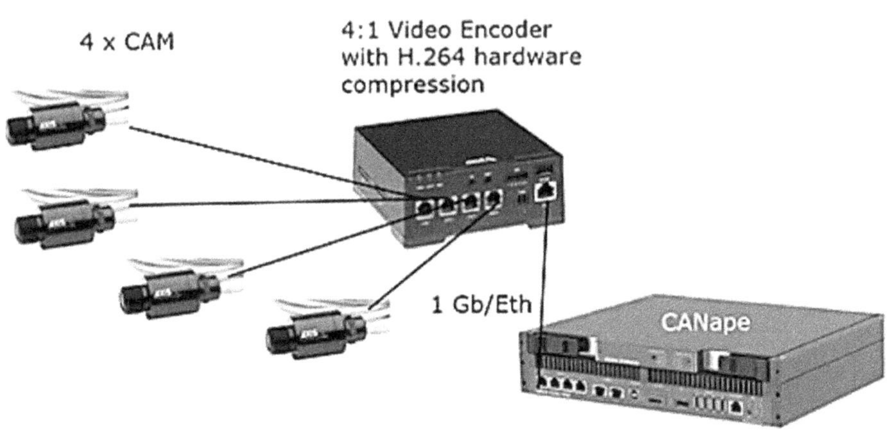

Figure 6.9 Data logging system.

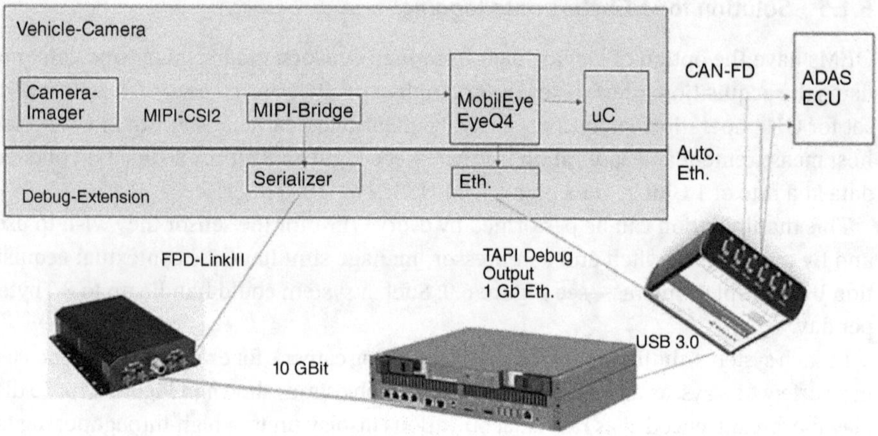

Figure 6.10 Use of a high-datarate video interface.

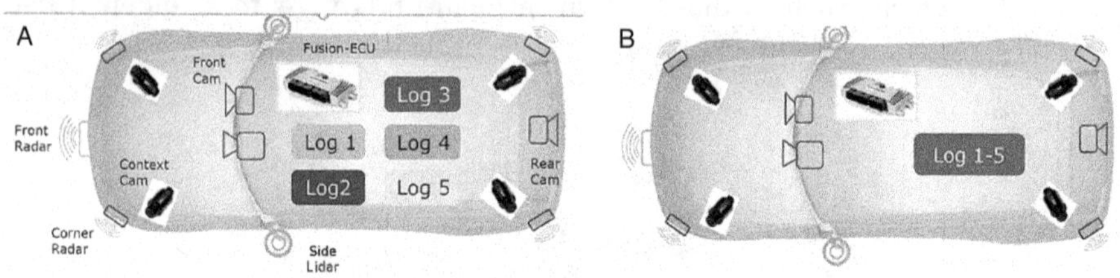

Figure 6.11 Installation and recording of ECUs.

streams need to be synchronized, it is preferable to fuse them and record them as a single dataset for the various sensors and ADAS – see Figure 6.11B.

For united recording of data from the whole system, enormous recording capacity is needed. The DHPR system – **D**istributed **H**igh-**P**erformance **R**ecording – consists of multiple PCs connected in a chain, all equipped with acquisition software such as CANape. This environment offers large storage capacity at a high datarate (over 1 GByte/s), and allows all incoming recordings to be synchronized. Such a setup ensures high performance and high disk capacity (see Figure 6.12).

To simplify vehicle instrumentation in a context requiring multiple VX1000 data-acquisition interfaces, Vector has developed a single module containing multiple acquisition interfaces (see Figure 6.13).

6.1.6 Solutions for transport of the Tbytes of data recorded

As Figure 6.14 shows, the data from the BrickPCs used as storage are compiled and sent to a server.

6.1.7 The three aspects of simulations

Having seen how the system is set up (an absolutely necessary first step in order to understand the subsequent discussion), let us now briefly look at three approaches to simulation, which are connected and complement one another.

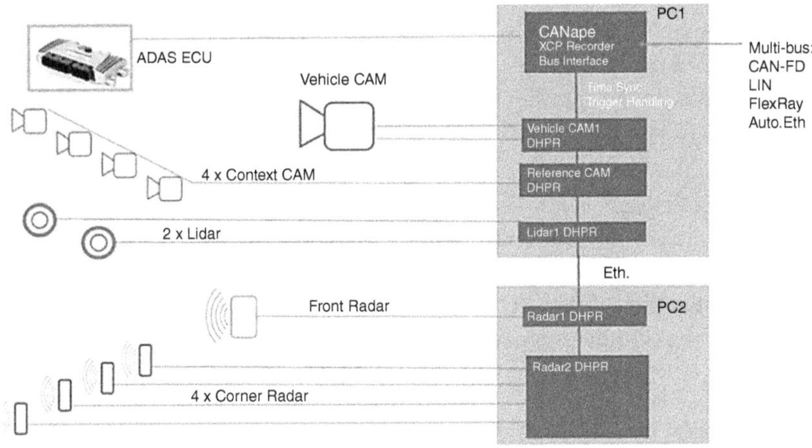

Figure 6.12 Synchronization of recordings.

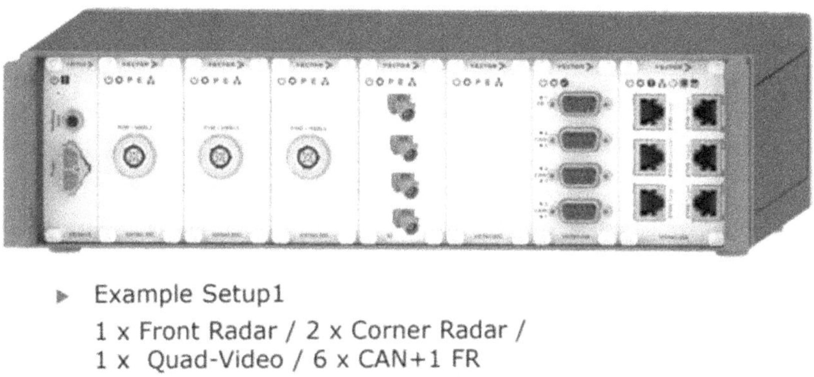

Figure 6.13 Data acquisition module.

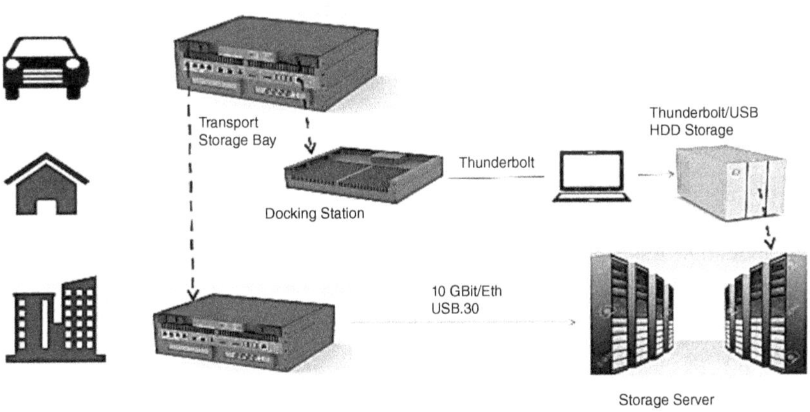

Figure 6.14 Overview of the whole data storage and transport system.

Functional simulations

Functional simulations focus on logic and the mathematical aspects of the functions involved in the different subsystems (for example, the precise and detailed operation of the ADAS computer, the networks, the communication systems, etc.).

Physical simulations

Physical simulations focus on the physics of the vehicle (for example, the mechanical characteristics of the shock-absorber springs, how the electronic suspension responds to the mechanical forces of those springs, how the chassis behaves, etc.).

Environmental simulations

Environmental simulations focus on the relationships between vehicle autonomy and the vehicle's environment as a whole (e.g. definitions of actions, obstacles such as pedestrians and other vehicles, the state of the road, and the quality of road signage – for example, how the system responds when signs are obscured by graffiti).

6.1.8 Physical and environmental simulations

There is a massive difference between the road testing of a handful of autonomous vehicles, navigating in controlled environments, and the millions of vehicles on the roads all over the world – were all these vehicles autonomous, they would have to be operable safely at all times, wherever they are. One of the major obstacles that must be overcome before fully autonomous "hands-free" vehicles (levels 3 and 4) can be rolled out on a large scale is to develop software to safely manage all conceivable driving scenarios without the intervention of a human driver (level 5 or higher on the open road).

- Example: identifying pedestrians, animals, vehicles, or other objects that may be encountered on the road, and envisaging the infinite number of scenarios that could affect the information received by the system's sensors – e.g. weather conditions, light levels, mist, condensation, obstructed view, and so on.

Today's autonomous driving systems are based on machine learning or deep learning, and software that can be designed and trained to recognize trends without having been specifically programmed for all possible situations that could occur. Of course, these systems cannot be released to the general public until automakers have proved that they meet today's exacting safety standards.

There is still a major problem to be addressed, though: how should such software be validated?

Safety simulations and validation

Today, traditional methods for validating a vehicle's safety software using the classic V-shaped model defined in ISO 26 262 are becoming outdated. In this approach, engineers specify the detailed requirements of the systems, define software architecture and create parallel subsystems to address each such requirement.

- The left-hand side of the V model shows how the system is divided into smaller and smaller subsystems;
- The right-hand side shows the process of validation of the software, beginning with the smallest of subsystems and ascending the V to finally validate the entire system.

Now, however, multiple autonomy levels (3, 4, and 5), defined by SAE J3016 (see Chapter 2) mean that the driver is out of the loop with regard to control of the vehicle, which creates a number of major challenges:

- Validation of the features of sensors that are now in charge of perceiving the driving environment;
- The deep learning systems used do not have the architecture, the requirements or the subsystems that are used to validate conventional safety systems;
- The machine learning and/or deep learning systems are probabilistic, and sometimes, even when presented with the same input data, they may behave differently;
- A vast array of driving tests can be conducted, but will only serve to validate one autonomous driving system. Indeed, if even a single line of machine code is modified (or the parameters of the experiment shift), if a single failure is encountered, we must go back to the drawing board and restart the whole process from scratch.

The autonomous vehicle industry must be able to use a closed-loop simulation, in which a virtual vehicle is placed in a realistic virtual world, using software that includes even the smallest details of autonomous driving. This requires precise simulation including an enormously broad range of elements – notably:

- Obstacles such as pedestrians, other vehicles, etc.;
- The road in different states, such as being wet with rain;
- The vehicle, including its mechanical characteristics, state, position, and behavior;
- The environment: the level of visibility, properties of the materials, weather conditions and road signage, etc.

The goal of all of this is to alleviate the need for billions of kilometers of road-testing, to validate the safety and reliability of autonomous vehicles so they can be brought to market sooner. To address this challenge, the simulation environment must be a closed loop, including a model of the virtual world in which the autonomous vehicle is to operate, and the virtual vehicle itself. The virtual vehicle must include physically accurate simulations of sensors, of the vehicle dynamics, and driving scenarios to allow for virtual tests of Software-in-the-Loop (SIL) and Hardware-in-the-Loop (HIL). Thus, the simulation platform can be used to safety-test complex autonomous driving systems integrating physics, electronics, onboard systems and software simulation. It models:

- Vehicle components that use input from actuators such as the steering wheel, brake, and accelerator;
- Vehicle dynamics, calculating the test vehicle's position, speed, and orientation;
- Driving scenarios, setting the test vehicle in motion, along with other vehicles and objects encountered in a road test;
- Sensors, which observe the surroundings in the virtual world, and the output signals from those sensors;
- Signal processing and deep learning, to identify objects, the position of the autonomous vehicle, and the driving conditions on the basis of sensor data;
- Decision control algorithms, which generate input for the actuators and display information and decisions to the passenger or operator.

These systems can be used to quickly and affordably simulate any vehicle, with any combination of sensors, any control system in any driving scenario. The interactive driving simulator considers all interactions between the autonomous vehicle, the environment, obstacles, and the state of the road, and provides the vehicle developers with indications to ensure compliance with the safety regulations and market expectations. Based on realistic driving conditions, the autonomous vehicle simulator makes the same reliable decisions as would be made by the future autonomous and/or connected vehicle.

Simulation of sensors

For a more concrete example, consider the optical and sonar sensors that are used in ever-increasing numbers in autonomous vehicles (infrared and visible-wavelength detection cameras, lidars for a 3D 360° view of the driving environment, radars, sonars, etc.). To produce a detailed simulation of such sensors:

- The software simulating the optical sensors must be able, in real time, to supply all the information needed for a realistic physics-based response, which precisely represents how those sensors would perform in the real world. This will help make the autonomous driving systems of the future safer. This requires a detailed physical simulation of these optical sensors, taking account of the optical lenses, mechanics, sensors, materials and optical properties, and fusing images captured by multiple cameras;
- Next, the software must be able to simulate the physical installation of those sensors in the vehicle, to determine the impact of that installation on the images produced (and the impact of optical screens, reflections from the bodywork, and metallic paint, etc.);
- Finally, a reduced-order model (ROM) of the models of the cameras and lidars must be integrated into a driving scenario, providing the requisite performances for real-time simulation.

Then, the simulation is able to precisely reproduce the images generated by the cameras and lidars, etc., and the path taken by the rays of light when they enter the sensors, to determine factors such as glare from the sun, reflections from buildings, from the road or from glass, when validating what the sensors are perceiving. A similar process, of course, applies to other sensors (radars, etc.), again with:

- Detailed simulations of the component;
- A simulation of its installation in the car;
- A simulation of the sensor's performance on the road to validate the safety of the autonomous vehicles.

Let us take a few examples.

- Sonars:
  - The propagation attenuation, atmospheric attenuation, absorption and reflection based on the object's geometry and the properties of the material, angle of aperture and tunable signal resolution, and the output from a histogram of intensity depth are all taken into account.

- Cameras:
  - Adjustable filters for distortion and color space, angle of aperture and tunable image resolution, and the output from an RGB image stream are taken into account.
- Lidars:
  - The simulation considers the intensity of reflection, based on the angle formed by the laser beam and the object's surface, the properties of the materials, whether or not it is a spinning lidar, the angle of aperture and tunable signal resolution, the output from a 3D point cloud as an ROS subject to DDS or UDP, in the format used by the specific lidar maker.
- Radars:
  - The simulation considers the diffusion of the radar waves as a function of the object's geometry and the properties of the material, the different features of the antenna, the angle of aperture and tunable signal resolution, output of raw data such as the relative velocity and distance from the object, and the intensity of the electrical field or the Fourier transform based on the graphic processor for the generation of remote Doppler plots.

In addition, we see pre-processing of the signals:

- Automated image segmentation for ideal sensor fusion, consideration and classification of all available objects in the object catalog, configurable object classes, angle of aperture and tunable signal resolution, output of fused data from relevant sensors, such as the object's relative velocity and distance, and also its category.

Safety validation process

The development of autonomous vehicles is (still) generally subject to ISO 26262, release 2, commonly called Safety of the Intended Functionality (SOTIF) – ISO 21448. This standard:

- Requires engineers to assess the safety of autonomous driving systems by identifying the relevant risks and modes of failure of the electronic components;
- Sets out broad guidelines on how to document driving scenarios and analyze the safety of those scenarios;
- Ensures that safety scenarios and triggering events are checked and the vehicle is validated in the environment in which the safety systems are to be used.

The developers of automotive systems must validate the functional safety of their electronic components, down to the silicon chips used, to achieve a high level of trust in the safety of their designs when autonomous vehicles are placed on the road.

- Note: ISO/PAS 21448:2019 – SOTIF

The absence of unreasonable risk due to hazards resulting from functional insufficiencies of the intended functionality or by reasonably foreseeable misuse by persons is referred to as the Safety Of The Intended Functionality (SOTIF). This document provides guidance on the applicable design, verification, and validation measures needed to achieve the SOTIF. This document does not apply to faults covered by the ISO 26262 series or to hazards directly caused by the system technology (e.g. eye damage from a laser sensor).

This document is intended to be applied to intended functionality where proper situational awareness is critical to safety, and where that situational awareness is derived from complex sensors and processing algorithms; especially emergency intervention systems (e.g. emergency braking systems) and Advanced Driver Assistance Systems (ADAS) with levels 1 and 2 on the OICA/SAE standard J3016 automation scales.

The vehicle's control and regulation loop

A combination of machine learning, deep learning, and control logic serves to implement a control loop in a fully automated/autonomous vehicle. This control and regulation loop comprises:

- Perception (what the vehicle sees);
- Motion planning (the vehicle's intended behavior);
- Motion execution (how the vehicle performs the maneuver).

This control loop is executed cyclically so that the vehicle can respond to continuous changes in its surroundings. In addition, in order to design and generate code for the control logic, it is necessary to automatically generate qualified code as output from an equivalent program in C, conforming to the safety standards of software models for autonomous vehicles, comprising a combination of solid-state machines and data streams.

Safe autonomous vehicle control software

To verify the safety of the complex algorithms used for perception, motion planning, and motion execution, developers often need to break down the overall architecture of an autonomous vehicle into a meaningful subset of components. The detailed architecture is generally made up of a "primary algorithm," which may be extremely complex, undergoing frequent updates, and very difficult to check. Generally, that algorithm is twinned with a corresponding safety barrier, which checks that the output from the primary algorithm is correct. If the safety barrier detects a problem, it transfers control to a "safety channel algorithm," which performs a brief function – e.g. guiding the vehicle to a kerbside stop. The primary algorithm performs a lengthy mission with no defined stop state, whereas the safety algorithm has only a short mission, which ends in a definite and safe state.

The detailed requirements of the safety barriers can be established, so that they are implemented in compliance with the highest safety requirements of ASIL D of ISO 26262, avoiding the need for the primary algorithm to meet these requirements. The primary algorithm may fail arbitrarily, and do things in the worst possible way, because the safety barrier makes the primary algorithm into a silent failure component, which is halted every time it produces erroneous data. If, say, the safety barrier detects a breakdown in planning or trajectory mapping – for example, if the planned trajectory collides with a parked car – then it halts the primary planning algorithm.

Perception validation

Autonomous driving systems depend on sensors to make decisions, on the basis of the vehicle's surroundings, and it is not difficult to create a safety barrier to check whether or not the output of those sensors is correct. On the other hand, when noise or interference are present in a perception signal, they can adversely affect the robustness of the function, and jeopardize the system's understanding of the situation, and hence safety. SOTIF, ISO 26262, addresses the dangerous behavior of systems with such limitations.

- Example: a simulation platform built into an autonomous vehicle with strong detection and safety checking may become extremely weak in the wake of barely perceptible environmental changes – for example, disguised attacks in the form of graffiti on a road sign can cause confusion and major fragility in the Convolutional Neural Network (CNN) processing of the signal.

With simulation, designers can deliberately introduce objects and events, simulate sensor readings, and determine whether the objects and events have been correctly detected, and whether the autonomous driving system has responded in the correct way. This is necessary to prove that the perception function is harmless, and to find troublesome scenarios that can interfere with autonomous driving systems and human drivers.

A simulation platform validates the safety of an autonomous vehicle's perception with a technique based on hardware-in-the-loop (HIL). For example, this is often used with a closed-loop simulation function that executes driving scenarios, while reduced order models based on physics simulate any combinations of sensors. The simulation of cameras can be connected to the electronic control unit (ECU) for hardware in the HIL simulation, in which perception software in the corresponding ECU interprets the image generated by the simulated camera. This simulation can be used to validate the autonomous driving simulation. The camera image and the image of the driver can be displayed in real time, so problems can quickly be identified and understood.

- Examples, among others: The companies TESIS/VECTOR with "ANSYS medini analyze" help manage the safety validation process by implementing safety analysis methods such as Failure Modes, Effects, and Criticality Analysis (FMECA). These tools help support design in accordance with ISO 26262 and safety analysis for the electrical/electronic and software-based control systems for safety-related functions. This integrated solution can offer end-to-end safety in deep learning and other autonomous driving systems.

This solution allows automakers and OEMs to virtually test and integrate their new generations of autonomous driving sensors before industrial products are actually made available and brought to market, by reproducing a 3D physical world and creating a virtual-reality driving experience in real time. In addition, it allows an autonomous vehicle to be tested in unlimited driving scenarios, day or night, on different roads and in different weather conditions, using vast libraries of content. These libraries include roads, crossroads, signs, signals, traffic, objects, pedestrians, and other human avatars, ensuring the pinnacle of realism.

The autonomous vehicle simulator thus makes the same decisions as would an autonomous and/or connected vehicle in the real world. This helps to eliminate the costly and risk real-world tests of sensor systems and reduce the time-to-market. Developers can be assured that the vehicle can accurately "see" traffic, pedestrians, signaling, and markings, and obeys the safety standards and driving regulations. These unique capabilities help speed up the engineering process by driving the autonomous vehicle, at an early stage, on digital test routes with realistic traffic conditions, including varied weather conditions, oncoming vehicles, and pedestrian scenarios to find out how your vehicle would respond in case of an emergency.

6.1.9 Conclusion of simulation

The automotive industry is well aware that it is far easier to build a handful of vehicles circulating in controlled conditions than build hundreds of thousands of vehicles and roll them out to be used anytime, anywhere. In addition, it is difficult to know how many road tests will be necessary and sufficient to ensure that such fleets of vehicles are safe. Thus, the greatest challenge in the large-scale rollout of autonomous driving systems is testing and debugging the machine-learning and deep-learning algorithms that function without defined requirements, and are designed to ensure the systems are sound and safe.

A multi-physical simulation and onboard software critical for safety to provide a complete closed circuit for virtual testing of autonomous driving systems – including (1) physically accurate models of the sensors, (2) a virtual environment, and (3) virtual SIL and HIL tests – allow the future autonomous vehicle to be driven on virtual test routes in realistic traffic conditions. The virtual world includes different weather conditions, oncoming vehicles, and scenarios involving pedestrians, which can be used to train machine learning software and validate its responses in any driving scenario. By providing a realistic, real-time simulation, this allows automotive engineers to create safer autonomous driving systems.

6.2 Evolutions of the electric/electronic (E/E) and software architectures of onboard systems

6.2.1 Evolutions of E/E architectures

Let us begin by looking at the history of development in this area. Over the past few decades, there have been huge leaps forward in the hardware and software architectures in the automobile market. The electronic architecture has shifted from "point-to-point" communicating systems to (mainly) "multiplexed networks," with distributed functions (see Figure 6.15).

As we showed in Section 4.5, numerous protocols have been implemented – primarily CAN, LIN, FlexRay, and MOST – to handle applications that require ever increasing datarates (see Figure 6.16).

Basic requirement for applications: bandwidth

For the new applications that are emerging, communication in E/E automotive architectures is growing rapidly, and therefore requiring burgeoning amounts of bandwidth. Applications such as multimedia infotainment, driving assistance using cameras, radars, lidars, etc. are growing in number.

Central gateway approach

As CAN uses a "bus" topology, the bandwidth is shared between the various ECUs. In itself, this is not hugely problematic, given the small payload that the CAN is able

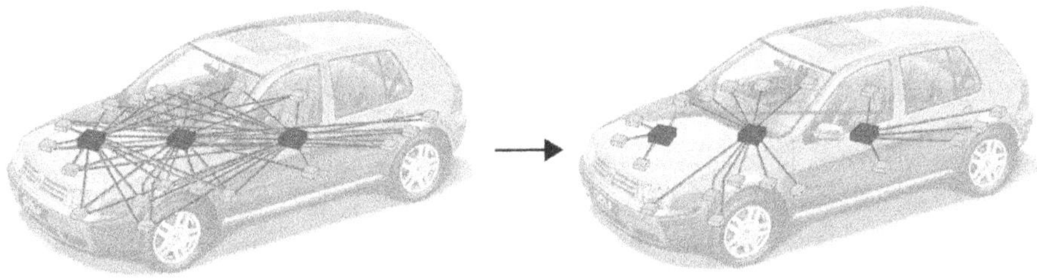

Figure 6.15 From point-to-point to multiplexed networks.

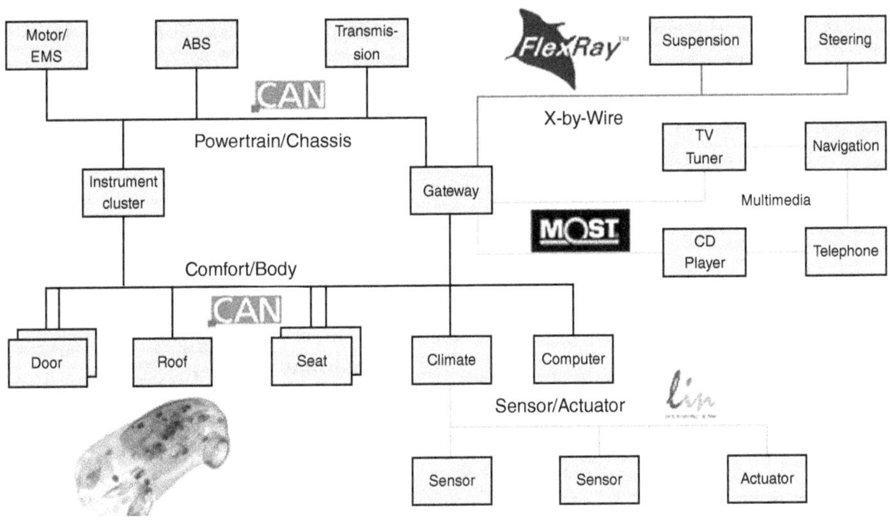

Figure 6.16 Architecture between 2010 and 2015.

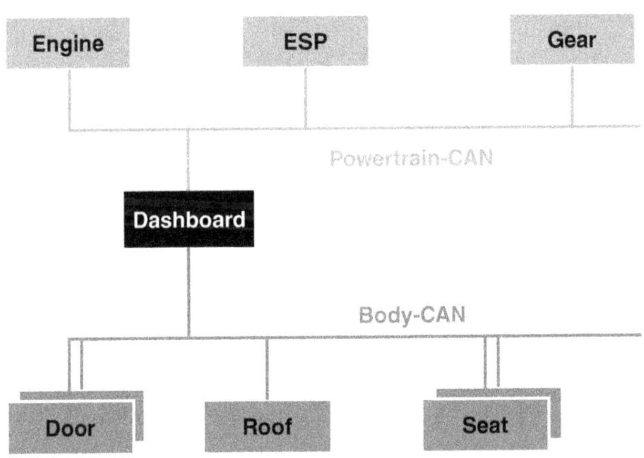

Figure 6.17 ECU gateway approach.

to handle (maximum 8 octets). To gain bandwidth, a CAN network is often split into multiple functional networks. The various CAN networks are linked to one another by a central gateway (one of the ECUs in the network – see Figure 6.17).

As we can see, each of the elements of a cluster can serve as a gateway for another subnetwork. For example, in Figure 6.18, the air conditioning is a gateway for the LIN.

Typically, all the clusters (the chassis, the vehicle body, etc.) are linked to a Central Gateway – see Figure 6.19.

Automotive software architecture

With E/E architectures at their current stage of development, software architecture is made up of two parts: the application codes and the software components making up the basic software – see Figure 6.20.

The software components, which are directly connected to the hardware, can provide a range of application services, and thus can handle low-level tasks. The application codes access these hardware services via the basic software.

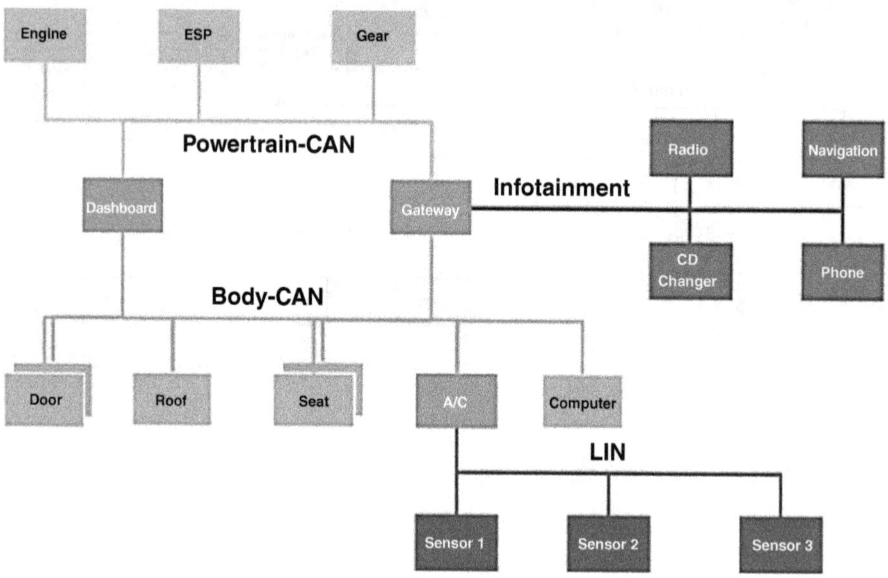

Figure 6.18 Cluster gateway approach.

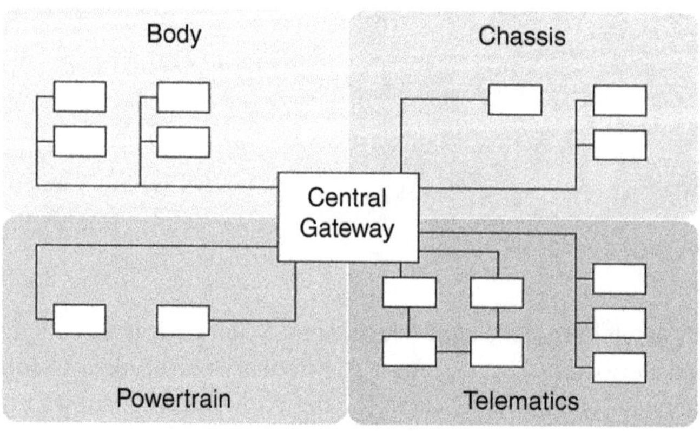

Figure 6.19 Central gateway architecture.

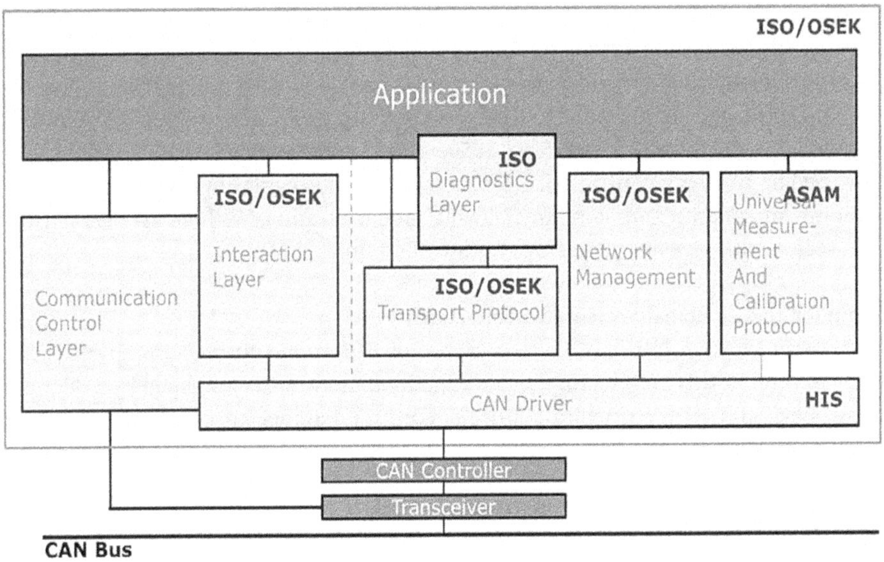

Figure 6.20 Software architecture.

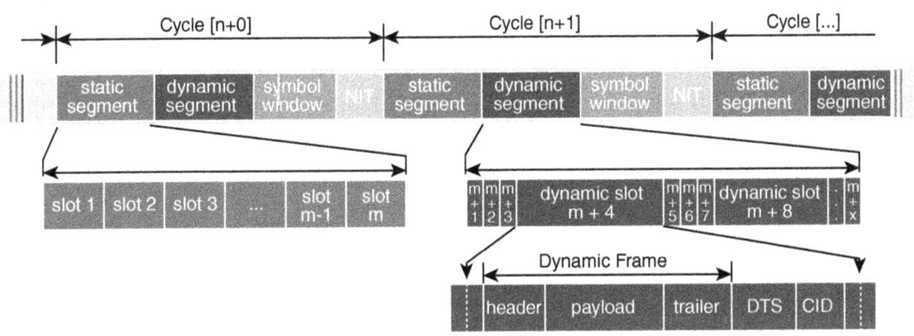

Figure 6.21 Examples of frames in the FlexRay protocol.

Security requirements and the dawn of new protocols

The advent of FlexRay

As indicated in Section 4.5.4, the FlexRay protocol was developed to handle new automotive technologies (X-by-wire), to provide communication security, and to increase bandwidth: for FlexRay applications, the payload may be up to 255 octets, with a maximum bitrate of 10 Mbit/s (see Figure 6.21).

To ensure that communication is deterministic, the FlexRay communication protocol is built around 64 repeating cycles. Each cycle (which might last for, say, 5 ms or so) is made up of "static slots" with guaranteed timing and "dynamic minislots" to cover the event-triggered part of the communication (see Section 4.5.4). The slot identifier ("slot-id") serves a similar function to CAN-ID, defining the content of the frame. Therefore, the central gateway approach is made more complicated every time a new network is added, using a new type of communication technology.

The dawn of CAN-FD

We saw in Section 4.5.3 that, in comparison to conventional CAN, the payload of CAN-FD is much higher (up to 64 octets) and can cope with a higher datarate. The CAN-FD protocol was designed to partly bridge the gap between ordinary CAN (which is generally used in vehicle networks at 500 kbit/s) and FlexRay (10 Mbit/s, but generally used at an average bitrate of 5 Mbit/s). CAN-FD is generally used at a bitrate of 2 Mbit/s for data transmission and 500 kbit/s for arbitration. With an average payload of 32 octets, this gives us an average bitrate of around 1.5 Mbit/s.

Evolution toward domain-oriented architectures

The use of FlexRay has led to a new model of E/E architecture: "domain-oriented" architecture. In this architecture, every functional network is managed by a "domain controller," and all the domain controllers are interconnected by a "backbone" – a high-datarate network such as FlexRay, facilitating communication between these different domains.

The four main domains typically found in a vehicle are Body, Chassis, Engine, and Infotainment. The advantage to this approach is that we can create a more flexible E/E architecture, in which the networks are managed by their domain controllers. The domain controller thus becomes a sort of router, which passes information from one domain to another in the required format. A backbone architecture reduces the need for bandwidth in all the slave networks, which use the backbone's higher bandwidth capabilities to communicate with one another (see Figure 6.22).

In automobiles, with the constantly growing numbers of protocols, and of computers, the embedded code is also growing rapidly. Every time the architecture evolves,

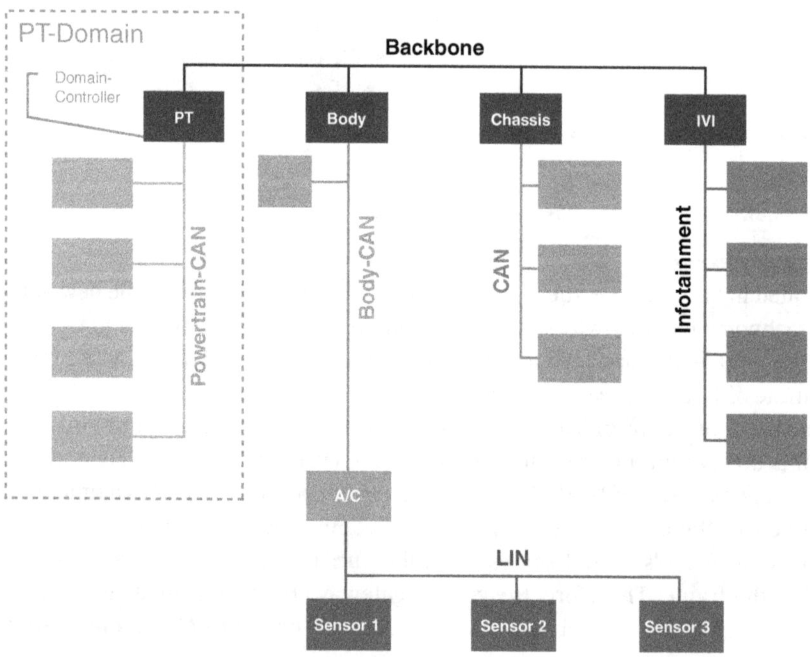

Figure 6.22 Backbone architecture.

automakers are faced with the need to adapt their codes to new types of microcontrollers, notably for the numerous sensors/actuators and, in particular, for critical applications with strict constraints and performance requirements, in which security and reliability are absolutely essential.

6.2.2 Evolution of software architecture in automobiles

Today, there are two major problems relating to software architecture in integration tests run by automakers:

- different software packages come from different sources, which can lead to issues with interoperability among the various ECUs in the integration phase;
- the software depends on the target microcontroller (proprietary layers), which means that the application code must be adapted every time the microcontroller is changed.

As the number of software packages used in vehicles is growing, a new strategy is needed, to allow this software to operate independently of which microcontroller is being used.

It is crucial that software evolve from an optimized design, closely tailored to the intended type of microcontroller, to a structured, standardized design that can serve any target hardware. With that goal in mind, the AUTOSAR standard (see Figure 6.23) is a standard, open-source automobile software architecture developed by a consortium of automakers, subcontractors, and tool suppliers, the aim being to open the door to innovation in electronic systems to achieve better performances, better operational security and better power management. From a technical point of view, this standardization offers a way to deal with the growing complexity of electronic systems. It lays the groundwork for future technologies and helps manage costs without compromising on quality. It also facilitates exchanges and updates to the software and hardware throughout the vehicle's life.

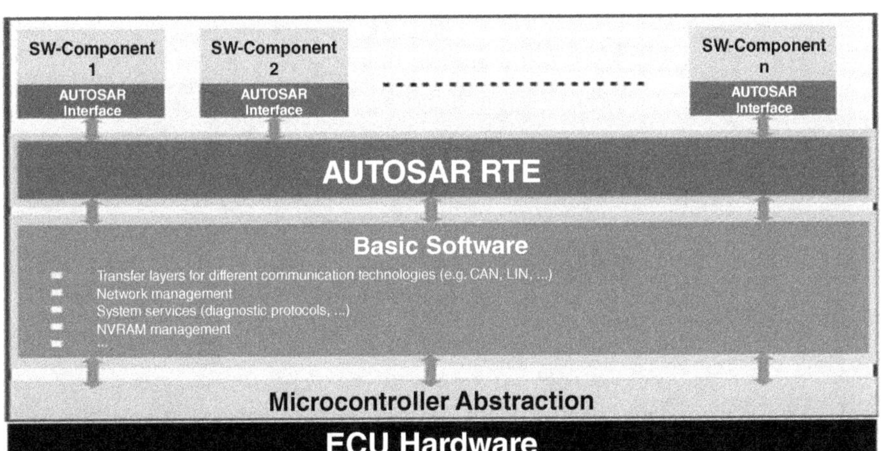

Figure 6.23 The AUTOSAR standard.

Migration from the data link layer to the PDU

Data Unit (DU)

In automotive networking, the data exchanged are generally signals, often very small. Sometimes, a signal comprises only a few bits – for example, the information about a button being pressed or released. These data can be processed in very little time. Therefore, for example, to optimize the payload within a single CAN frame, multiple signals can be carried together (Figure 6.24). The full content of the CAN frame can then be defined as a single "Data Unit" (DU). The CAN-ID of the frame defines its content. In this system, it is often helpful to split one or more signals across multiple frames, and thus across multiple ECUs.

Protocol Data Unit (PDU)

In view of the maximum data payloads that can be supported by CAN-FD (64 octets) and FlexRay (up to 254 octets), it is urgent to improve data transport and data sharing between computers.

A new concept has been developed: the Protocol Data Unit (PDU), with which multiple DU signals can be grouped into PDUs, and those PDUs can be encapsulated into frames (see Figure 6.25). CAN-FD and FlexRay are well suited to the PDU design. In

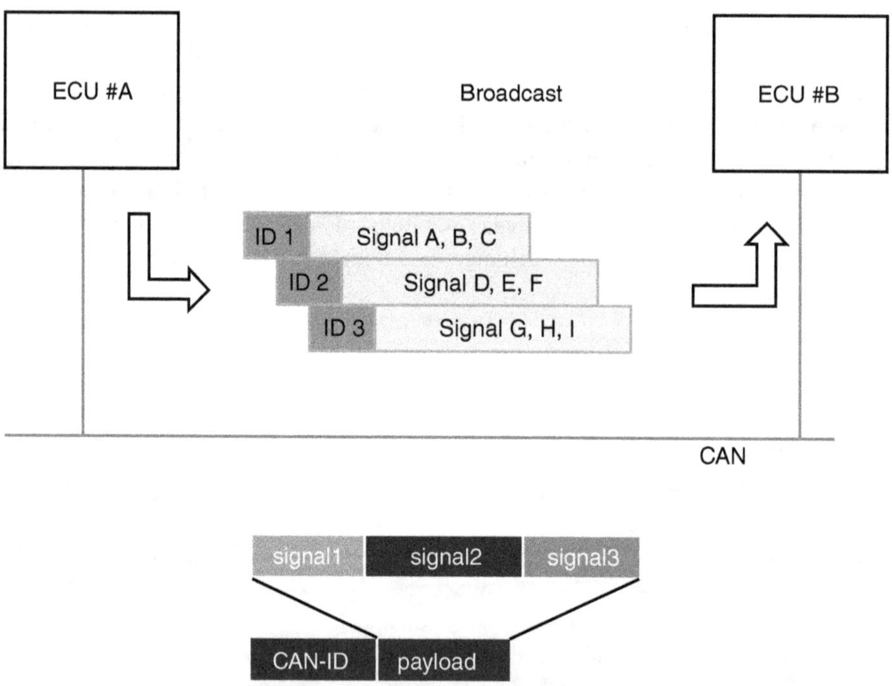

Figure 6.24 Signal mapping.

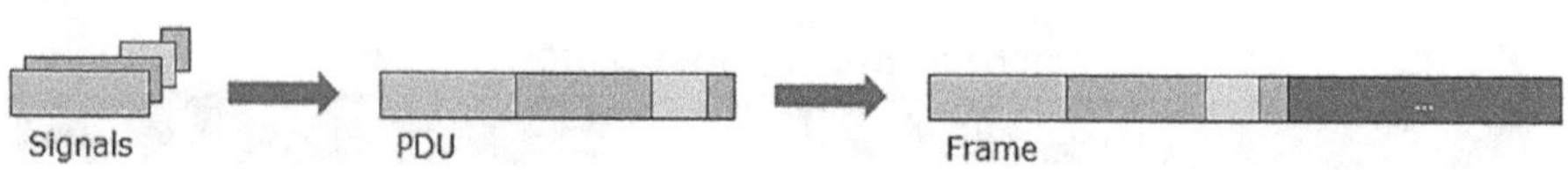

Figure 6.25 Encapsulation of PDUs.

light of its small payload of 8 octets, conventional CAN does not lend itself particularly well to this new approach.

In this system, each PDU could be said to replace a CAN frame. Ultimately, this is tantamount to sending multiple CAN frames in one go (see Figure 6.26). In addition, these PDUs can be transmitted by different frames and therefore by various ECUs.

Software tools, which are also affected by this new concept. Testing and analysis require new functions (see Section 6.4).

With respect to the gateway

A higher bandwidth leads to higher workloads in the gateways, given the increased amount of data that needs to be processed. PDU helps take some of the strain off the gateway by copying signals into a PDU packet as a contiguous whole, which is easier to handle than with conventional signal routing (see Figure 6.27).

Static PDU mapping: FlexRay

For FlexRay, PDU mapping in the data link layer is significant. A FlexRay frame can contain multiple PDUs, with each PDU corresponding to a fixed position in a given slot (Figure 6.28).

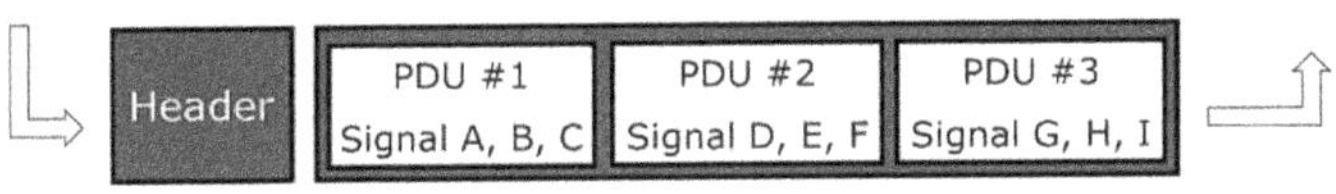

Figure 6.26 PDU frame.

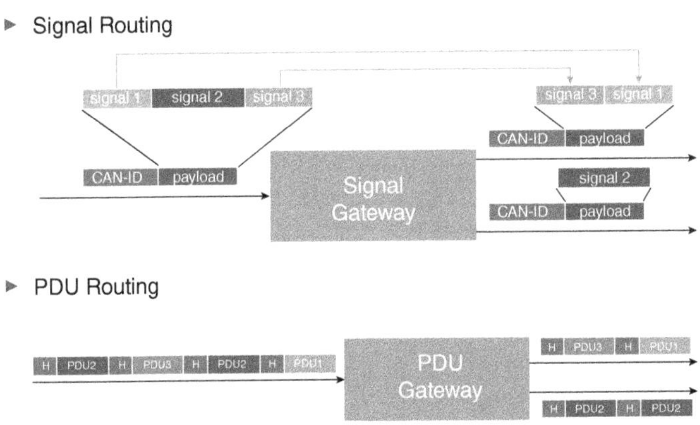

Figure 6.27 Signal routing mode.

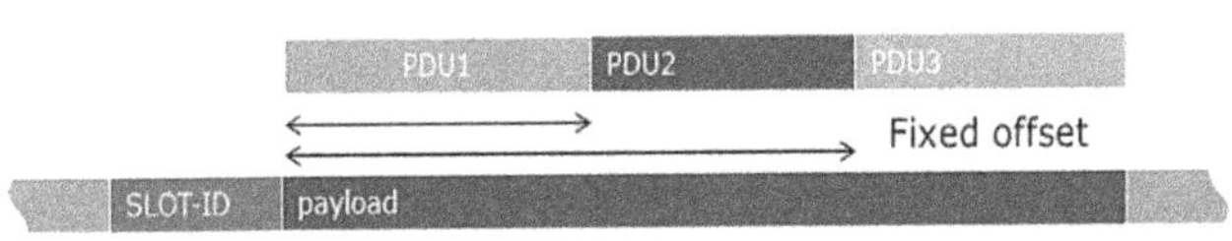

Figure 6.28 Static PDU mapping.

Initially, the basic software processes the application signal in the data link layer. The data link layer (or communication layer) compiles signals with a definite identifier and fixed position. This assembly of signals is mapped onto a frame of a given ECU to be transmitted over the bus.

Moving forward to the concept of a PDU, AUTOSAR defines a new data link layer to process the signals in PDUs. The signals are not mapped directly to a frame, but to an "I-PDU" by the DLL. Then, the I-PDUs are mapped onto a frame by the data link layer, which defines an L-PDU (see Figure 6.29). This means that the same I-PDU can be used in different environments and can be propagated via various frames on different networks. For conventional CAN, AUTOSAR defines the CAN frame as a single PDU, with the CAN-ID defining its content.

As stated earlier, multiple CAN frames represented in the form of a PDU may, if we wish, be mapped onto a single FlexRay frame. Unfortunately, time-triggered behavior in FlexRay is not very compatible with the event-triggered function of CAN. It may happen that updates to the cycles of CAN PDUs are different from FlexRay cycles and, consequently, the same data are transmitted twice or a PDU update is lost.

Dynamic PDU mapping: CAN-FD

As is the case with FlexRay, CAN-FD is able to improve the sending of large packets in a single transmission (multiple CAN frames in a single CAN-FD frame). However, unlike with FlexRay, CAN-FD has event-triggered behavior just like CAN, which means it is easy to migrate from existing CAN architectures to CAN-FD with a higher datarate.

The backbone approach and domain architecture, which we looked at in Section 6.2.1, adds a new constraint on data routing. Indeed, the data in each domain, in the form of a PDU packet, is routed to the central gateway. The backbone compiles all the PDU packets from the different domains into a container. The problem that arises here, obviously, is the size of that container and the position of each PDU within it. As data arrive in an event-triggered manner, for CAN and CAN-FD, it is not possible to predict what the container will contain. A dynamic notion is then required for the backbone and for event-triggered networks such as CAN-FD.

Unlike FlexRay, dynamic frame construction during runtime is used for CAN-FD. This means that a PDU's position in the frame is no longer static. The length of a PDU is also dynamic.

Given that the PDU's position in the frame is no longer static, how can a receiver (which typically knows the exact position of the relevant data in the frame) extract it? With dynamic mapping, it is crucial to attach some information to each PDU to distinguish them. This is the role of the new AUTOSAR layer, which inserts a header into each PDU. This header contains a PDU identifier (PDU-ID) and its data length code (DLC) (see Figure 6.30).

A CAN-FD frame, then, is a container PDU, with PDUs at changeable positions, identifiable by their ID and their DLC. Each PDU is identical to a true CAN frame (see Figure 6.31A). The container has neither its own identifier nor a DLC, which is what makes it dynamic. The routing of CAN-FD packets to the backbone, and vice versa, is therefore simplified (see Figure 6.31B).

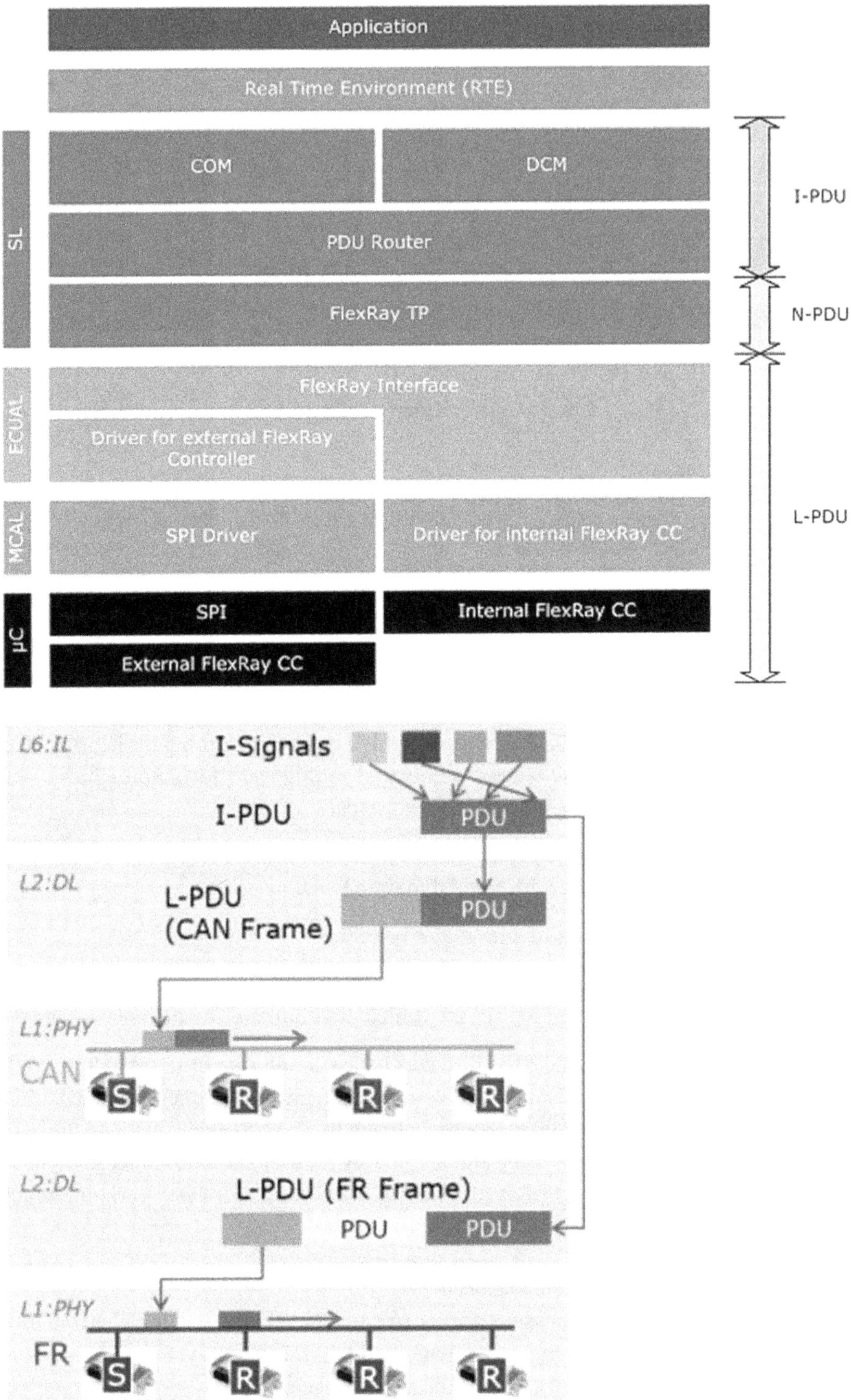

Figure 6.29 L-PDUs and I-PDUs.

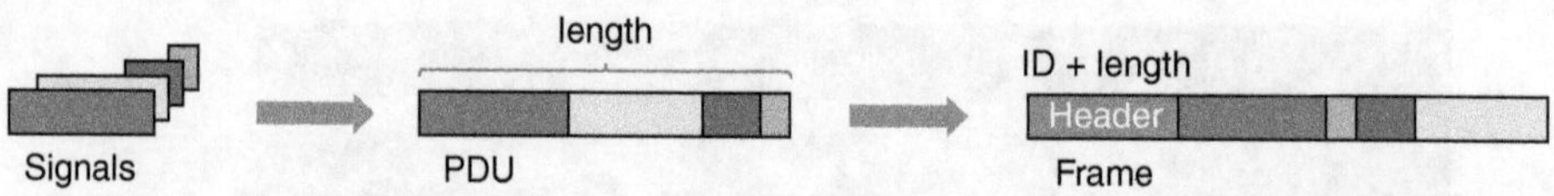

Figure 6.30 PDU header.

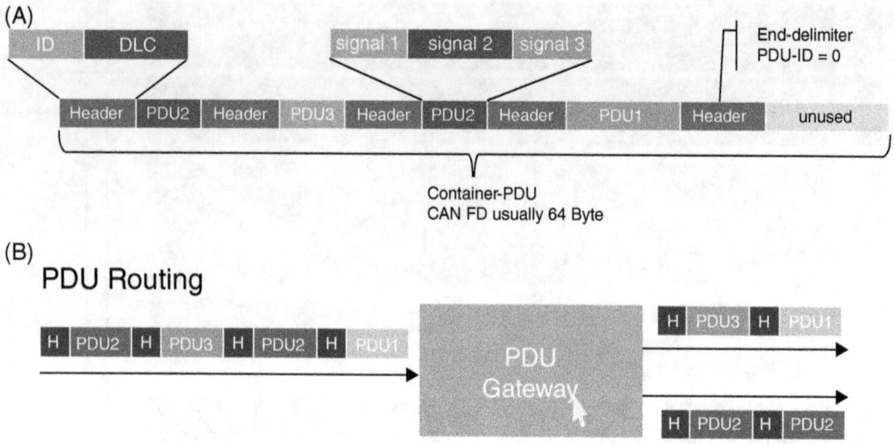

Figure 6.31 Container PDU and PDU gateway.

The work of the AUTOSAR consortium – which has been working for nearly a decade on defining standard exchange formats and middleware for automobiles – has facilitated the introduction of CAN-FD in two steps:

- AUTOSAR 4.1.1: data acceleration;
- AUTOSAR 4.2: increased payload and dynamic PDU.

This dynamic approach also provides major benefits:

- Communication becomes independent of the architecture and of the CAN-ID;
- The functions can be shifted within the vehicle without altering the receiver nodes.

The event-triggered aspect of CAN-FD makes it easy to migrate from CAN. The cyclic broadcasting of CAN frames is not a problem in itself, given the small size of data they carry. On the other hand, the large payload carried by CAN-FD may overload the various receivers. Whether or not they are affected by all or some of the incoming data, these receivers will have to search for the PDUs that are addressed to them, which creates work for the CPU.

Sending of CAN-FD frames or "Container PDU"

A dynamic strategy of transmission during runtime is necessary, as well as dynamic PDU mapping. Various triggers are implemented in AUTOSAR to reduce the work of receiving. Figure 6.32 shows the various possible triggers:

- After a container time-out;
- After a PDU time-out;
- When a PDU is added;

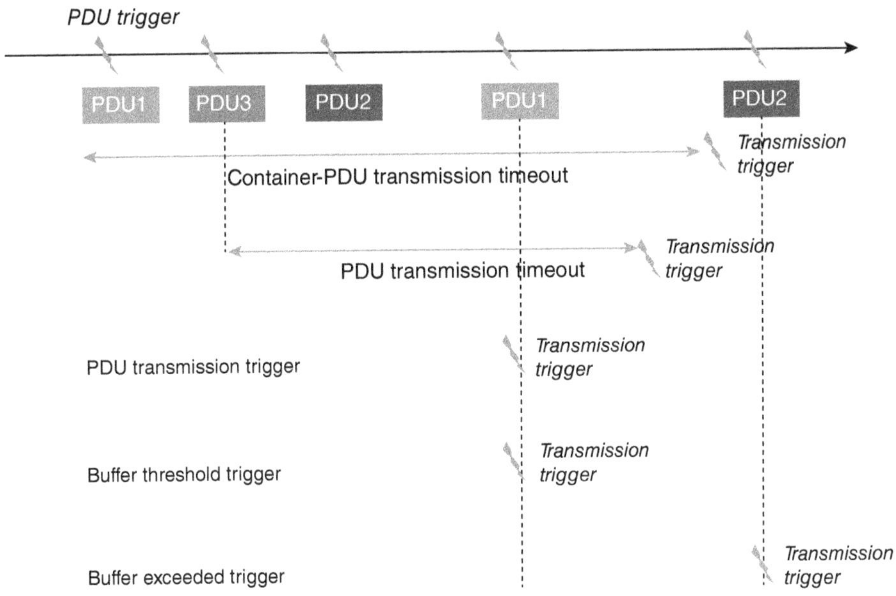

Figure 6.32 PDU transmission triggers.

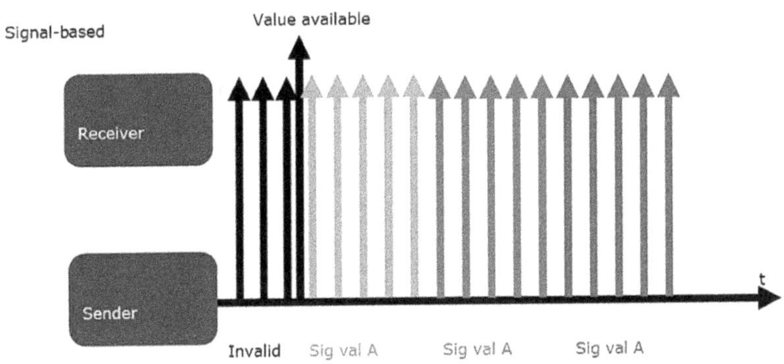

Figure 6.33 Receipt of useful information.

- When the buffer threshold is reached;
- When the buffer is exceeded.

Extensive CPU resources required

CAN-FD improved the sending of dynamic size packets in a single transmission. In addition, the transmission, which is controlled by triggers, avoids the need for cyclic transmissions and takes some of the strain off the receivers. Hence, this dynamic approach offers great flexibility for future E/E architectures, but the requirement for CPU resources for each domain controller is still high: although the PDUs are dynamic and their transmissions are controlled, transmission takes place in broadcast mode, which overloads receivers that are not the intended addressees of the transmission, and leads to numerous CPU interrupts. Figure 6.33 shows the useful information in relation to everything that can be received. Hence, although this solution is attractive, it is insufficient – particularly for handling the huge volumes of data that will be needed for future applications.

For each received frame, the receiver algorithm must also check whether relevant PDUs are present. To do so, it analyzes whether the first PDU is pertinent for receipt, then hops to the next PDU to check whether that is also relevant, and so on until it reaches the end of the frame.

This approach creates massive complexity for the new generations of E/E architectures for future vehicles. Therefore, a range of ideas have emerged for architecture designs that could satisfy requirements over the coming decades, and are a fundamental reshaping of the way in which communication works.

6.2.3 Automotive Ethernet

The vehicles of the future are intended to be carbon free, autonomous, and connected. New applications with even stricter requirements therefore need to be specified and developed (see examples in Figure 6.34).

Owing to the higher datarates (up to 1–2 Gbit/s) and large payloads (1500 octets) that Ethernet can deliver, this protocol is becoming a promising candidate for use in automobiles. In addition, new technological advances have meant that Ethernet (see Chapter 5 for information on the physical layer) can operate on a single unshielded twisted pair of wires (UTP). In view of these three qualities – datarate, payload, and medium – Ethernet is well suited to applications in ADAS, infotainment, Diagnostics over IP (DoIP), and ECU flashing (see Figure 6.35).

In addition, for networks carrying a high load such as Ethernet (1500 octets), it is essential that signals be grouped into PDUs. AUTOSAR 3.0 is the first version that supports Ethernet. Since this version, the following standards are referred to as AUTOSAR Classic.

Evolution from signal to service

The possible payload that can be carried by a 1500-octet Ethernet frame places an even greater strain on the CPU than with CAN-FD or FlexRay. For this reason, it is even more important for Ethernet to transmit as much data as possible within a single frame, with a dynamic transmission system, as mentioned for CAN-FD. This means a PDU with a tailored size and mapping can be used for each receiver. However, there is still a need to improve the CPU load in relation to reception, by avoiding the use of broadcast mode if it is not absolutely necessary.

Ethernet is employed in switched network mode (switch topology) with a communication architecture that offers great flexibility, such as point-to-point links between

Figure 6.34 Emerging applications.

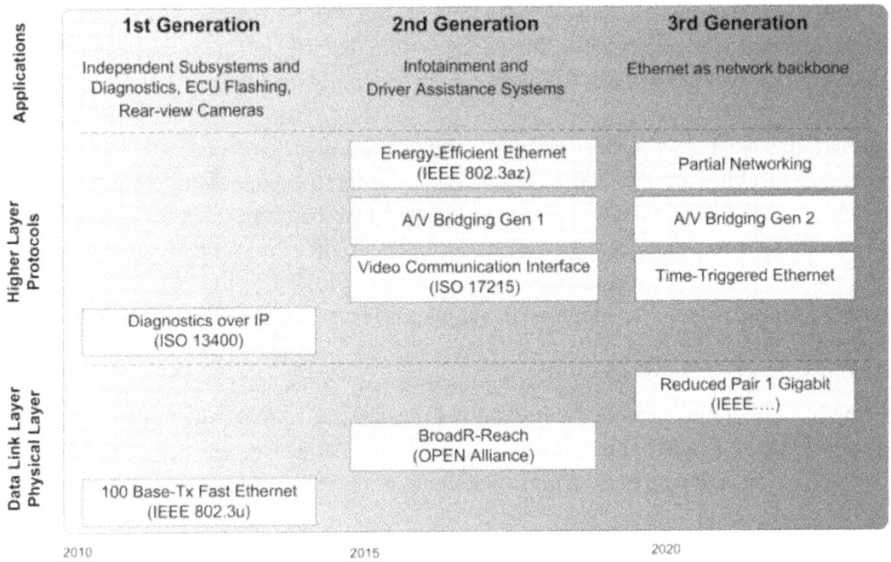

Figure 6.35 Position of Ethernet applications in the near future.

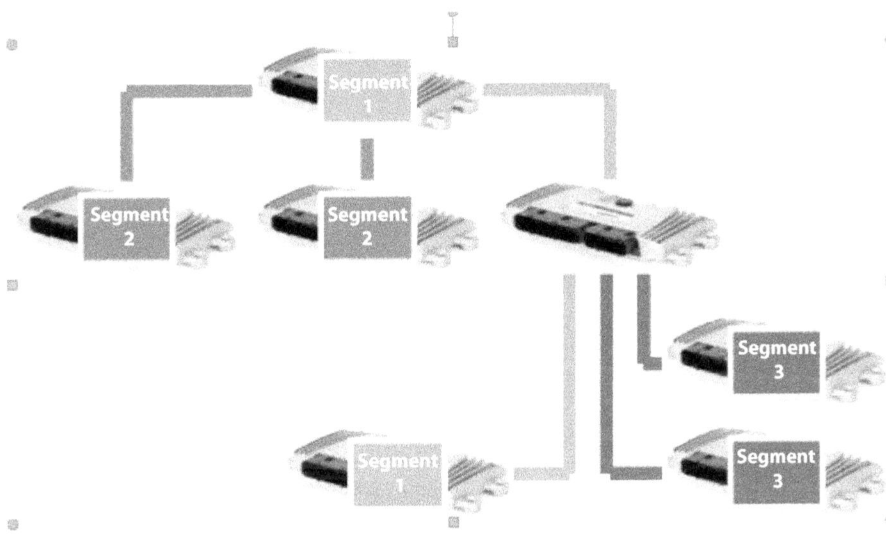

Figure 6.36 Casting modes.

segments. The data are generally not sent in broadcast mode to all network nodes, but, instead, are sent individually to each receiver (in unicast mode) or to a group of receivers (in multicast mode) – see Figure 6.36.

This means that the concept of a dynamic frame as used in CAN-FD can be employed separately for each receiver. In other words, the frame containing dynamic PDUs can only be sent on request from the receiver, which is the transmission trigger. This new topology, again, opens the door to a reorganization in E/E architectures in automobiles and a new concept of "service-oriented" communication. The objective is to take advantage of the "switch/unicast" topology of Ethernet to further reduce the amount of data received by each receiver to just those data that are useful for that receiver, and also to reduce the bandwidth devoted to each segment (see Figure 6.37).

Evolution to a dynamic service-oriented architecture

In a static architecture, the various sensors/actuators (the camera, in Figure 6.38) cannot be shared between devices. The transmitter and receivers are set in advance when the system is being designed. The exchange between the communicating elements is a transmitter/receiver relationship.

The number of cameras included in vehicles is increasing rapidly, so this fixed structure is no longer suitable.

With service-oriented design, the architecture is dynamic and therefore more flexible. The sensors/actuators can be shared among the various devices. No longer is it a transmitter/receiver relationship, but a client/server one. However, this does require that a server (i.e. the data supplier) be informed if one or more receivers need shared data (see Figure 6.39).

Evolution of E/E architecture with Ethernet and IP

The penetration and integration of Ethernet/IP serving the needs of the future has brought the concept of a backbone from the earlier FlexRay system back to light. This "service-oriented" approach means Ethernet is a good candidate to take the place of "Backbone" in the network (see Figures 6.40 and 6.41).

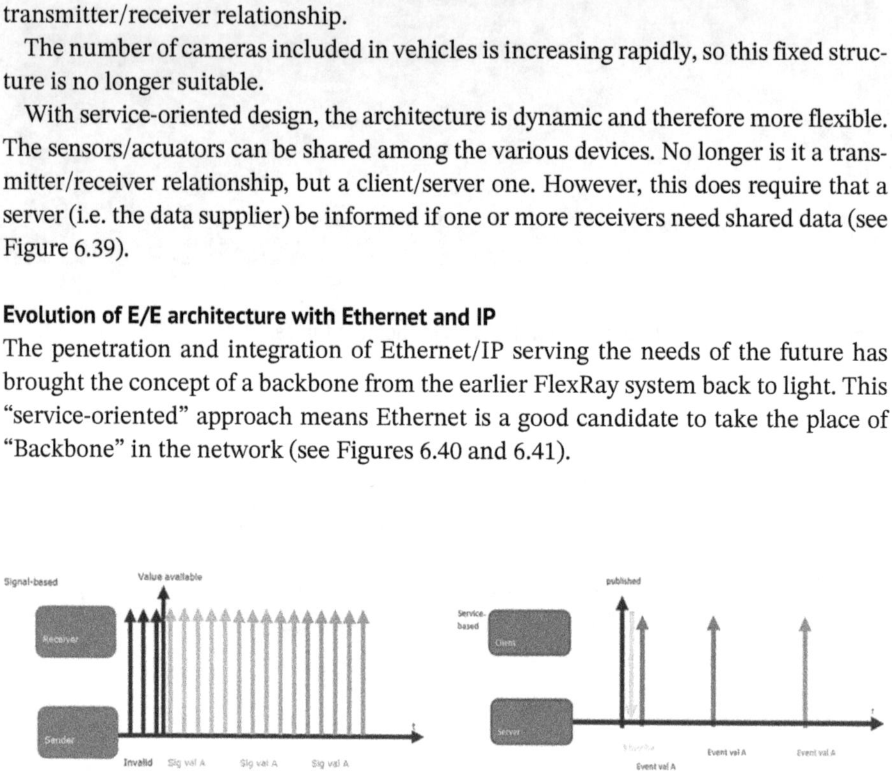

Figure 6.37 Reduction of data stream for a receiver.

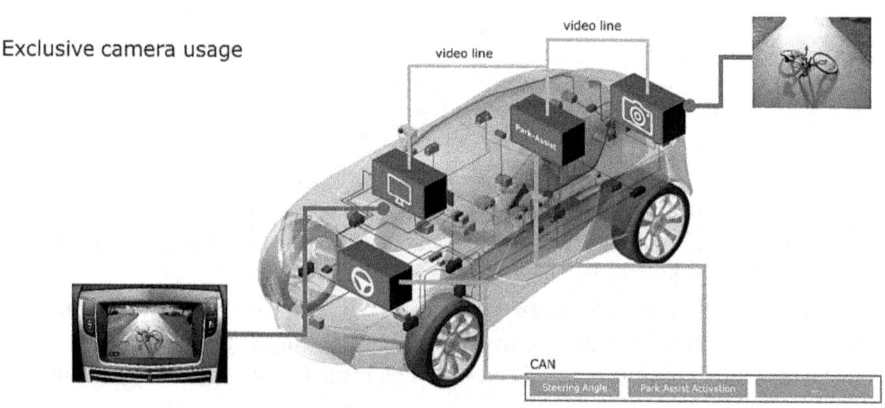

Figure 6.38 Static architecture "before".

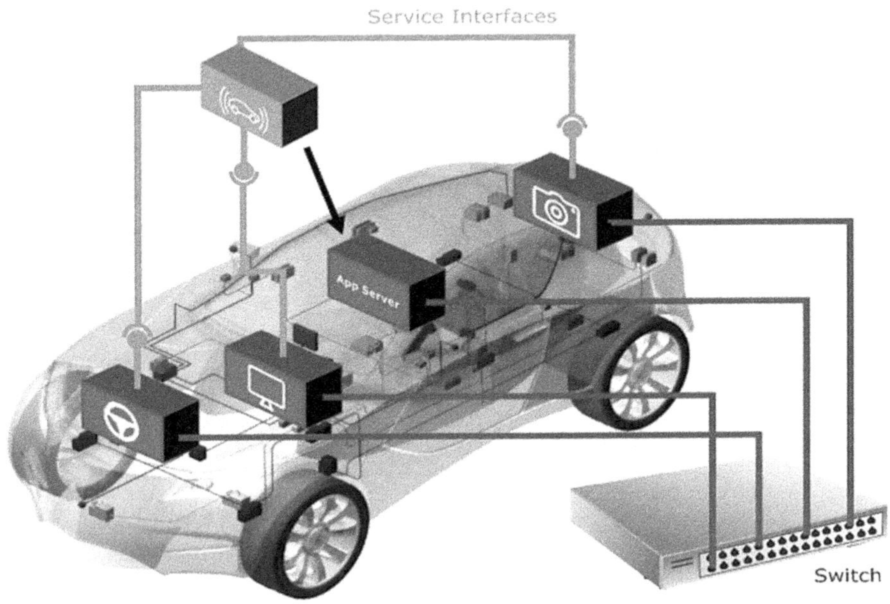

Figure 6.39 Dynamic architecture "after".

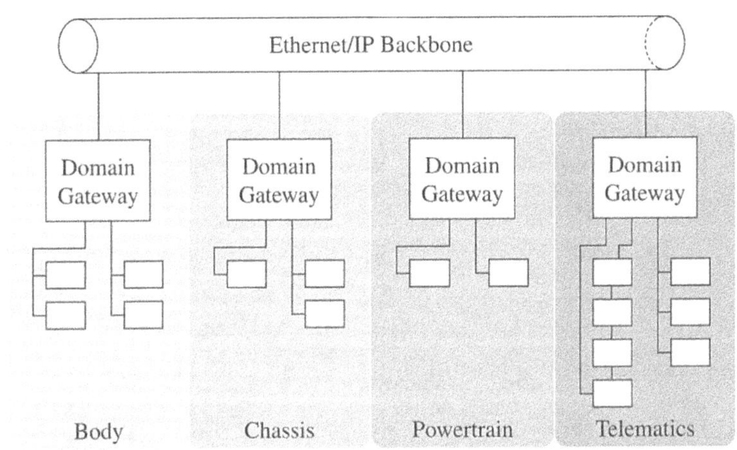

Figure 6.40 "Backbone" architecture.

Thanks to dynamic communication, the "domain controller" architecture can forward pertinent messages, which relieves the load on the network.

Evolution of AUTOSAR software architecture

The communication approach implemented in AUTOSAR 3.0 does not fully satisfy the requirements in terms of load on the CPU on the receiving side. In view of the reasons mentioned above, new protocols introduced by BMW and supported by AUTOSAR 4.3 have been created to define a new concept of "service-oriented" communication by introducing the notion of "service" (see Figure 6.42):

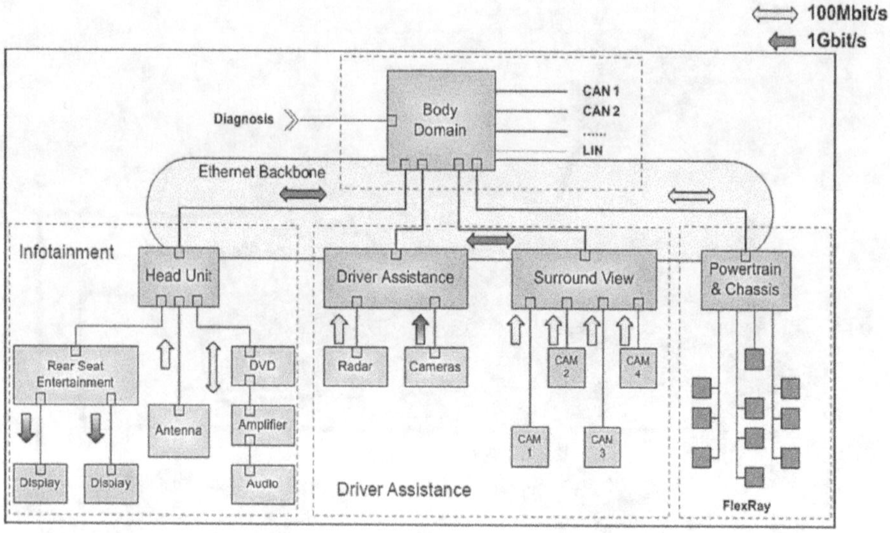

Figure 6.41 "Backbone" architecture.

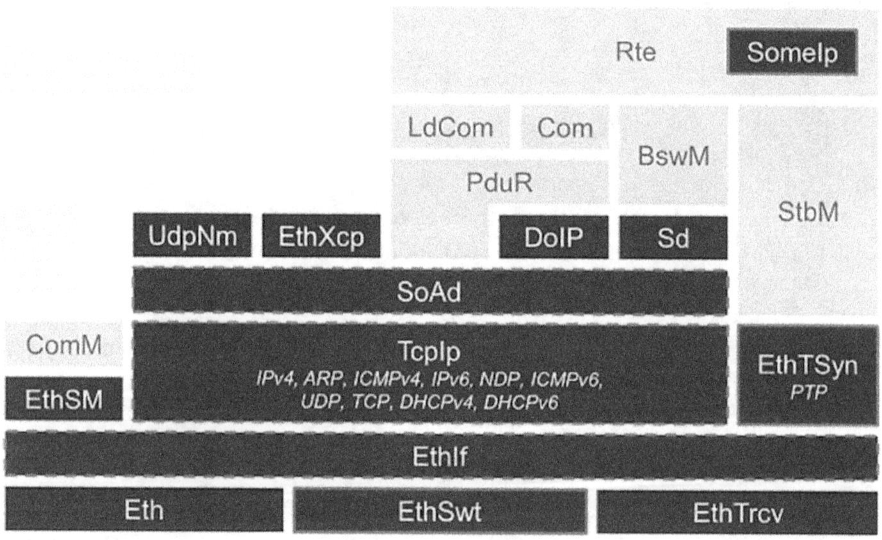

Figure 6.42 Evolution of AUTOSAR for Ethernet.

- SOME/IP (Scalable service-Oriented MiddlewarE over IP): each receiver must dynamically define its needs, so that it only receives useful information from the service providers. This consists of activation or deactivation of the data paths during runtime. The active "data paths" allow data to be routed by SOME/IP frames;
- SOME/IP/SD (Service Discovery) or "Control Path": controls the data paths. These protocols are detailed later on (see Section 6.3.5 on SOME IP).

In updates of the Ethernet stack in the AUTOSAR 4.3 versions, there is also a "Socket Adaptor" – a component that will be discussed in detail later on (see section 6.3.1

"Model of Automobile Communication") – and new building blocks of the SOME/IP and SOME/IP SD protocols are introduced. These protocols, along with DoIP and AVB/TSN, will be detailed in the next section on "Model of Automobile Communication."

6.3 Functions

6.3.1 Model of Automobile Communication

Remember that the ISO's general model of communication is based on the well-known seven layers of the OSI model. The model used for communication in automobiles is presented, in overview, in Figure 6.43.

This figure clearly shows a basic socket formed by the first four layers: Ethernet/PHY, MAC/Ethernet, and TCP/IP. This socket, or basic stack, posed a fundamental problem of data compatibility in the AUTOSAR stack. Indeed, Ethernet is a socket-oriented communication protocol (TCP/IP), while other protocols used today are PDU-oriented (see Figure 6.44). The integration of Ethernet into AUTOSAR requires an adaptation from socket to PDU orientation.

To make this adaptation, an additional layer (Socket Adaptor, or SOAD) has been introduced into the AUTOSAR layers, as shown by Figure 6.45. The SOAD interprets the IPDU header, made up of an ID and a DLC and, if necessary, adds the TCP/IP headers (Port and IP addresses).

Above this basic socket are the application layers – namely:

- The three application entities: SOME/IP, DoIP, and XCP;
- Directly above layer 2, AVB/TSN;
- Unrelated, smart charging (for batteries).

We shall examine these application parts one by one to briefly explain what they contain and how they work. Let us start with XCP.

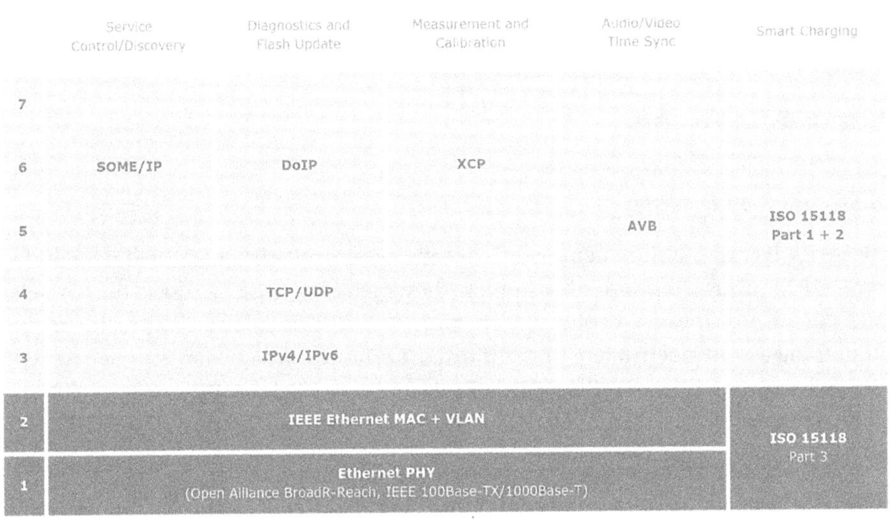

Figure 6.43 Adaptation of the OSI model to the needs of modern automobiles.

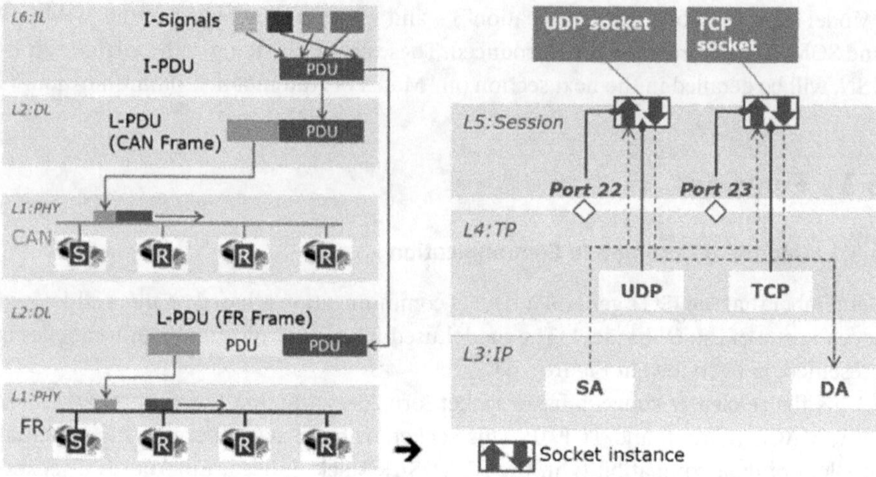

Figure 6.44 Adaptation from socket orientation to PDU orientation.

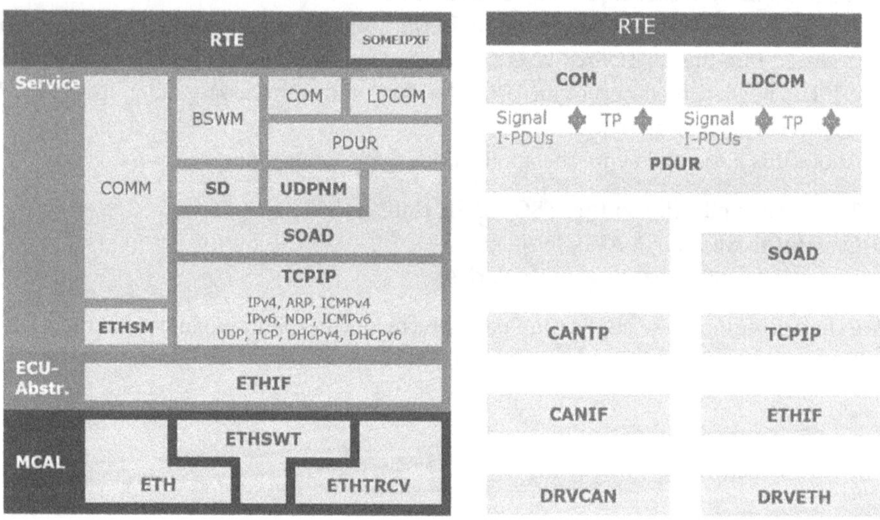

Figure 6.45 Socket adaptor.

6.3.2 XCP on Ethernet

XCP – Universal Measurement and Calibration Protocol – is a dedicated networking protocol for measuring and calibration. It allows read and write access to the variables and contents of the microcontroller memories. XCP also allows flash memory programming and is independent of the transport protocol, which was not the case with its predecessor, CCP (CAN Calibration Protocol) – see Figure 6.46.

This protocol is based on TCP/UDP to transport data (see Figure 6.47).

- UDP: is used for communication in broadcast or unicast mode;
- TCP: is used for point-to-point communication.

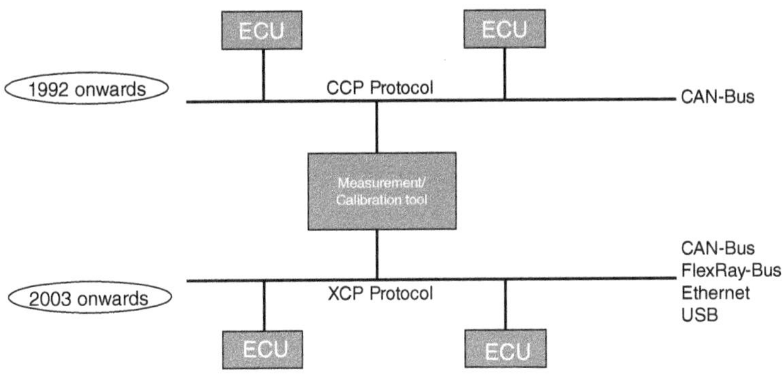

Figure 6.46 XCP.

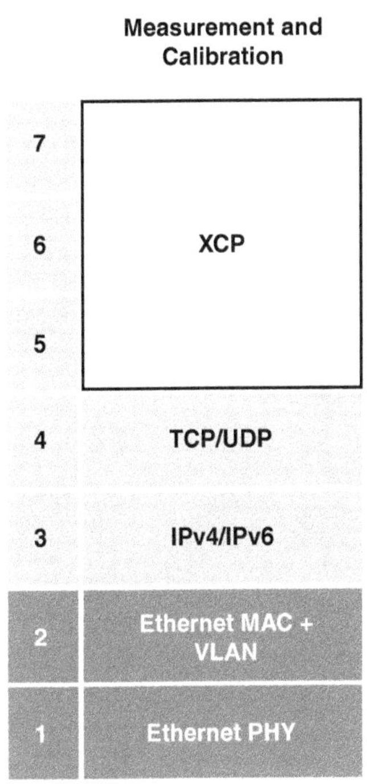

Figure 6.47 XCP on Ethernet.

Applications

- Measurement of data in the computer's memory, synchronously or asynchronously, with events triggered by timers or by other conditions (see Figure 6.48);
- Calibration: modification of the computer's internal parameters in order to fine-tune the applications.

To access data stored in memory addresses, a standardized database from ASAM (Association for Standardization of Automation and Measurement Systems) is used by XCP. The filename extension for this file, in AML format, is ".a2L", and contains the memory addresses for the various parameters (see Figure 6.49).

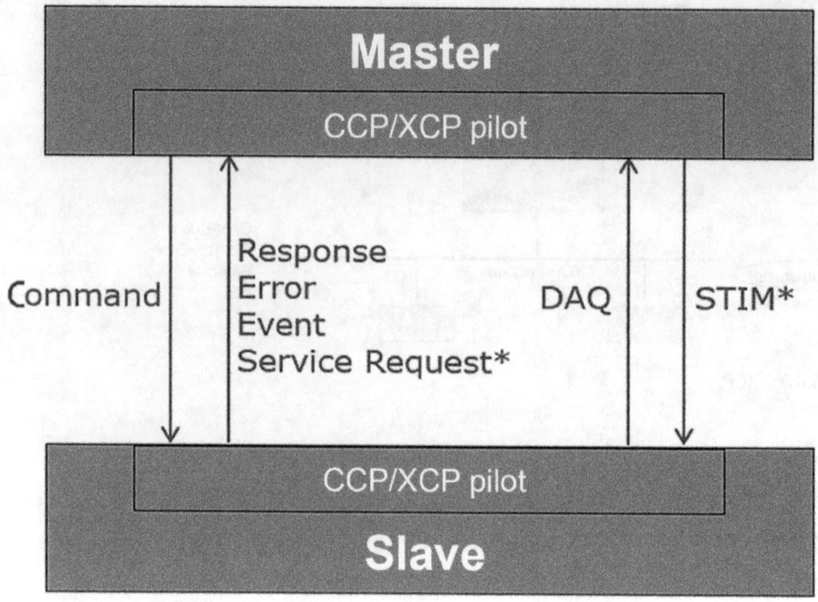

Figure 6.48 Various modes of measurement.

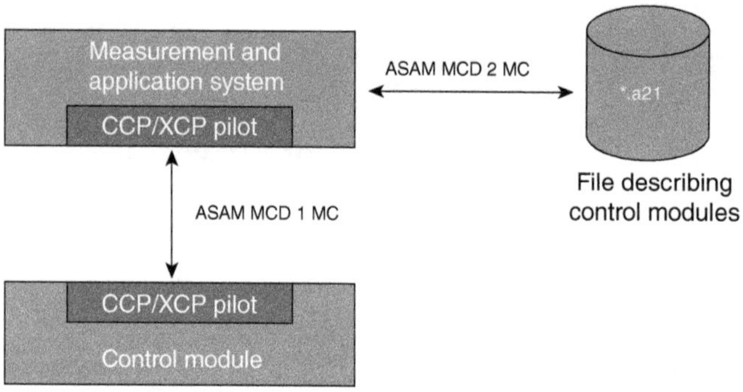

Figure 6.49 ASAM database – A2L.

The XCP frame transporting the memory data is encapsulated in the DATA field of the transport protocol (e.g. for CAN) – see Figure 6.50.

6.3.3 DoIP – Diagnostics over IP

The ISO, seeking to develop a standard for communicating diagnostics information over IP in the automotive sector, set up the working group "Diagnostics over IP" (DoIP). This group's aim is to design a standard interface separating a device's onboard communication technologies from an external testing device. The implementation of that interface is intended to switch data between different onboard networks in the

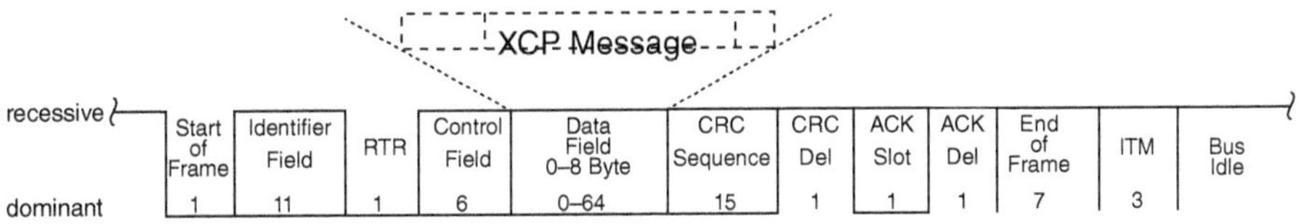

Figure 6.50 XCP on CAN frame.

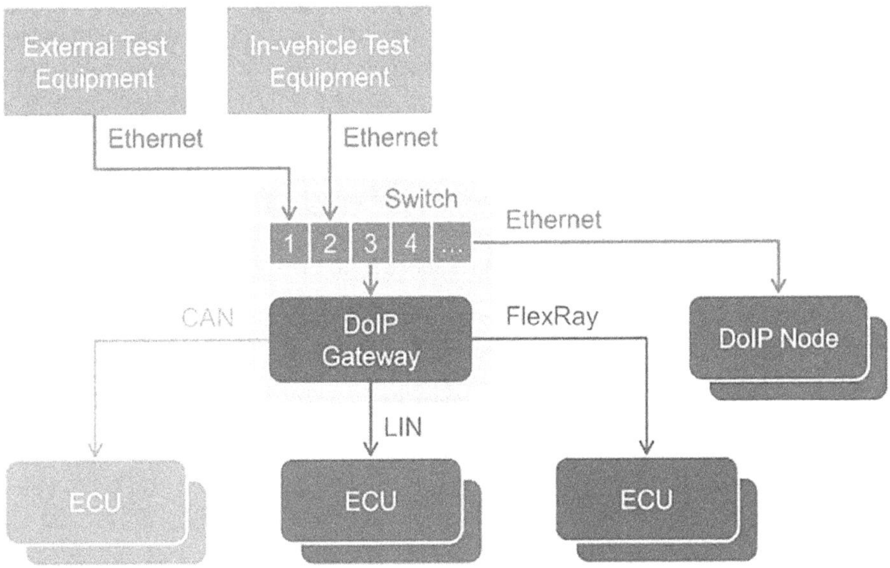

Figure 6.51 DoIP gateway.

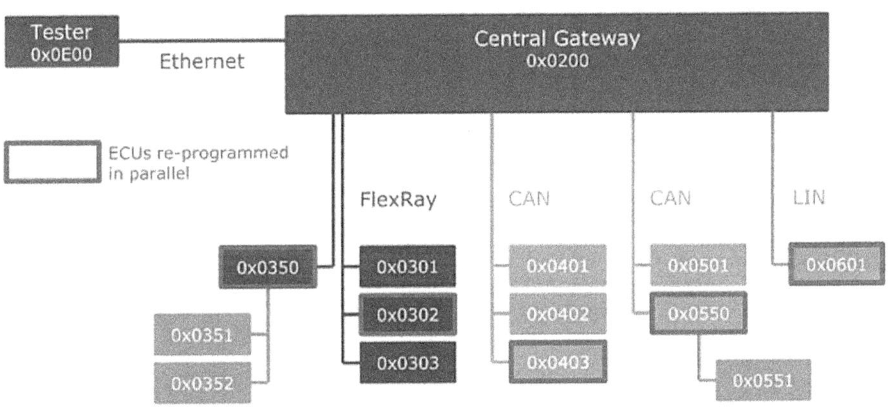

Figure 6.52 Logical addresses in DoIP.

vehicle toward the IP network of the test device, and thus works as a DoIP Gateway (see Figure 6.51).

- The tester, whether or not connected via the Ethernet gateway, uses DoIP for diagnostics;
- The addressing of ECUs is based on logical DoIP addresses, which are indicated in Figure 6.52;

- The gateway maintains the address mapping table and transfers messages under the Unified Diagnostic System (UDS).

How the system works

Let us now examine the stack needed for DoIP (see Figure 6.53):

- DoIP can use either UDP or TCP depending on the requirements – UDP for communication in broadcast mode and TCP for point-to-point. At the start of the diagnostics process is a Network Management step (see Figure 6.54) to connect the tester to the DoIP gateway. DoIP network management frames are then transmitted (we shall discuss their format later on). The first phase in this network management is identification, comprising the following exchanges:
- Vehicle announcement: the gateway sends three DoIP packets via UDP in broadcast mode. These packets serve to identify the target vehicle to the tester.
- Vehicle identification: the tester then asks for confirmation, by means of a “vehicle identification request”, followed by a response from the vehicle identified.

Once identification has been established, the tester sends a message via TCP (point-to-point) to activate the DoIP gateway (see Figure 6.55):

- Routing Activation and Diagnostic Request/Response.

If activation is successful, a UDS frame transporting the required service is then sent via the DoIP frame.

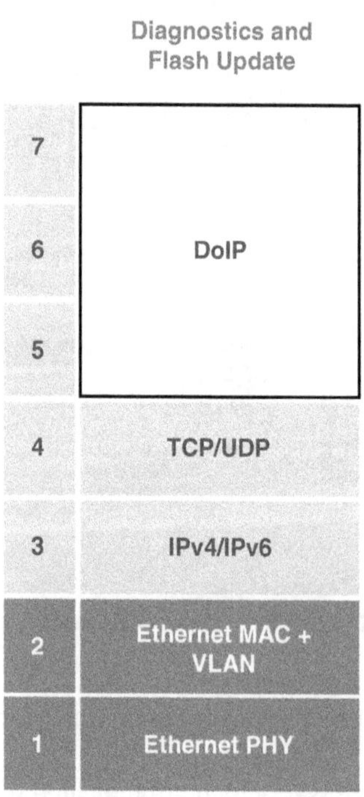

Figure 6.53 DoIP on Ethernet.

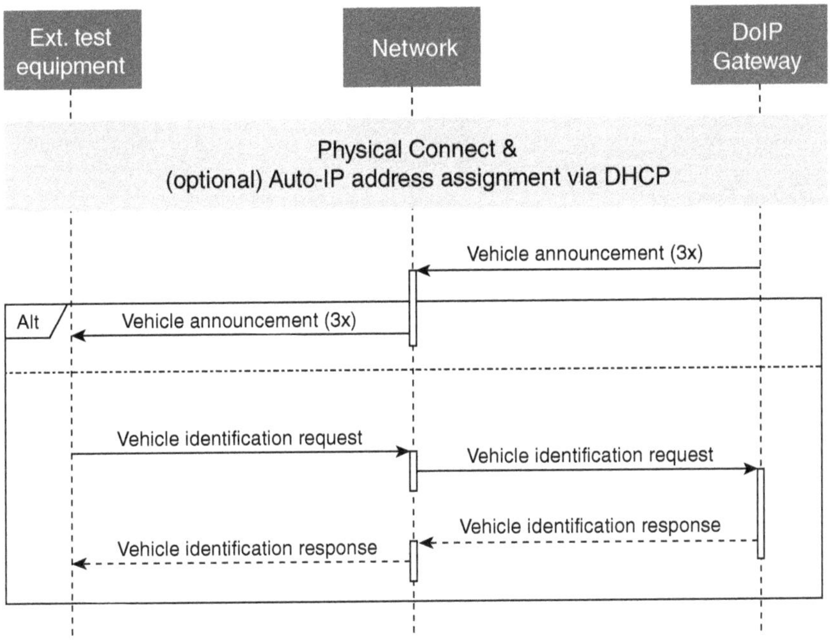

Figure 6.54 Process of identification in DoIP.

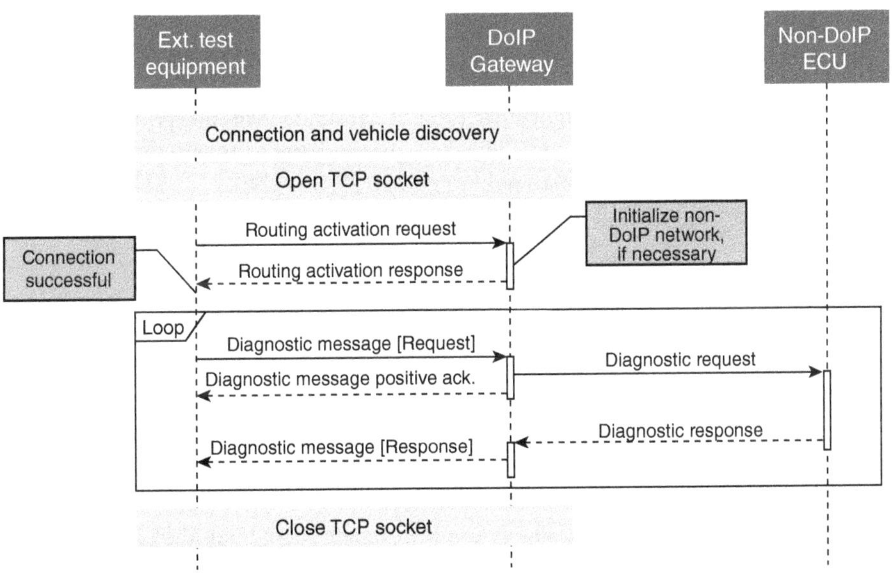

Figure 6.55 Activation of DoIP gateway.

DoIP message

In the various phases, the tester or the vehicle (gateway) transmit DoIP frames over the network – see Figure 6.56.

The "Payload Type" field indicates whether that frame is an announcement frame or identification frame, transmitted during the process of connection between the tester and the vehicle. Given that data transport takes place in UDP (broadcast) or TCP

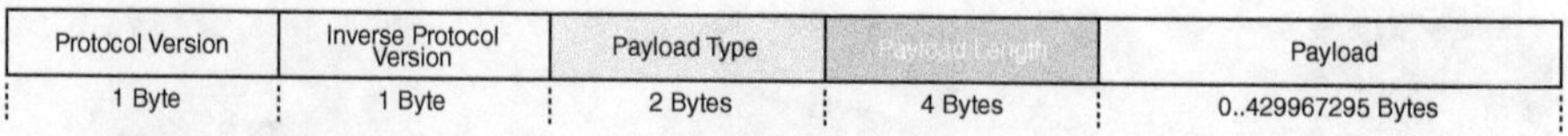

Figure 6.56 DoIP frame.

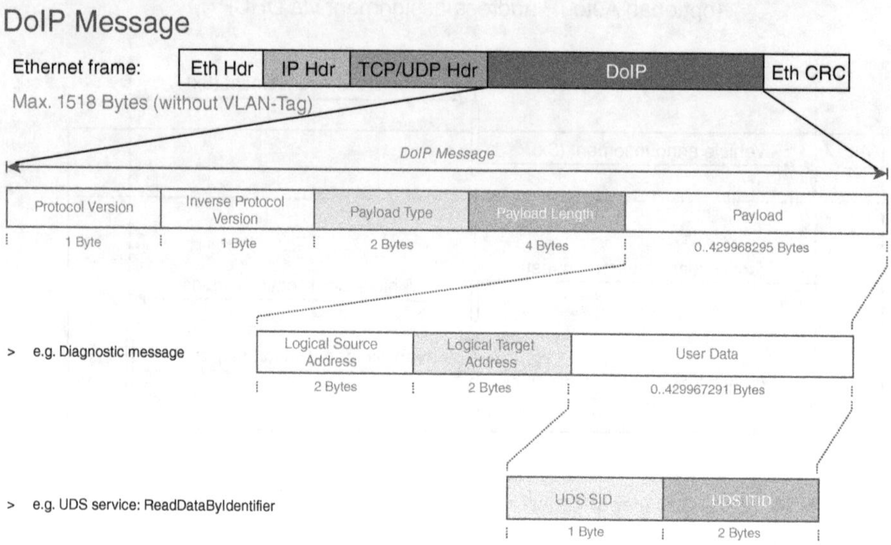

Figure 6.57 Loading and encapsulation of the DoIP message.

(point-to-point), we see the diagnostics frame encapsulated via the various upper layers – see Figure 6.57. The “payload” field contains the service transmitted by the tester or the vehicle’s response, as illustrated in Figure 6.57.

Applications

DoIP applications serve mainly:

- To quickly access the vehicle’s ECUs, with or without a hardwired connection (WLAN);
- To access the systems of existing networks and buses, such as CAN, LIN, FlexRay, and MOST, and to download or program flash memories and carry out diagnostics on the vehicle’s ECUs on assembly lines. In view of Ethernet’s high bandwidth, this protocol can also be used to reprogram the ECUs in parallel;
- Ethernet is well known and widely used, in repair shops that already have infrastructure in place. A diagnostic tester can easily be built into the dealership’s garage or repair shop;
- In the workshop, using wireless technology, a vehicle can automatically connect to a network via a hotspot.

6.3.4 AVB – Audio Video Bridging (IEEE 802.1)

As pointed out in Chapter 5, in 1980, the IEEE launched the 802 project to create standards for applications on a local area network (LAN). In 1985, IEEE 802.3 became the official Ethernet standard, as a data transmission system using a single Ethernet cable.

Designed by the "Video Bridging Task Group" of the IEEE Committee 802.1, "Audio Video Bridging" is the name of a set of technical standards pertaining to the transport of audio and video streams via a standard Ethernet network. The standards were designed to transmit audio and video streams at high speed, with a single cable, in real time and synchronized, with the concepts of priority and quality of service (low latency).

AVNU Alliance

For its part, the AVNU Alliance, set up in 2009 by Xilinx, Harman International, Broadcom, Intel, and Cisco Systems, is a consortium of professionals in the automotive and electronics industries and industrial manufacturers. It has over 60 members from all over the world, representing a broad sample of enterprises. This Alliance has worked with its members to establish certification procedures and ensure interoperability of the open-source standards Audio Video Bridging (AVB) and Time-Sensitive Networking (TSN) (see Section 5.4.2).

Applications

- Driver assistance systems with multiple video streams;
- Infotainment/multimedia applications.

AVB nodes

An AVB system contains a switch, referred to as "AVB-Bridging," an AVB-Talker (the node which is communicating), and a Listener. The bridges have more than two ports. Switches have the ability to make multiple copies of incoming images, for transmission via multiple outgoing ports (that is, multicasting) (see Figure 6.58).

AVB domain – Ethernet-compatible zone of AVB

In the AVB version, the audio/video data are not transported beyond the boundaries of the network. Therefore, the IP, TCP, and UDP use an LAN. In this type of system, a single transport layer is needed to transport the streams to the Ethernet layer. AVTP (Audio/Video Transport Protocol) is used (see Figure 6.59). Other transport protocols

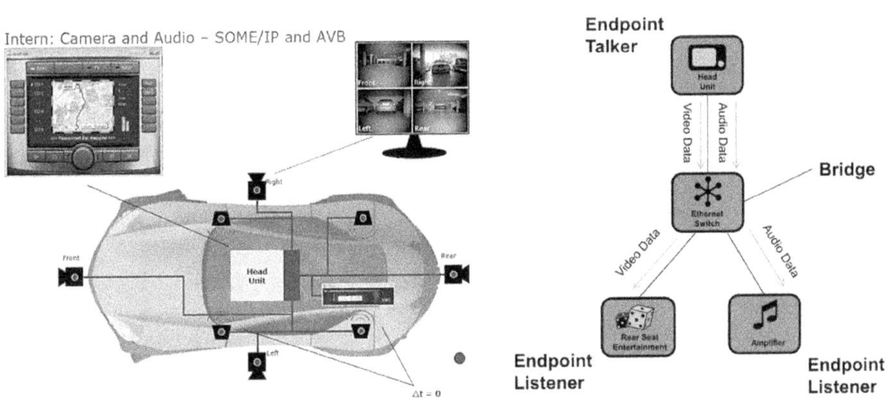

Figure 6.58 Examples of AVB networks.

are also used, such as RTP/RTCP (Real-time Transport/Control Protocol), which is based on the UDP/IP layers.

Audio/Video Transport Protocol – AVTP

The specifications for this transport layer are also set out in IEEE 1722(a). This transport standard supports raw video and raw audio, and compressed data conforming to IEC 61883, parts 1 to 8.

- 61883 – 2 SD – DVCR
- 61883 – 4 MPEG2 – TS compressed Video H264
- 61883 – 6 Uncompressed Audio
- 61883 – 7 Satellite TV MPEG
- 61883 – 8 Bt.601/656 Video

The earliest version of an AVB node contained only that transport layer, as illustrated in Figure 6.60.

This configuration did not take account of synchronization issues and, consequently, the applications were vulnerable to losses, stream interruptions, and random latency.

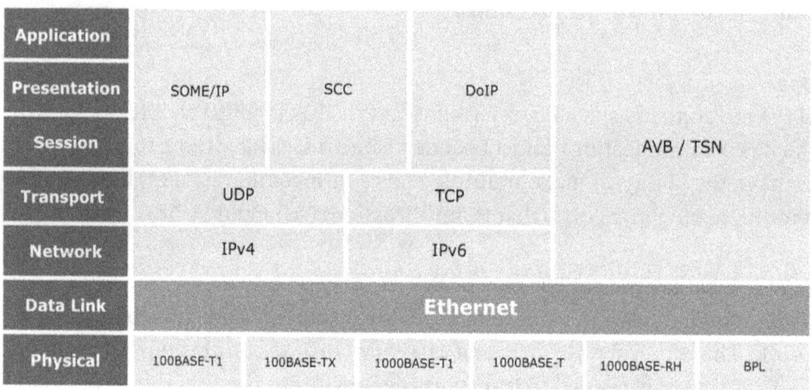

Figure 6.59 Position of AVB/TSN in the OSI model.

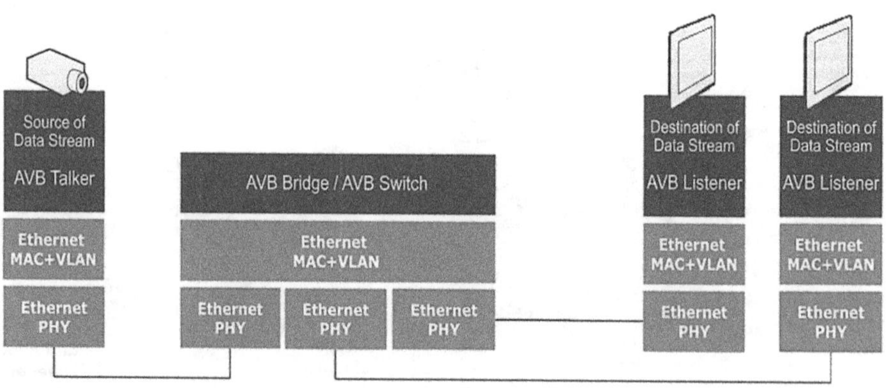

Figure 6.60 The earliest version of an AVB node.

For automotive applications where the safety constraints are most stringent, such as in ADASs, these limitations are unacceptable, and end-to-end transmission times must be calculated deterministically. In the interests of better quality of service (QoS), extremely precise synchronization is required to avoid any problems with stream losses and interruptions. In this context, the latency of critical, high-priority streams across a maximum of seven "hops" on the network must be no more than 2 ms at 100 Mb/s and 1 ms at 1 Gb/s between a transmitter and a receiver.

Quality of service – QoS and AVB/TSN

With the implementation of QoS, AVB gave rise to the generation AVB/TSN – time-sensitive networking (see Section 5.4.2). A range of additional protocols have also been produced to ensure uninterrupted audio and video synchronization among multiple Talkers and Listeners on a switched network. The extensions needed for AVB to comply with IEEE 802.1 are as follows:

- IEEE 802.1BA: Audio Video Bridging (AVB) Systems;
- IEEE 802.1AS: All clocks are permanently synchronized to within 1 μs: PTP/gPTP;
- IEEE 802.1Qat: Stream Reservation Protocol (SRP) to prevent cutoff;
- IEEE 802.1Qav: Forwarding and Queuing for Time-Sensitive Streams (FQTSS).

Software components, which were initially designed for an AVB system that was not real-time critical, as shown in Figure 6.61, are evolving towards a new form of distribution, and providing distributed QoS, across each node (see Figure 6.61A). Figure 6.61B shows the positions of each layer in the software stack.

Synchronization process – PTP/gPTP (Precise Time Protocol)

- PTP/gPTP ensures that the clocks of all nodes within an AVB domain are synchronized at all times. The first step in this process is to define a GrandMaster node – that is, a point of reference for all the nodes in the domain. The algorithm BMCA (Best Master Clock Algorithm) is used to select the GrandMaster. In automotive applications, the GrandMaster is often defined in advance and PTP distributes that clock's time across the whole of the AVB domain – see Figure 6.62.

PTP synchronization frames are then sent cyclically:

- Every second, to measure the signal propagation time through the various network segments. This calculated time is transmitted with each audio/video stream to inform the addressees of the critical delay. As Figure 6.63 shows, this information can be used to synchronize the sound with the images, transmitted along different paths but needing to be displayed or broadcast at the same time;
- Every 125 ms to adjust the clock offset in relation to the GrandMaster.

Stream reservation protocol – SRP The talker in an audio/video stream must ensure that sufficient resources are available on the bridge and the listeners (buffers, bandwidth, etc.). It is necessary to have this absolute guarantee, to avoid any loss or cutoff of the stream. SRP uses specific frames to request information from all listeners to which the talker needs to transmit, to ensure resources are available. No stream will be sent without first being accepted by the listener, and without confirmation that the

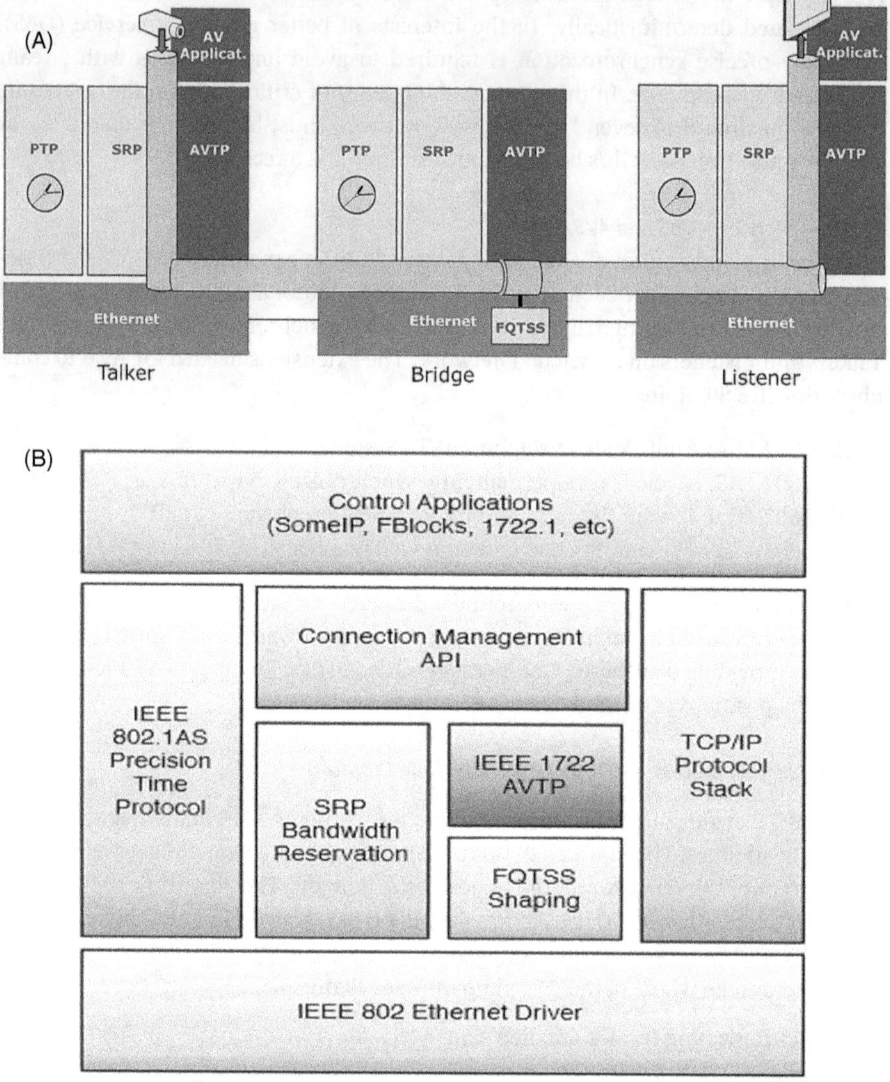

Figure 6.61 Exchange between a "talker" and a "listener", and the structure of the software stack.

space needed to receive it is available at the other end. A maximum of 75% bandwidth is allocated to high-priority streams (see Figure 6.64).

FQTSS Once the resources have been reserved, the various streams are transmitted via the various channels. At that point, the bridge needs to decide on the scheduling of the streams of differing nature and levels of priority (see Figure 6.65).

FQTSS classifies the streams into three categories (see Figure 6.66):

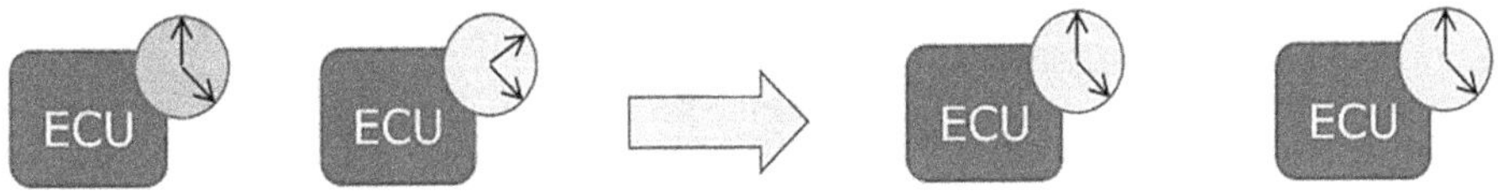

Figure 6.62 Clock offset in relation to the grandmaster.

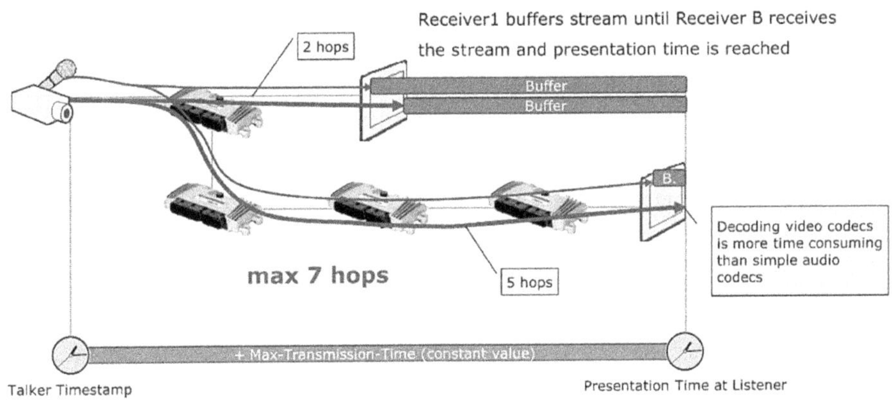

Figure 6.63 Sound and video synchronization.

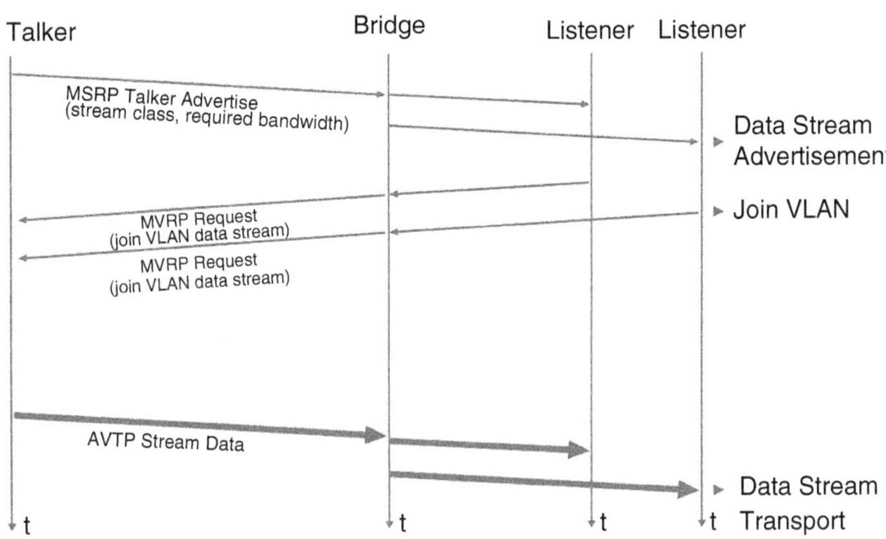

Figure 6.64 Checking resource availability and stream reservation.

- High-priority streams (classes A and B) are transmitted without any condition (Time-Aware transmission);
- Lower-priority streams are transmitted within an allotted transmission window (Credit-Based transmission). That credit diminishes as the frames are transmitted;

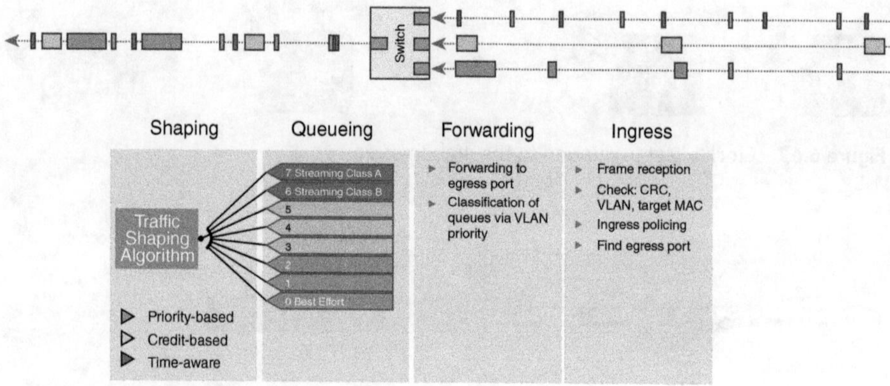

Figure 6.65 Scheduling of feeds with differing levels of priority.

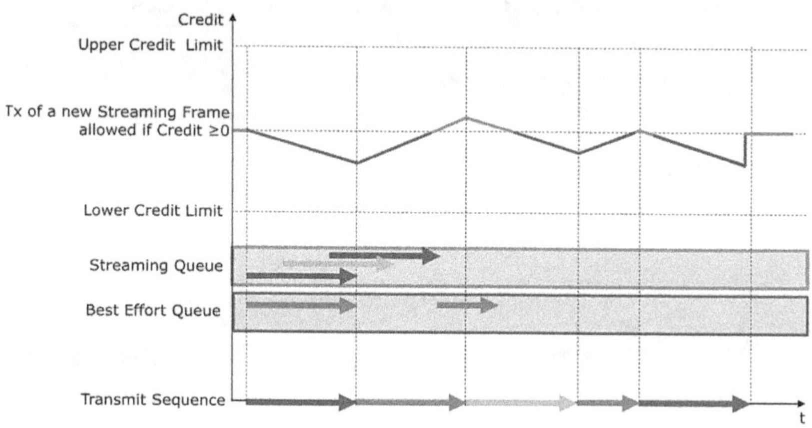

Figure 6.66 FQTSS and category-based scheduling of streams.

- Non-priority frames, referred to as "Best Effort" frames, are transmitted upon expiry of the credit allotted to the previous types of frames. This approach helps prevent significant latency in the delivery of these frames.

AVTP Ethernet frames

The AVB stream is a VLAN frame that defines its level of priority (see Figure 6.67):

- Class A VLAN Priority = 5 or 6, sampling 125 μs
- Class B VLAN Priority = 4, sampling 250 μs
- Class C VLAN Priority < 4, sampling 1 ms

The Ethernet frame, with the MAC address of the Talker and the Multicast address of the Listeners, transports the AVTP stream complying with IEC 61883 (see Figure 6.68).

The header of this AVTP packet contains (see Figure 6.69):

- The Stream ID: the identifier of the audio/video stream;
- The "Present Time" calculated by PTP, which indicates to the listeners the exact time at which the stream was broadcast;
- The stream in question, the type of which (IEC 61883) is defined by the "subtype" field in the header.

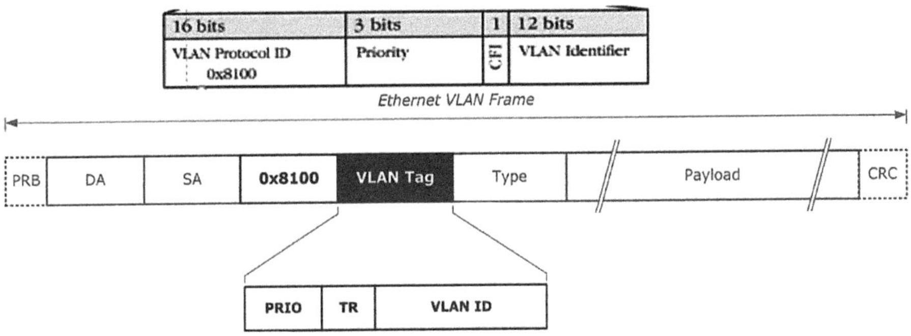

Figure 6.67 Format of a VLAN frame.

Figure 6.68 Detailed view of a VLAN frame.

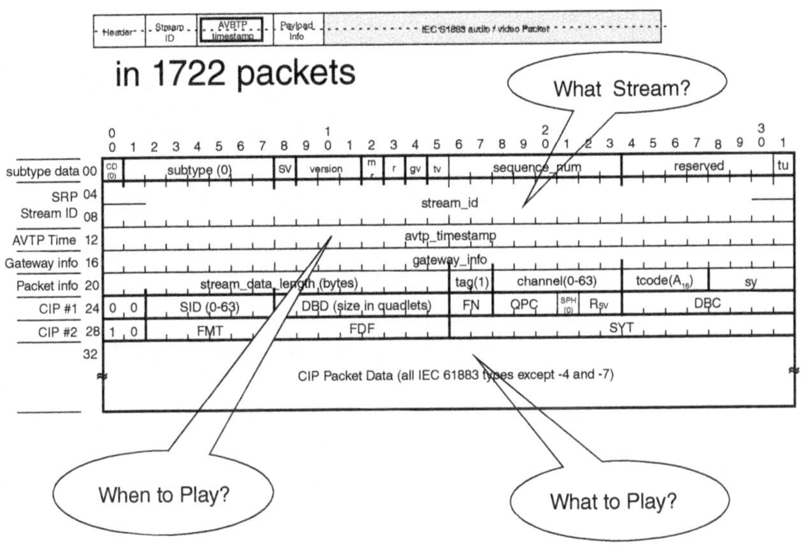

Figure 6.69 Header of AVTP packet.

6.3.5 SOME/IP – Scalable service-Oriented MiddlewarE over IP

Designed by BMW in 2011, the concept of service-oriented communication – SOME/IP – was created as lightweight middleware to cater for the needs of the automotive industry. In particular, it was intended to serve for ADAS and infotainment applications, replacing MOST. It can be implemented with a range of different operating systems (AUTOSAR, GENIVI Linux, and OSEK).

How it works

Figure 6.70 shows the position of SOME/IP within the OSI layered model.

In this design, it is important to distinguish between server nodes (or service providers) and client nodes (or service consumers).

- Any server node offering a service announces itself to a network group (using a multi-cast IP address);
- Any client node seeking to avail itself of a service will first join that multi-cast group, and can then access the services offered by the group.

The obvious advantage of service-oriented communication is that it prevents any one communication channel from being inundated unnecessarily with data. This method significantly reduces the bandwidth on the segments and the receiver-side CPU workload. Hence, two protocols are needed:

- SOME/IP, which provides services and transports data: this is known as Data Path;
- SOME/IP-SD, which dynamically activates or deactivates communication between various nodes during runtime: this is known as Control Path.

In order to properly understand this principle, it is important to define the following components:

- The protocol SOME/IP;
- How the data are sent over the network: serialization/deserialization;
- The protocol SOME/IP-SD.

SOME/IP and services

Let us begin by defining what a service is and how it is transmitted:

- In SOME/IP, a service sent by a server is simply an exchange interface between the different nodes, containing information such as the transmitter ID, the service being provided, and the service data;
- A client wanting to use this service initially requests to subscribe, by sending a SOME/IP-SD message corresponding to its request.

Service-oriented communication model

In order for a client to receive a service, it must first ask for that service, either asynchronously or synchronously (see Figure 6.71).

- Asynchronous request or method: no need for subscription to the service. As subsets of this method, we can distinguish between:
  - **RPC** – Remote Procedure Call (Request/Response): the client queries the server for the result of a function and awaits a response. An example of this would be a mathematical addition;
  - **Fire and Forget**: the client "fires off" a request to the server to activate a particular function and does not await a response. An example would be activating the vehicle's air conditioning;
  - **Field Get/Set**: the client requests the server to either "Set" or "Get" the value of a given field, and awaits a response. An example would be to set or get the camera frame rate.

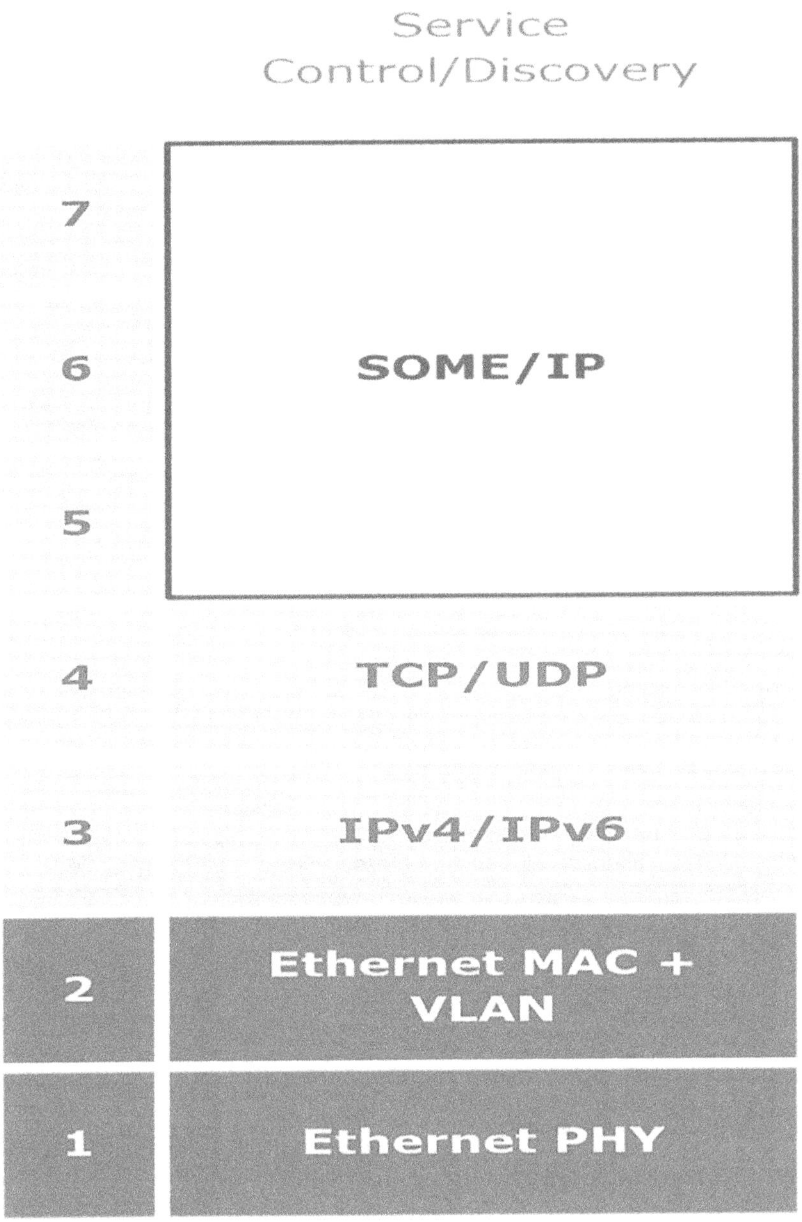

Figure 6.70 SOME/IP over Ethernet.

- Synchronous request: the client must subscribe to the service. We can distinguish between:
  - Event Notification: the client receives a notification when an event corresponding to the service occurs: e.g. the vehicle encounters a road sign;
  - Field Notification: when a Set/Field is performed, the client receives a notification of the change in the value of the Field. For example, a client requests to change the "Frame Rate" field by means of a Set/Field operation; if a client is subscribed to Notifications for that Field, it will be sent the new value of the "Frame Rate".

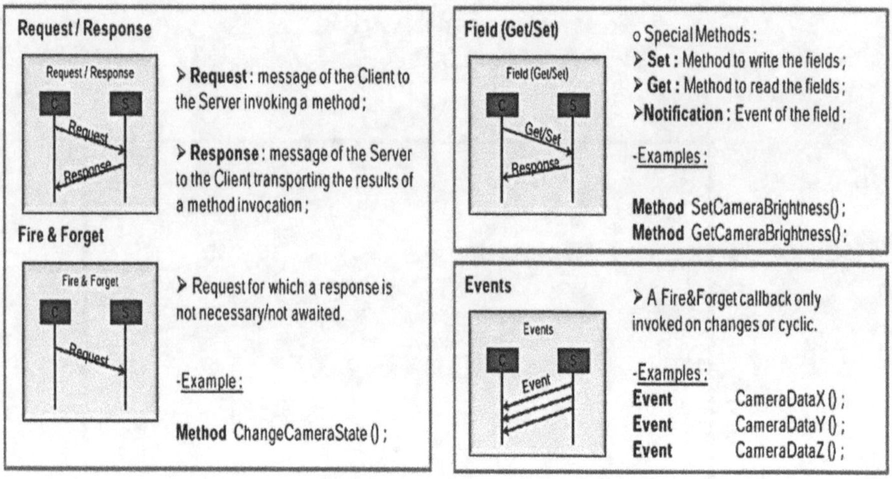

Figure 6.71 Synchronous and asynchronous service exchanges.

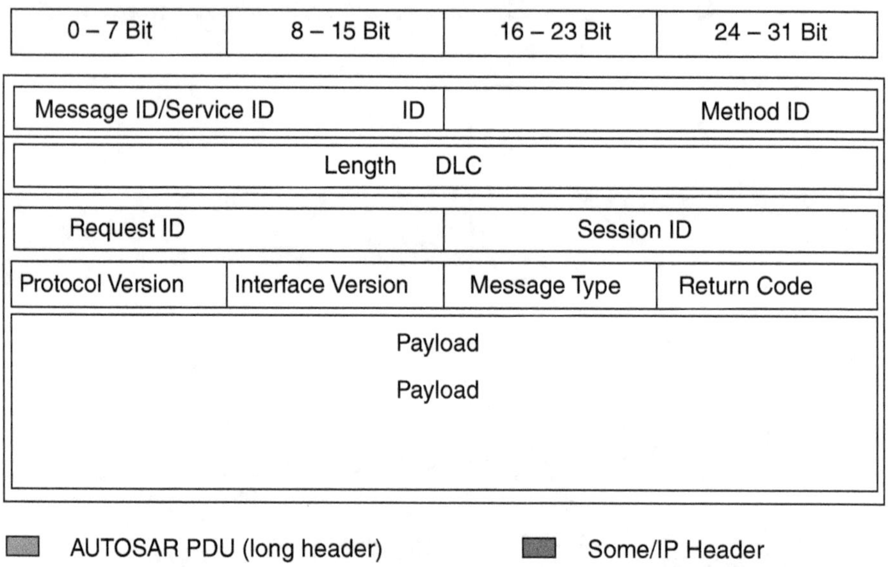

Figure 6.72 AUTOSAR PDU and PDU header.

Header and data field of a service

For the purposes of interoperability (see Figure 6.72), this interface was designed to match the format of the header of an AUTOSAR PDU. The header in the interface contains the ID and the DLC, as does any conventional PDU, produced by the AUTOSAR communication layer. The ID is composed of the service's unique number and the method ID. A given service ID may have multiple corresponding method IDs. The DLC indicates the total packet payload.

One part of the header (the Client ID) is extracted by an AUTOSAR layer "SOMEIPXF" – see Figure 6.73. The PDU "Socket adaptor" layer switches between a PDU socket and an Ethernet socket, as required.

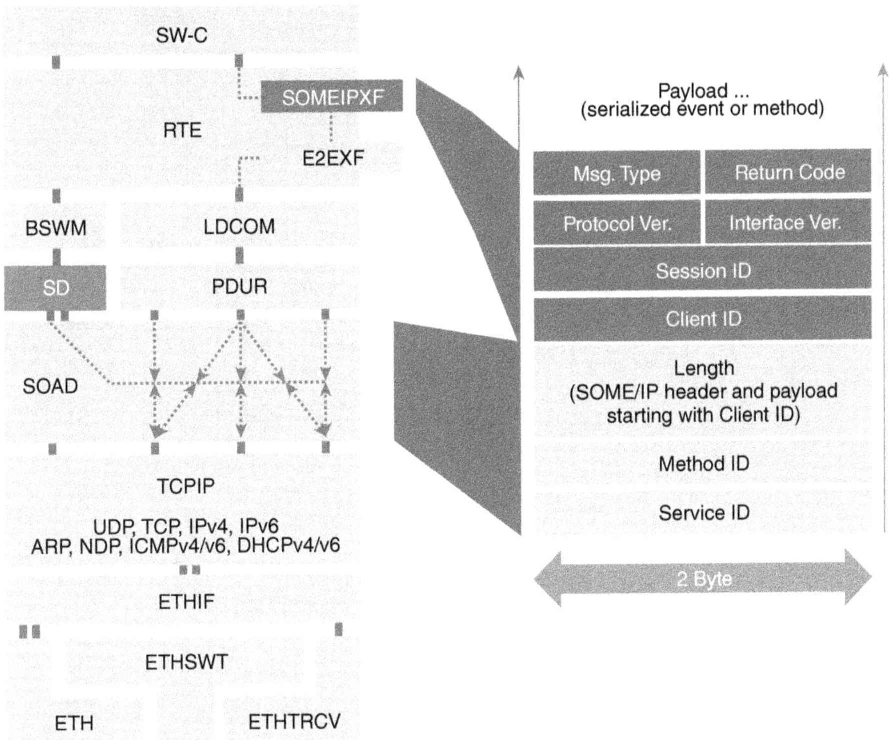

Figure 6.73 The SOME/IPIP header.

- The "Request ID" or client ID is unique to each transmitter or client;
- The "session ID" corresponds to the request for the service and is incremented each time a request is made;
- The "Message type" field represents the method used to send the service: Notification, Request/Response;
- The "payload" defines the useable data in the service transmitted by the server. The data field is not fixed in advance – it depends on the data size, as a camera may detect, for example, 10 or 20 road signs.

Serialization/deserialization of a service

Let us now examine how the structures and contents of the header and payload fields in SOME/IP are transmitted over the network. As shown in Figure 6.74, serialization is a means of flattening structures that may otherwise be highly complex.

A SOME/IP message is dynamic and may contain one or multiple parameters. In order for the "deserializer" to know the exact position and length of the parameters in the payload received, it is important to maintain some form of understanding between the serializer and the deserializer – see Figure 6.75.

One of the solutions to establish understanding between the serializer and deserializer is to use a "database file" containing the description of the configuration of the computers. All applications wishing to access a computer begin by reading that database and update the structures in order to access that computer (see Section 6.4.1).

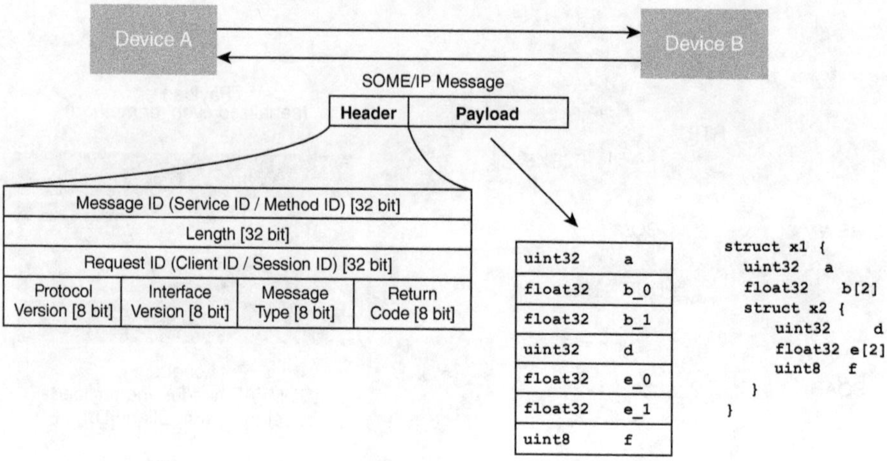

Figure 6.74 SOME/IP frame serialization.

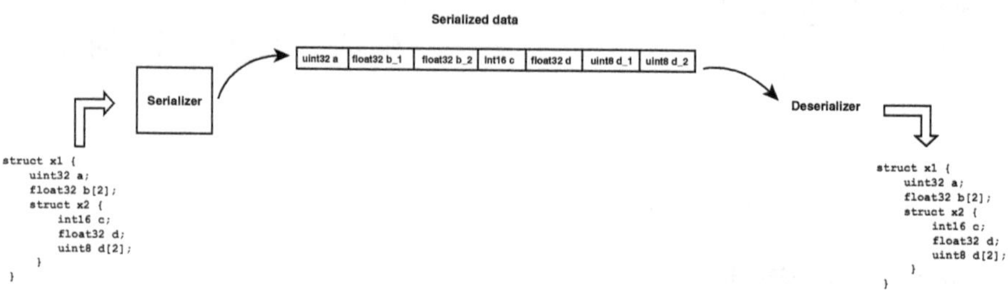

Figure 6.75 Detailed view of serialization.

SOME/IP Service Discovery – SD

The Service Discovery protocol creates links of communication during runtime. Notification events are only transmitted to receivers that have subscribed to event-triggered services.

Type of messages in SOME/IP-SD

Let us begin by defining the various types of SOME/IP-SD messages. As Figure 6.76 shows, there are:

- Offer (Publish) messages: multicast over the network by any server on the network of the group to which it belongs;
- Find messages: multicast by any client that has not seen an Offer on the network of the group to which it belongs;
- Subscribe message: unicast by a client to subscribe to an offered service;
- Subscribe Ack: unicast by the server to acknowledge the subscription.

Communication is a client/server relationship. It includes (Figure 6.76):

- Service Provider: a server, provides information about the existence of services to the Broker;
- Service Broker: a network of groups of ECUs, registers the services and enables all clients in the group to find those services;

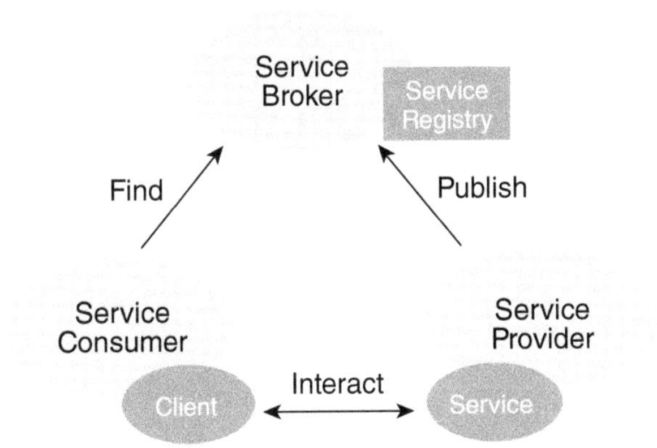

Figure 6.76 Messages between the client/server and the group.

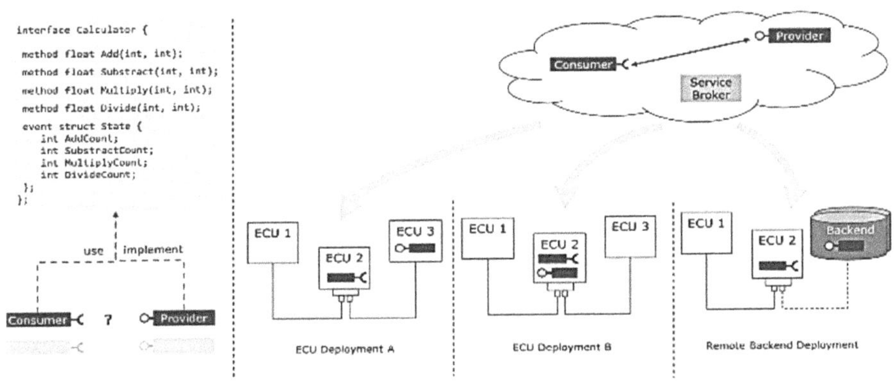

Figure 6.77 Multiple options for deployment of a service.

- Service Consumer: is helped by the Broker to locate the services and thereafter use a Service Provider. A service consumer may access multiple service providers or servers.

Service localization

Figure 6.77 shows the multiple options for deployment of a service. The providers and consumers may be hosted on the same machine, but may also be moved onto a back-end machine.

Communication model for SOME/IP-SD messages

When a network starts up, the SOME/IP nodes, both servers and clients, go through a number of phases. We can distinguish three main phases, as shown in Figure 6.78:

- Init Phase: there is a certain period during which the servers and clients become available.
- Repetition Phase: the repetition phase lasts for a certain period, which is specified at the design stage. During this phase, any client declares itself with a "Find" message, which allows it to find the services it needs. If the corresponding offer is found

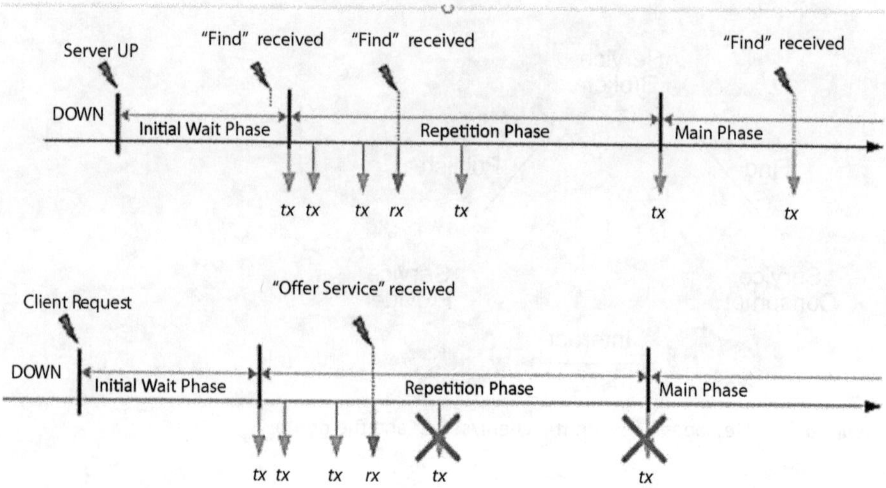

Figure 6.78 Phases of operation of the service discovery protocol.

during this phase, the client registers it or subscribes to an event-triggered service, and immediately enters the main phase (see below). The server completes its repetition phase, defined at the design stage, and, in turn, goes into the main phase.

- Main Phase: this is the phase of normal operation in which any server will broadcast offers in cyclical fashion. The Find messages no longer need to be sent in this phase, unless the offer transmitted in this phase has expired.

One of the problems with this transmission model lies in the fact that the cyclical offer is not a guarantee of non-latency for the client. Indeed, a client can only subscribe to a service if it perceives the offer broadcast over the network at the right time. Although this system exhibits a number of disadvantages, many users have adopted it because of its excellent flexibility – particularly in event-triggered areas.

The SOME/IP-SD header and data field

The header of the SOME/IP-SD PDU is the same as that for SOME/IP. The Service ID/Method ID field contains the fixed value: 0xFFFF8100 (see Figure 6.79).

In a data field in SOME/IP-SD, the various types of SD messages pertinent to the ECU are stacked: Find, Offer, Subscribe, Subscribe ACK, Stop Offer, etc. Another data field contains the service addresses (the UDP/TCP port number and the IPV4/IPV6 address).

IGMP – Internet Group Management Protocol

IGMP is used to manage groups, allowing services to be provided to the members of a group using the IP address and Multicast MAC address (see Figure 6.80).

As explained above, the Offer/Find messages are multicast over the group network. Multicast IP addresses are reserved by the class D group IANA: a multicast address always begins with "0xE..." (see http://www.iana.org/assignments/multicast-addresses).

Below are a few examples of the IP fields in multicast addresses:

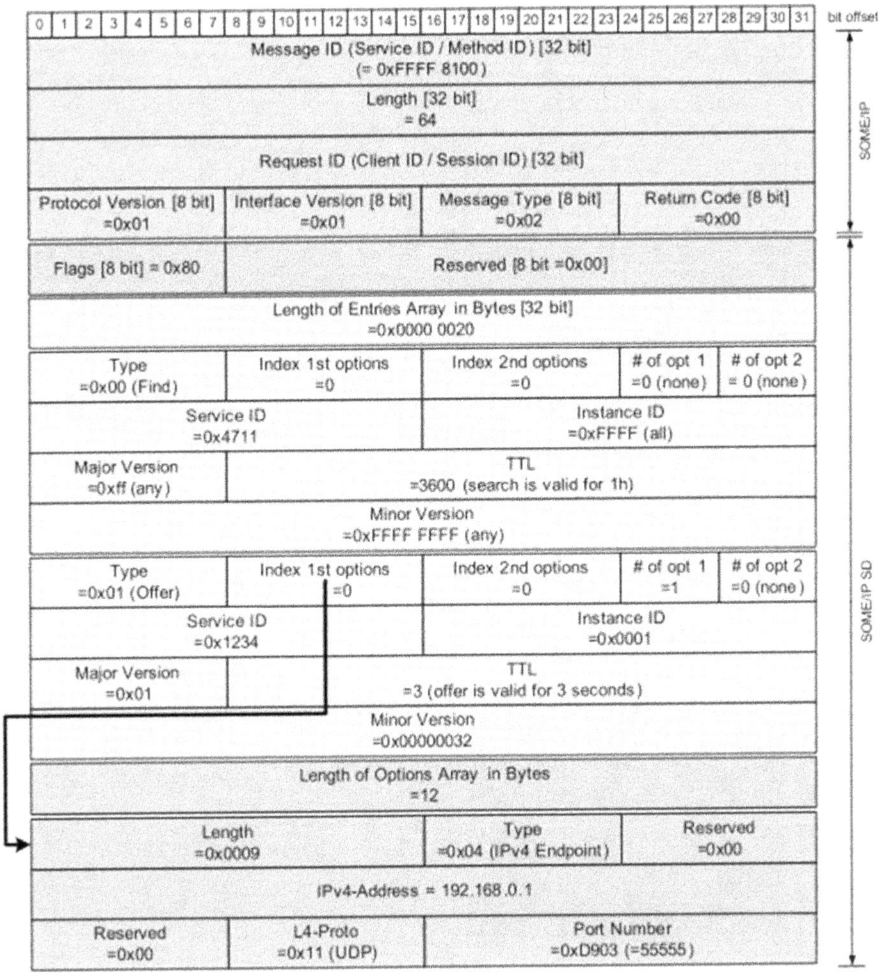

Figure 6.79 Example of a PDU in SOME IP SD.

- 224.0.0.0 – 224.0.0.255
- 232.0.0.0 – 232.255.255.255
- 239.0.0.0 – 239.255.255.255

Each multicast IP address has a matching multicast MAC address (see the example in Figure 6.81).

Evolution of software architecture: Classic and Adaptive AUTOSAR

The future applications mentioned above have slightly different requirements. For example, with regard to the vehicle's internal data from multiple sensors – cameras, radars and lidars – CPU performance and real-time operation are essential safety criteria. This means that critical services such as emergency braking can be provided in a

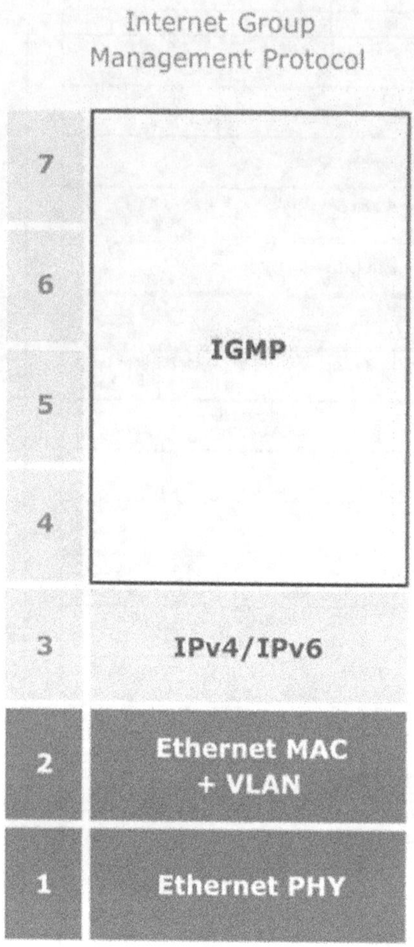

Figure 6.80 IGMP on Ethernet.

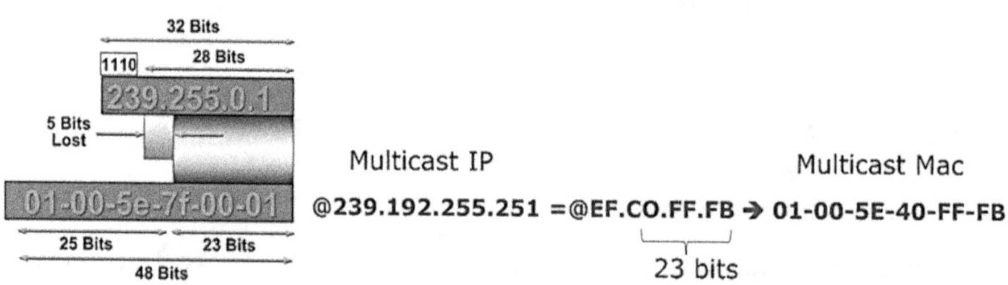

Figure 6.81 Multicast IP and MAC addresses.

timely fashion. The Classic AUTOSAR standard describes how the ECU software should be configured to deliver these critical services. The functional configuration is set in stone after the development phase, so the execution model is static.

Today, with ADAS, the data may actually be transmitted beyond the confines of the vehicle, for communication with infrastructures or other interlocutors. Therefore, security is needed – in the sense of cybersecurity – to guard against hacking. These data must be encrypted with strict management of multiple keys. The intention, in opening up to the outside world, is to make the ECU configuration dynamic rather than static. Adaptive AUTOSAR, which was launched in March 2017, manages the evolution of applications over time – notably by means of over-the-air (OTA) updates. Thus, this platform interfaces with the outside world (a Cloud service) and, on the various ECUs, hosts dynamic applications (or services), which can be flashed or downloaded depending on what the end user needs, as can be done with smartphones such as the iPhone (see Figure 6.82). In this platform, there is not a single stack, and the communication paths are established during runtime rather than being set statically

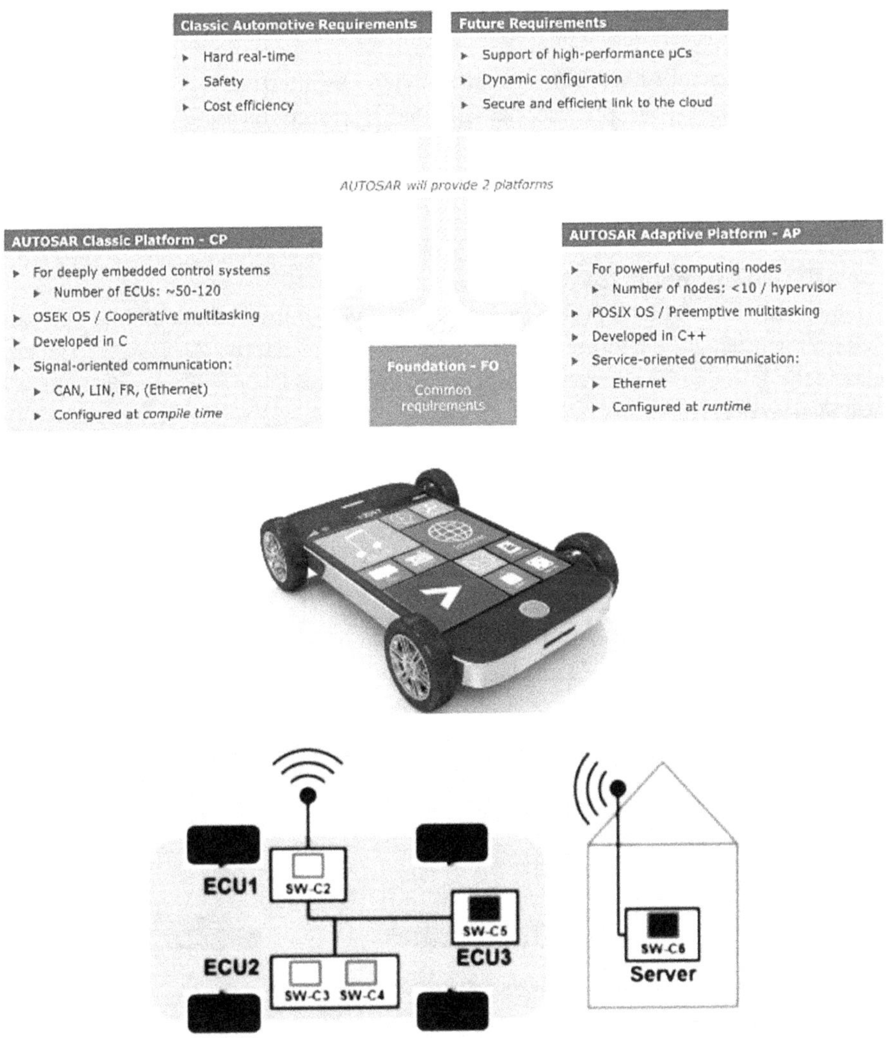

Figure 6.82 AUTOSAR: adaptive platform (AP) and classic platform (CP).

at the system design stage. This Adaptive version also provides better handling of DoIP (Diagnostics over Internet Protocol), based on compliance with ISO 13400-x. Here, basic DoIP services are built in to provide routing operations and are available to provide the vehicle identification number (VIN) and the group ID (GID).

Version 4 of the Classic platform includes extended serialization of data structures for SOME/IP (Scalable service-Oriented MiddlewarE over IP). This evolution simplifies data communication and improves compatibility between the two AUTOSAR platforms.

In a vehicle, the number of dynamic nodes included in Adaptive AUTOSAR is between 1 and 5, whereas the number of conventional nodes is around 100. Different levels of security and insulation are applied. Adaptive AUTOSAR uses C++ as its main programming language and POSIX (Linux) as its operating system.

At present, a European consortium of automakers, OEMs, software solution supplies, and semiconductor suppliers are working on service-management middleware to manage both local and remote services (cloud services) – see Figure 6.83.

Evolution of E/E architecture: SOA – Service-Oriented Architecture

Thanks to the new concept of service-oriented IP communication and the software concept of coexistence of a static, "signal"-oriented configuration and a dynamic, "service"-oriented one, a new service-oriented architecture (SOA) has been developed. SOAs are hybrid E/E architectures that allow various networks – CAN/CAN-FD, LIN, IP and so forth – to coexist, thus solving the challenge of compatibility between the two AUTOSAR platforms (see Figure 6.84).

Let us take a closer look at what a future E/E architecture could look like – in particular, the interaction between Classic AUTOSAR and Adaptive AUTOSAR. The key idea is the introduction of two basic parts: an execution layer and a connected layer (see Figure 6.84).

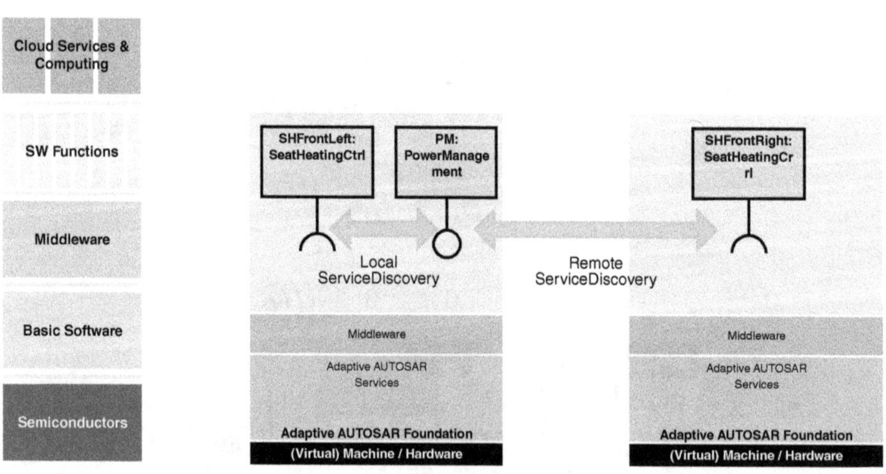

Figure 6.83 Middleware: standard service-management platform.

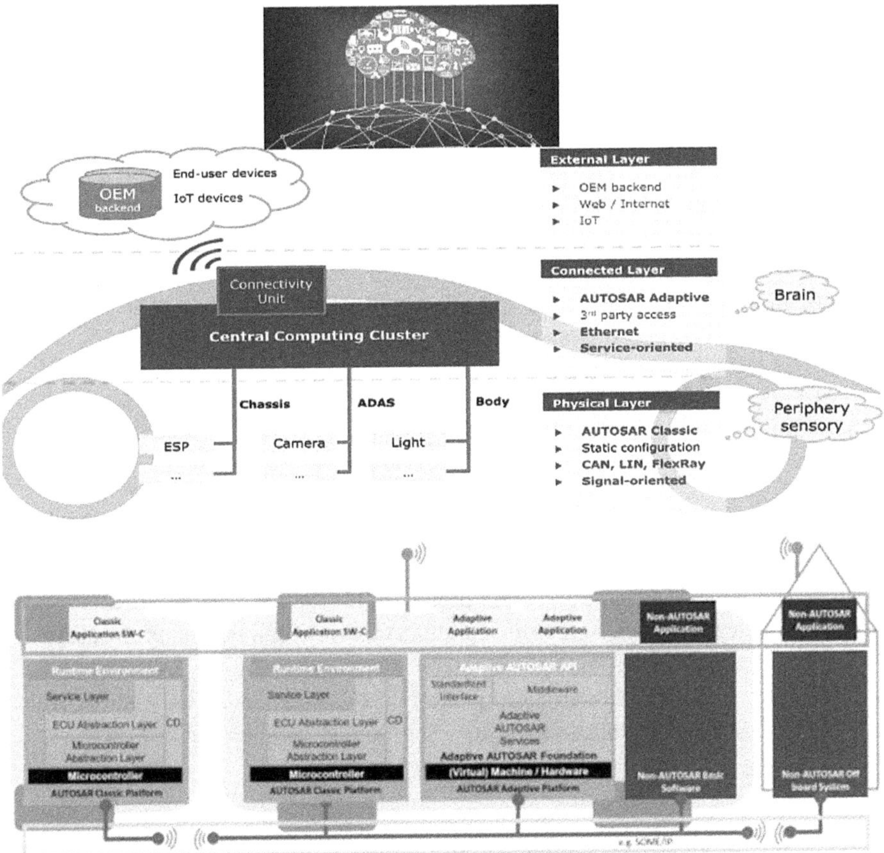

Figure 6.84 The two fundamental parts of SOA: the execution layer and the connected layer.

- The execution layer is used to provide basic functions using series of sensors and actuators. It is based mainly on Classic AUTOSAR. Safety is a more critical concern than security, because the data serve to execute critical tasks such as braking and steering. For the execution layer, CAN/CAN-FD, FlexRay, and LIN dominate the bus system. Classic AUTOSAR is the more suitable platform;
- The connected layer essentially performs all new functions beyond those of a conventional automobile. These functions are not required for the critical tasks of basic driving. Cybersecurity is the dominant concern over operational safety, because the information is being sent outside of the vehicle. For the connected layer, switched Ethernet networks with Unicast, Multicast, and Broadcast options are the preferred solution for fast, service-oriented communication. Adaptive AUTOSAR is the more suitable platform.

6.4 Testing and analysis tools

6.4.1 Tools needed for these new architectures

In this section, we present a number of real-world examples to concretely illustrate what we are talking about. We have drawn these examples from the catalog of the company VECTOR, which offers numerous software packages and basic services for the implementation of new generations of applications. VECTOR provides all that is needed to analyze, simulate, and test computers and IP networks (e.g. AVB, DoIP, SOME/IP), and a whole network implementing a range of protocols – CAN, CAN-FD, and LIN. Of these solutions, we can cite the following examples:

- Interfaces supporting BroadR-Reach® and other types of physical layers;
- A string of multi-bus tools facilitating the integration of these new technologies into the architectures of existing vehicles.

This string of tools can be summarized by the V-shaped cycle shown in Figure 6.85.
This software chain includes numerous elements:

- CANoe – for various communication protocols, this tool can be used to analyze the bus, simulate the vehicle's electronic architectures (virtual, hybrid or physical bus), test, and run diagnostics on the computers;
- VT System – a modular test environment for functional tests on ECUs. With CANoe, VT System forms a flexible and powerful solution for creating test arrays;

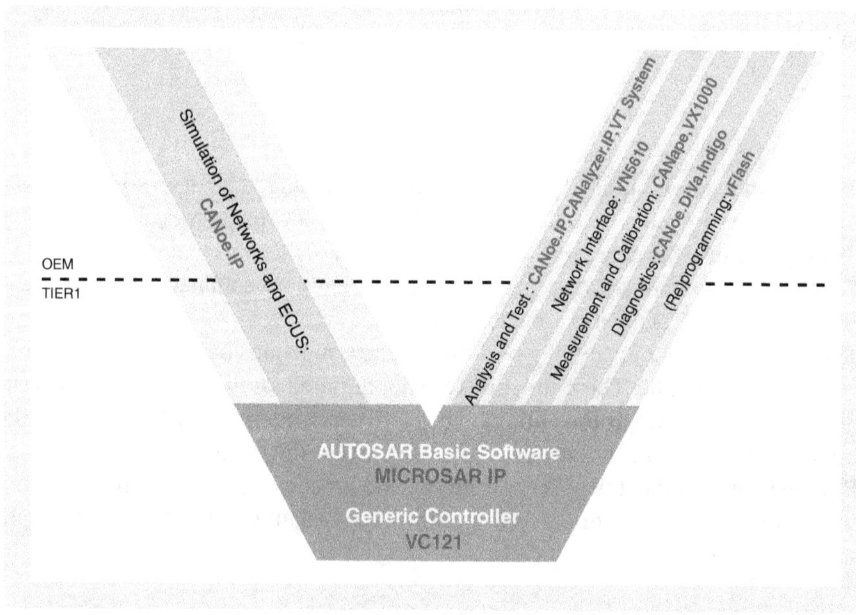

The IP solution from Vector covers the entire development process for networks and ECUs.

Figure 6.85 V-shaped cycle.

- CANape and VX1000 – high-datarate measurement and calibration using XCP, resulting in a minimum of runtime effect on the ECU;
- VFlash – flashing of one or more computers, also using DoIP.

6.4.2 Evolution of development tools

As previously indicated, there are already complete solutions in place for analysis of architecture simulation and development of test arrays for the CAN, LIN, FlexRay, and Ethernet protocols (see Figure 6.86).

The starting point for these tools is the database or messaging system, which characterizes communication in the computer network. Multiple database editors are available with these solutions, for the various formats dedicated to each protocol. For example, for CAN, the format .dbc is edited by the tool CANDB++, which was designed for CAN payloads of 8 octets, and CAN-FD payloads of up to 64 octets.

It is not our intention to linger over the description of each tool – this has already been done in an earlier work. Rather, we shall focus on what allowed CANoe to evolve to cater for Ethernet systems, and then on the evolution of the tool with respect to the evolutions of SOA.

Architecture of signal-oriented tools

Sticking with CANoe, we shall begin by looking at the initial architecture of this tool, with its different functions – message-oriented and signal-oriented, for analysis, simulation, and testing for the various communication automobile networks – see Figure 6.87.

Starting with the bottom layer, we can distinguish the different physical layers for the various buses, then the communication layers, the transport and network management layers, and finally the application layers, which describe the behavior of the various computers on the network in a proprietary language such as CAPL (Communication Access Programming Language), or C++. Alternatively, that behavior can be modeled by external tools such as MATLAB or Simulink. The input for this architecture is a

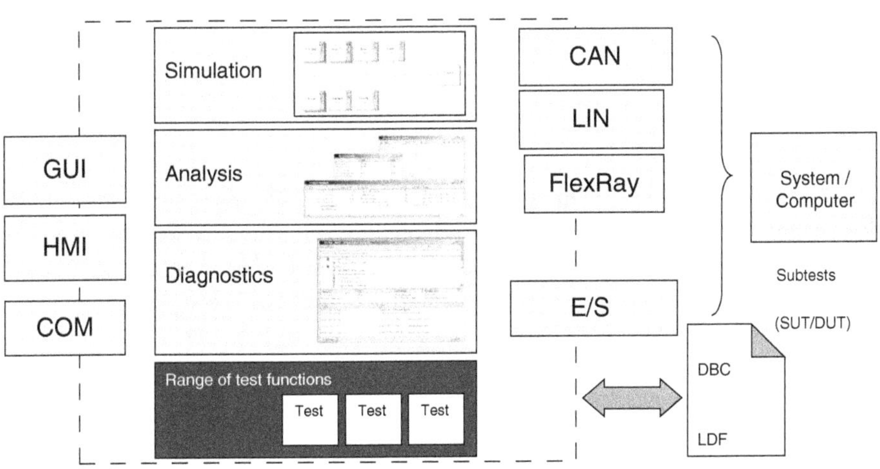

Figure 6.86 Example of VECTOR's platform.

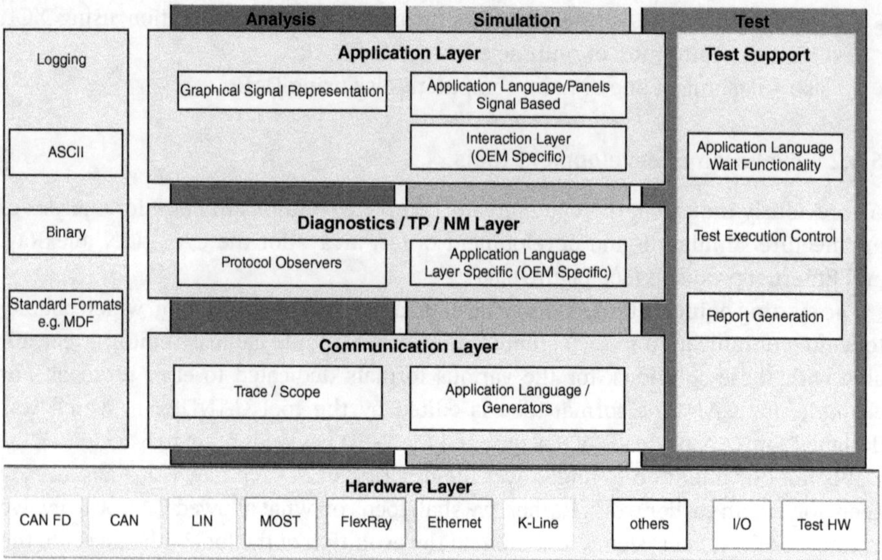

Figure 6.87 Initial architecture of VECTOR's CANoe.

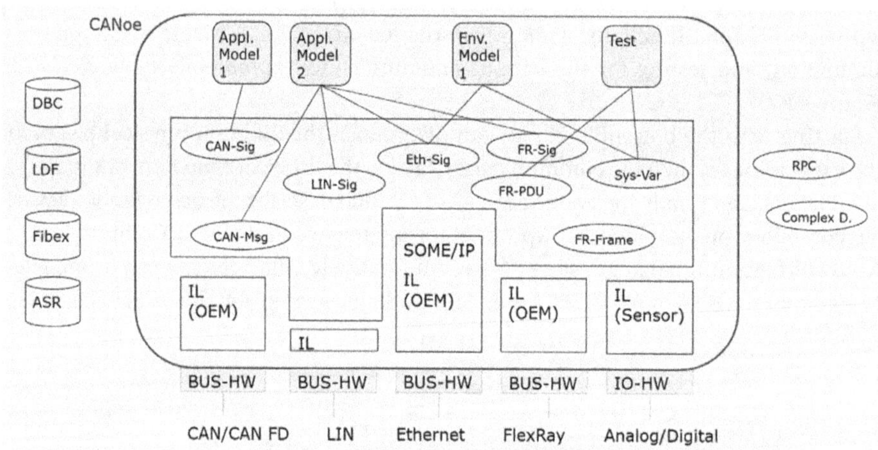

Figure 6.88 IL of protocols.

signal-oriented or message-oriented database such as DBC for CAN or Arxml for Ethernet or FlexRay. To simplify the development of application functions, a range of application programming interfaces (APIs) have been deployed in the Interaction Layer (IL) for each type of protocol. The main role of these APIs is to send and receive messages and signals (see Figure 6.88).

Architecture of PDU-oriented tools

In view of the payload that can be supported by certain protocols such as FlexRay (up to 254 octets) and CAN-FD (64 octets), data transport urgently needs to be improved. As explained earlier, AUTOSAR introduced a new concept: the "Protocol Data Unit"

(PDU), where data units (DUs) are clustered together to form PDUs and those PDUs are then encapsulated in frames. The database format has been completely overhauled in order to integrate PDUs.

Two of the database files used in the automotive industry are as follows:

- The ASAM standard (ASAM-MCD-2) introduced Fibex (.xml) format;
- AUTOSAR introduced .arxml format. This description is a better alternative to describe dynamic networks with wide data fields, such as Ethernet. However, it can also apply to any other network, such as CAN/CAN-FD or FlexRay, which means that a unified descriptive format can be used for the entire vehicle.

As the connection protocols – CAN-FD, FlexRay, and Ethernet – differ widely in their frame formats and communication principles (e.g. event-triggered, time-sensitive, switched, or service-oriented), an initial solution is to differentiate the architecture on the basis of each protocol (see Figure 6.89).

In this architecture, a test script, for example, would be dedicated for a CAN/CAN-FD, Ethernet, or FlexRay bus. To enable users to focus on their applications whatever protocol is being used, a new communication layer, independent of the data link layer, has been implemented. This independent layer of abstraction between the signal (application) and the data link layer (protocol) would operate in the same way for CAN-FD, FlexRay, and Ethernet. The concept is similar to that of AUTOSAR. This has given rise to a new communication architecture that masks the data link layers from the user by strict separation between the communicating components and the medium used for transmission. The layer that provides this separation is middleware, known as "PDU Extraction" (see Figure 6.90).

This concept means that test scripts and application modeling scripts for CAN/CAN-FD, FlexRay, and even Ethernet will be similar and compatible.

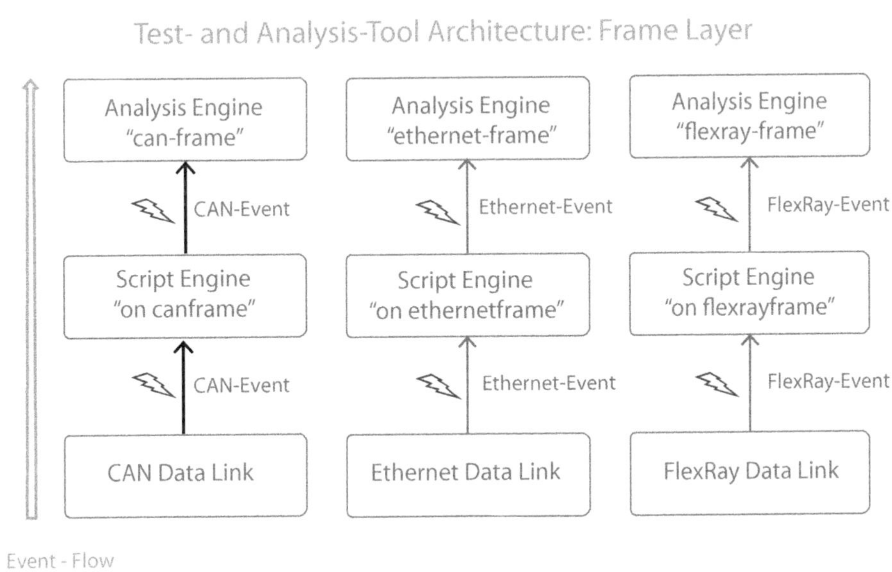

Figure 6.89 Protocol-dependent architecture.

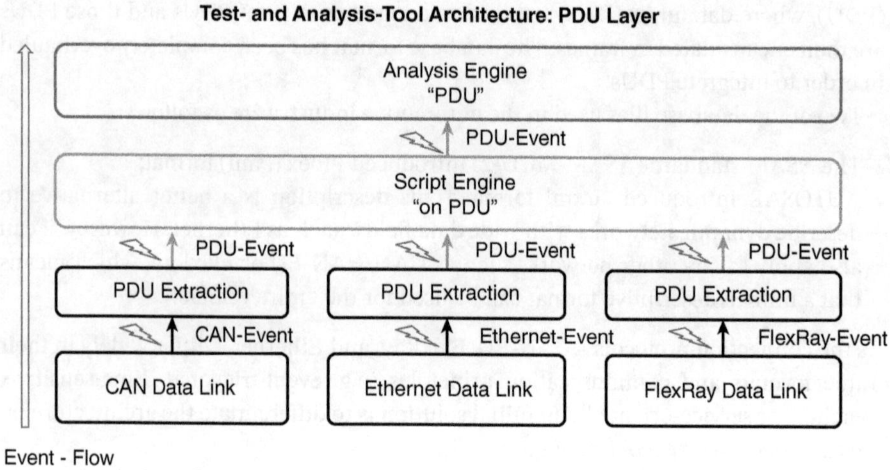

Figure 6.90 Protocol-independent architecture.

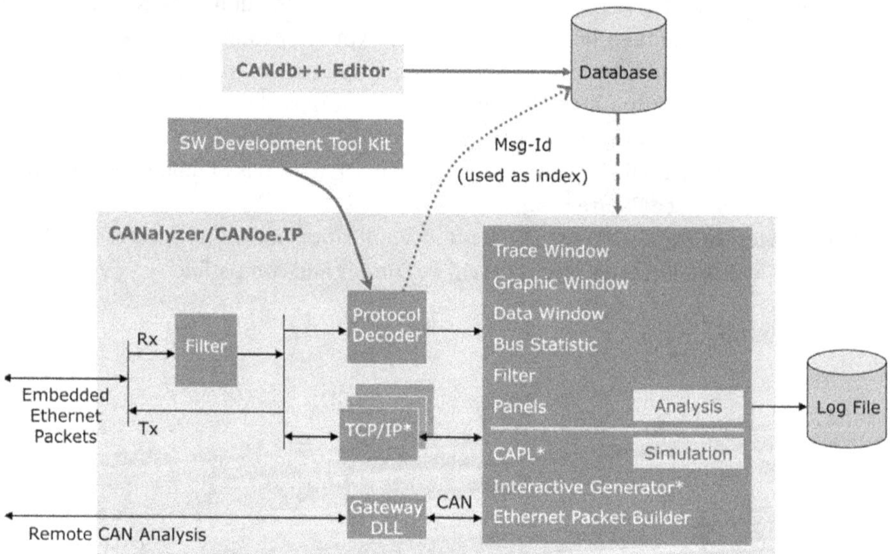

Figure 6.91 Individual TCP/IP stack.

At this stage of development, implementations for IP have extended the scope of this tool for simulating and testing computers and IP networks.

Let us examine these implementations in CANoe, facilitating the rollout of Ethernet systems.

Evolution for IP integration: TCP/IP stack

To simulate or test Ethernet architectures or computers, it is essential that the tool be able to simulate models of the TCP/IP stack on each computer. In fact, every simulated node must contain a TCP/IP stack. CANoe is able to emulate an individual TCP/IP stack for each ECU (see Figure 6.91).

Thanks to this concept of an individual TCP/IP stack configuration associated with each ECU, users are able to simulate an Ethernet node that is missing from their system (see Figure 6.92).

Ethernet packet generators supplement the simulation, and also offer the possibility of interfering with the network in order to test its response – by sending an incorrect checksum, for example (see Figure 6.93).

TCP/IP packet simulation and automatic testing

To facilitate the simulation of TCP/IP packets, a set of APIs have been developed (the directory Ethernet_IL.dll) – see the example in Figure 6.94.

Architecture of the service-oriented tool

By way of example, we shall now examine the principle used in CANoe for the deployment of SOA.

As we saw earlier, CANoe ensures signal-oriented communication by the interposition of the PDU layer (middleware) between the applications and the IL data link layer (linked to the protocol). To deal with the issue of service, the tool uses an

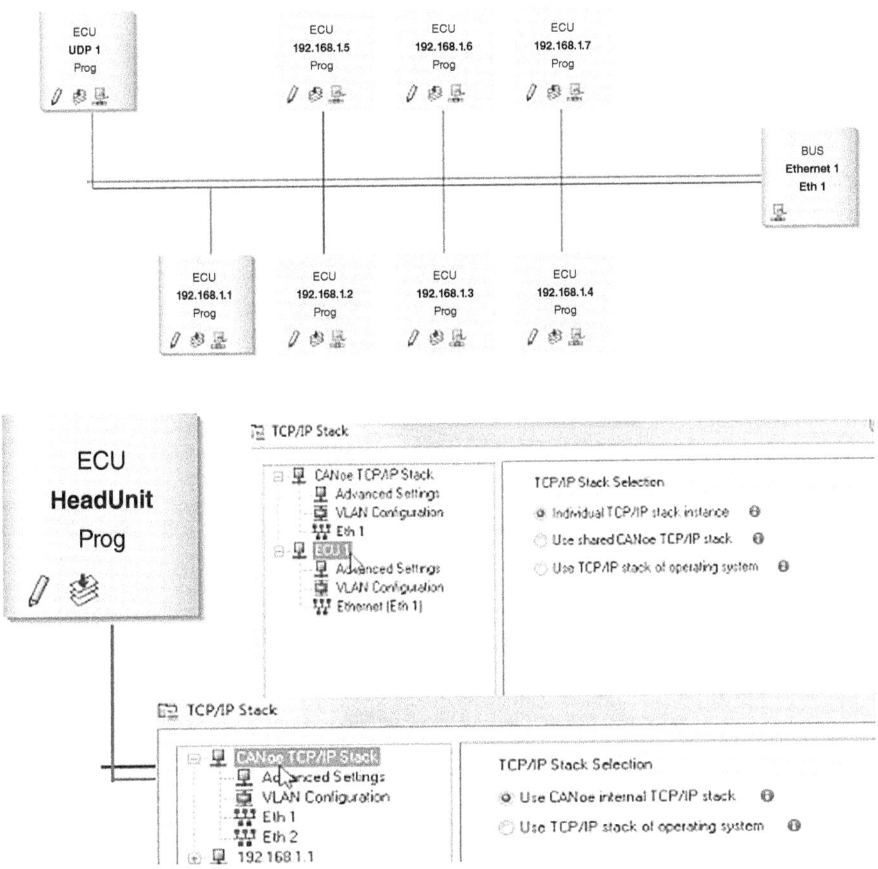

Figure 6.92 Simulation of a missing Ethernet node.

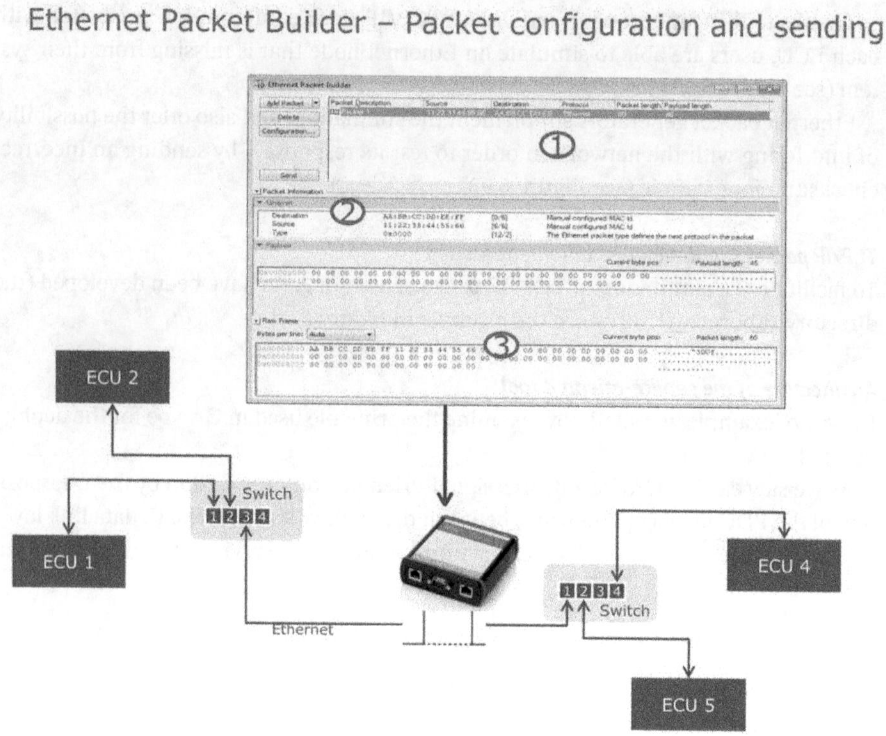

Figure 6.93 Configuration and transmission of IP packets.

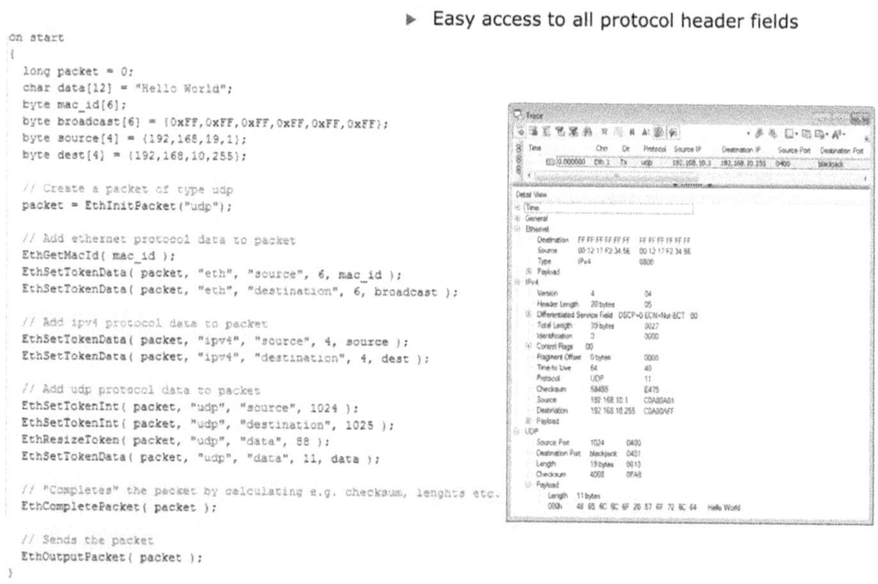

Figure 6.94 Example of a script.

object-oriented communication approach, rather than a message-, signal-, or PDU-oriented approach. A new intermediary layer, or middleware, between the application and the data link layer is therefore introduced (see Figure 6.95).

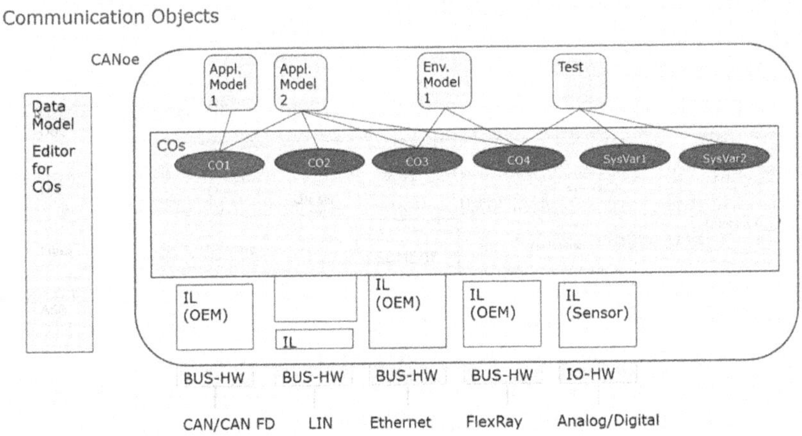

Figure 6.95 Middleware between the application layer and the data link layer.

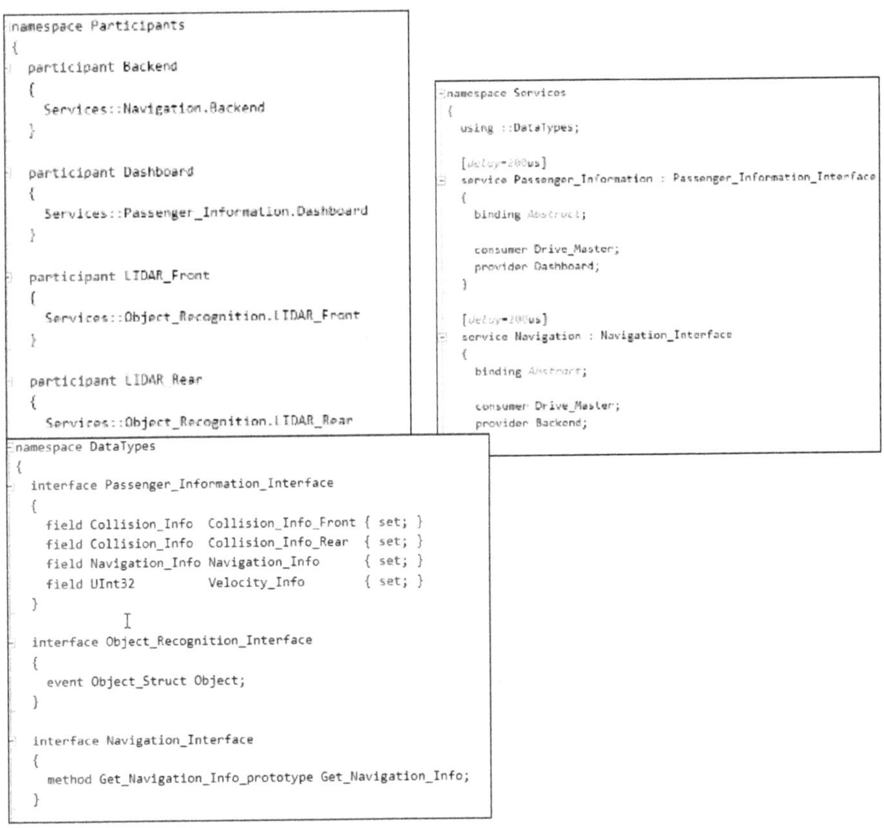

Figure 6.96 Service-oriented data model.

A new object-oriented database format (or Data Model) is therefore necessary. ARXML, which is PDU-/service-oriented, is also supported by the new model. This database describes the different participants and the services provided and received. An example is shown in Figure 6.96.

Let us recap the principle behind SOA and services.

Service-oriented architecture

The fundamental principle behind SOA is that intelligence is shifted away from the network core, into the extremities (see Figure 6.97).

This approach requires two types of communication: one service-oriented and the other signal-oriented. Two types of middleware – Adaptive AUTOSAR and Classic AUTOSAR – are also needed.

- The signal-oriented approach with Classic AUTOSAR applies to the outer parts of the network, for which real-time requirements are of prime importance. Typically, these domains use the protocols CAN-FD, FlexRay, LIN, and IP;
- The service-oriented approach with Adaptive AUTOSAR applies to the IP network architecture, the "backbone and Domain Controller", for which dynamicity and bandwidth are crucial.

The interface services are exchanged between the backbone and the multiple Domain Controllers. Each Domain Controller routes the signal service to the appropriate low-level network.

As an example of an SOA model, consider the vehicle represented in Figure 6.98A, with service-exchange interfaces.

Each participant constantly reads the database file described above. Any modification, addition, or withdrawal of a service is updated during runtime.

Service interfaces

- Represent the elements exchanged. They are imported into CANoe from the database (in vector format, .vCDL) or from the AUTOSAR standard descriptions (.arxml) – see Figure 6.99;
- Can be statically defined, but may also be deployed flexibly across the whole of the network, in relation to supply or to demand. In other words, a computer may be extended dynamically, during runtime, to offer or receive new services. Thus, the

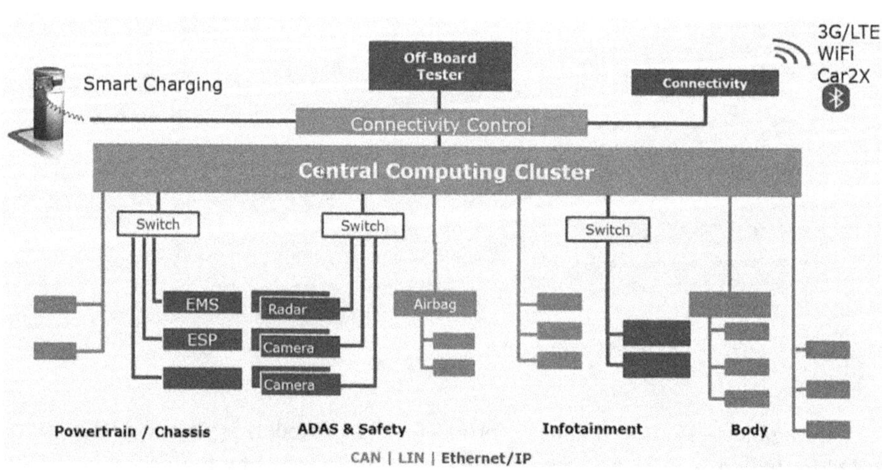

Figure 6.97 Fundamental principle behind SOA.

input files may be modified and extended as necessary. An OEM can deploy an application doing a reconfiguration of the entire system.

Communication Objects

Communication Objects (COs) are the key elements in this new concept of communication in CANoe. These imported objects are configured or personalized in the CANoe environment by means of a configuration interface (Figure 6.100A). They represent the participants or nodes in the network. With each participant is associated an application model that handles the network services. On the basis of the configuration, the communication object provides communication channels to its partners at runtime (see Figure 6.100B).

The service interfaces, and the associated methods and events, can be implemented in model form in CANoe (in CAPL or .net language) – see Figure 6.98B. Users access these models to view or modify the content directly using test and simulation scripts.

Thus, the user can dynamically create new communication objects during the simulation, if necessary. At the same time, it is possible to modulate the availability of a service. This makes it possible to simulate dynamic topologies with configurable communication identical to what is used in connectivity applications.

(A)

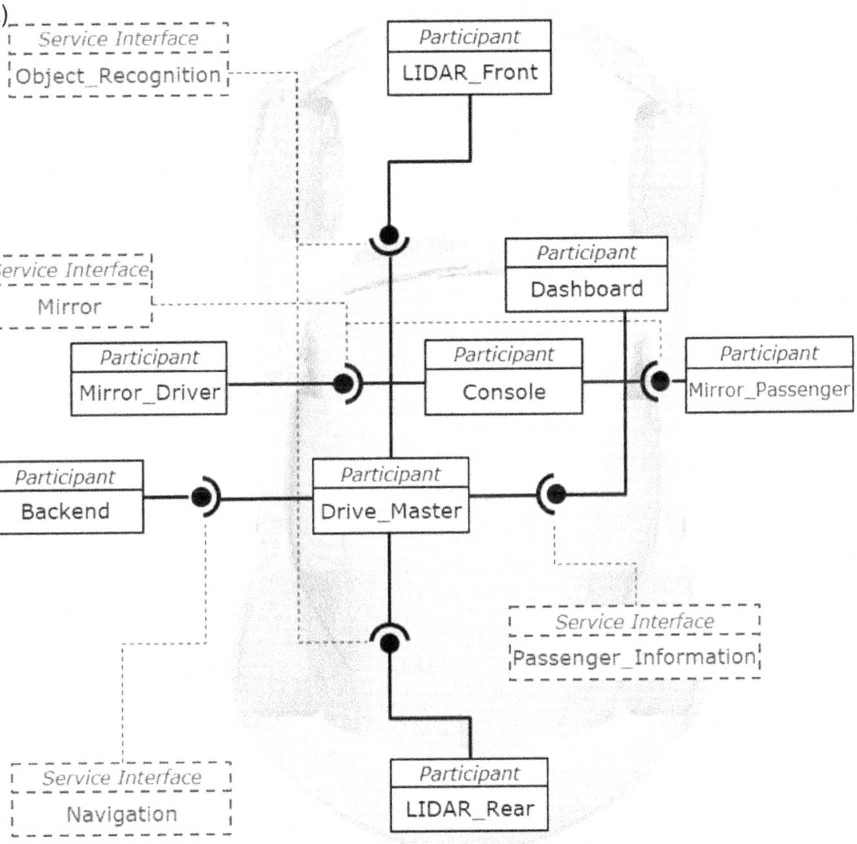

Figure 6.98 Service interface and service model.

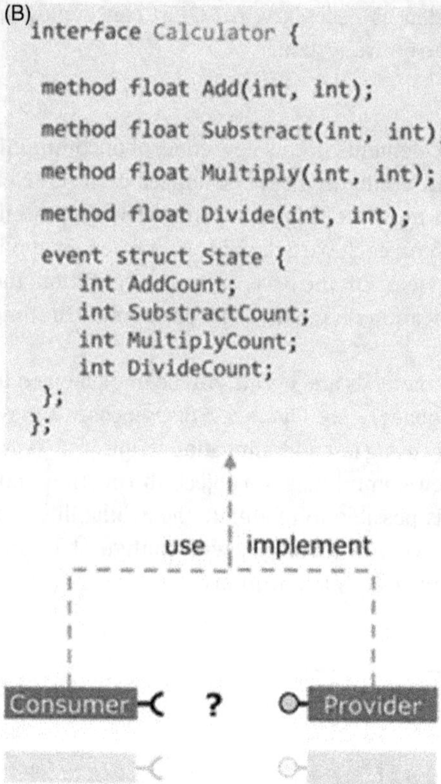

Figure 6.98 (*Continued*)

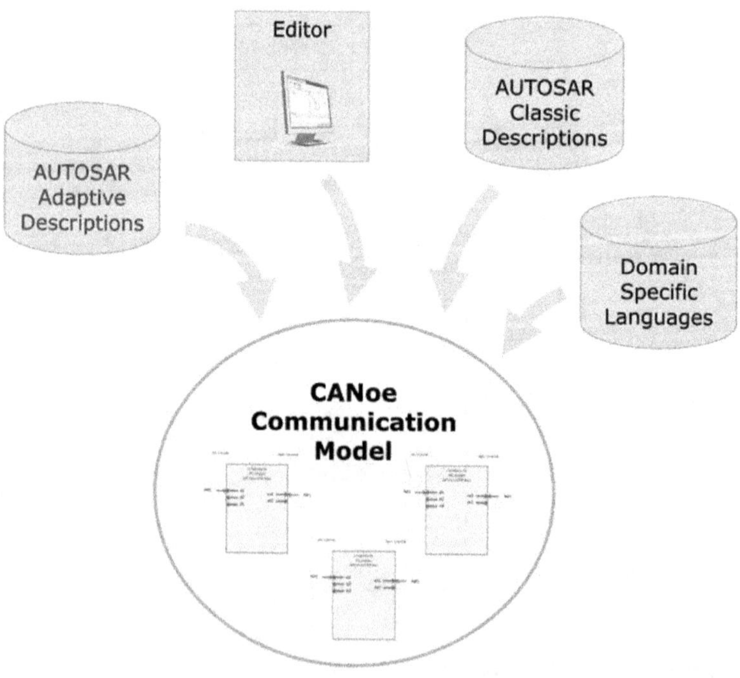

Figure 6.99 Importation of service interfaces.

Figure 6.100 CO configuration interface.

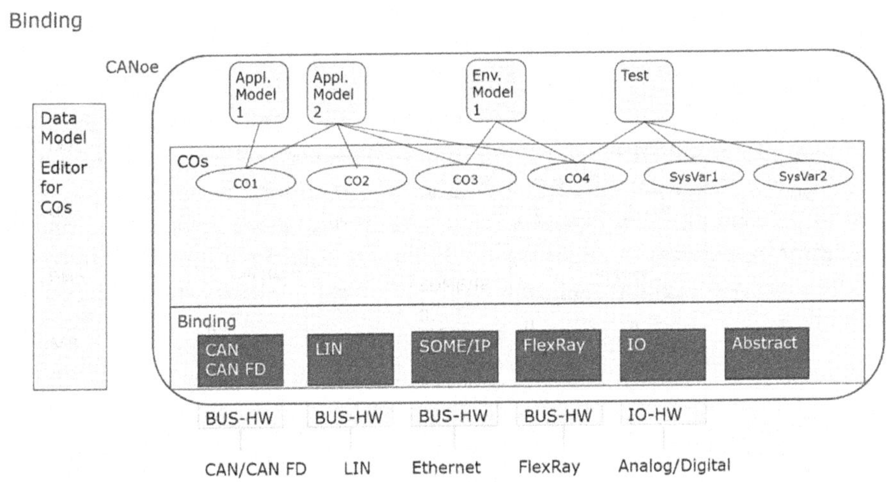

Figure 6.101 Links between COs and different network protocols.

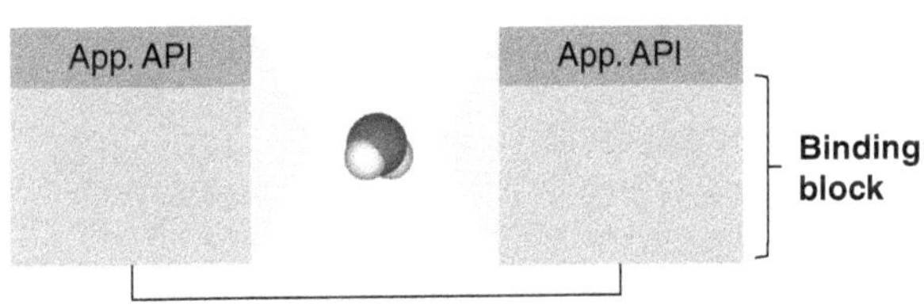

Figure 6.102 Communication between COs depending on protocol.

Links from COs to communication protocols

Communication objects may be linked to different network protocols for transmission. The configuration of the links defines the properties of transmission between the different connection points of a CO. This means that the configuration of the link will determine which protocol and which network are used for the transmission. Thus, the link creates a transition between the virtual world of CANoe and the real world (see Figure 6.101).

In CANoe, a library of predefined binding blocks can be used to make objects communicate via the protocol to which they are connected. These blocks encapsulate the complexity of the network-level layers and provide an interface for the application models and test scripts – see the example in Figure 6.102.

The complexity of the transmission varies depending on the protocols and the underlying networks. As a general rule, the transmission for an Ethernet network is complex. Therefore, the protocol parameters for the TCP/IP layers used must be configured. CANoe offers two types of binding blocks.

Abstract transmission

Abstract binding blocks allow for transmission between terminals simulated by CANoe. Thus, abstract transmission facilitates communication between components simulated in CANoe without needing to describe and simulate lower-level protocols. Abstract transmission is appropriate, for example, for virtual prototyping – see Figure 6.103.

SOME/IP binding

During the development of a vehicle, certain nodes tend to be simulated while others tend to be real. SOME/IP binding allows transmission between terminals on the basis of the SOME/IP protocol via TCP/UDP/IP and Ethernet (see Figure 6.104).

SOME/IP binding takes care of transmitting events, calling methods and finding and acquiring services through Service Discovery. In addition, the transmission of I-PDU signals with end-to-end protection and onboard secure communication are supported.

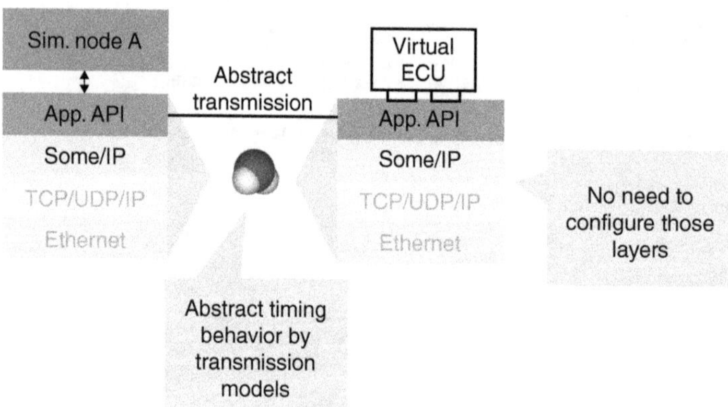

Figure 6.103 Abstract transmission for virtual prototyping.

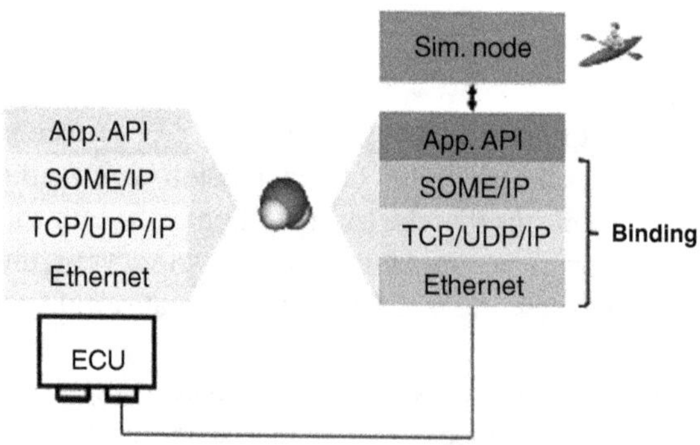

Figure 6.104 Binding between COs via SOME/IP.

Other types of binding

In today's world, there are multiple platforms for development of Ethernet applications to serve a range of functions, in the automotive field (SOME/IP), in robotics (ROS), in Web services (HTTP/REST), and the IoT (MQTT), and so forth. In view of the new directions in automobile applications (connected vehicles, importantly), future versions of CANoe will also need to support HTTP, REST, and MQTT connections to facilitate the development of these applications in automobiles (see Figure 6.105).

To cover the next decade, new binding interfaces are under development (see Figure 6.106).

Let us now turn to the implementations that have facilitated the simulation of ECUs and typical networks (SOME/IP, AVB, DoIP) in the CANoe environment.

API development

To test and simulate increasingly complex environments with new application protocols such as AVB, SOME/IP, DoIP, and others, APIs (Application Programming Interfaces) are indispensable to designers, allowing them to simulate Ethernet nodes that are missing from their system without needing to manage the transmission of the various packets.

API for SOME/IP CANoe is one of the tools that can handle service-oriented tests. To decode the SOME/IP packets, we need the database FIBEX4.1 or ARXML4.3 for the description of the SOME/IP services. For simulation of SOME/IP packets, a directory

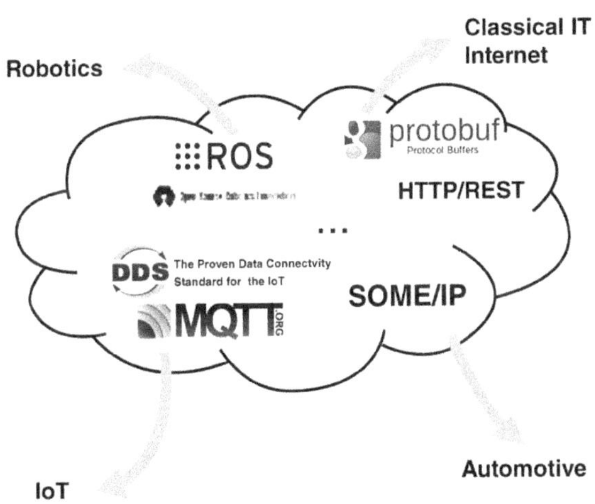

Figure 6.105 Examples of other technologies.

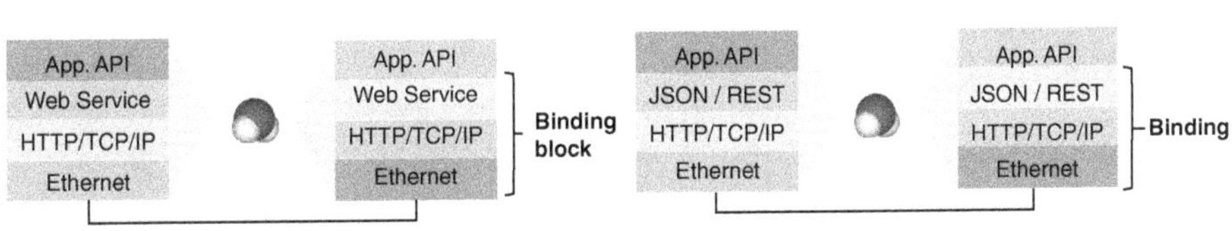

Figure 6.106 Examples of future applications.

of functions, SomeIp_IL.dll, has been developed, which allows interactive simulation via HMIs (Human–Machine Interfaces) – see Figure 6.107.

Windows such as Trace are also able to easily decode this protocol (see Figure 6.108).

APIs for DoIP CANoe offers the possibility of simulating the tester or the SUT (System Under Test), or both. For this purpose, it is necessary to have the database describing the diagnostic services (in ODX or equivalent format). APIs (DoIP.dll) have also been developed, to facilitate the simulation and handling of DoIP packets. Windows such as Trace are also able to track the stages of the protocol (see Figure 6.109).

APIs for AVB For the simulation and modeling of AVB nodes, a function directory AVB_IL.dll has been developed, which simplifies the simulation of a camera (Talker) or a (Listener) via HMIs (see Figure 6.110).

CANoe decodes the various protocols employed in RTP and AVB/TSN (PTP, AVTP) for a Talker or Listener. CANoe supports various media formats (including CVF – Compressed Video File) – see Figure 6.111.

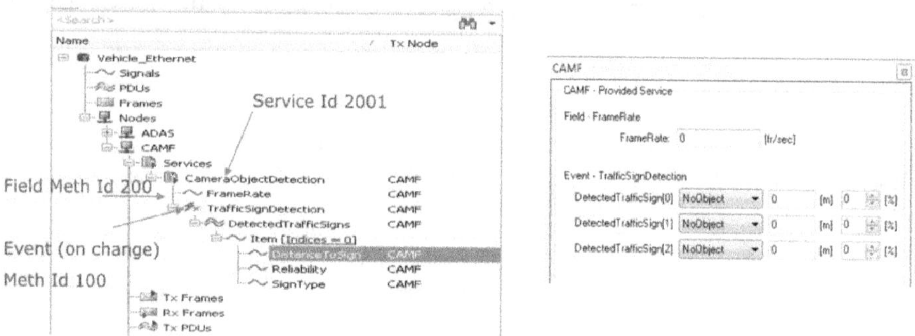

Figure 6.107 Example of a service-oriented HMI.

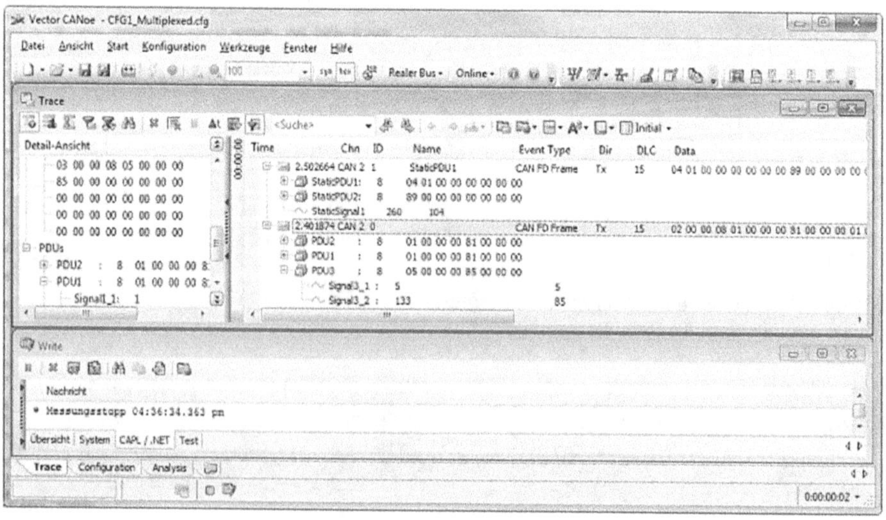

Figure 6.108 Trace window for SOME/IP.

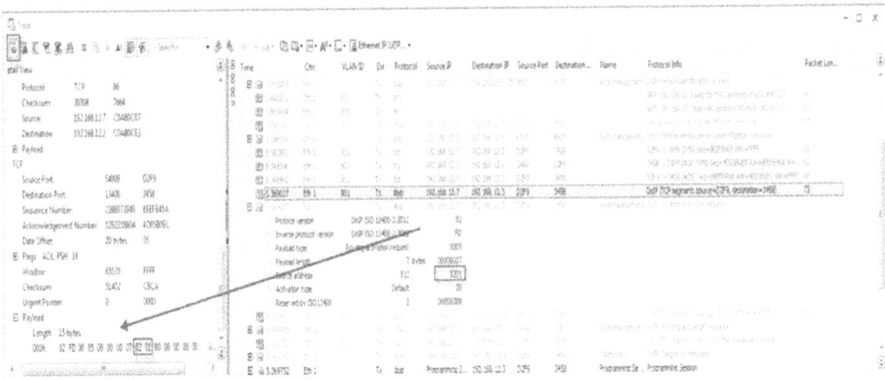

Figure 6.109 Trace window for DoIP.

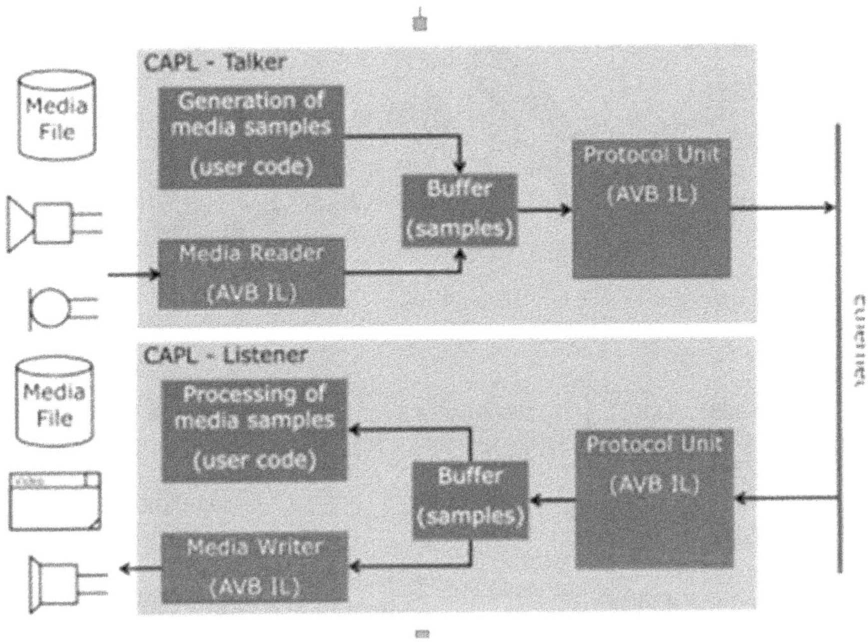

Figure 6.110 API for AVB.

Network communication interfaces

It is also necessary to have hardware interfaces to connect to physical Ethernet networks. These interfaces support the different automobile hardware layers, BroadCom and conventional Ethernet – that is, Ethernet 100 and 1000 BASE T1 and 100 and 1000 BASE TX. With this hardware, the isolation of the Ethernet Windows network is ensured. These interfaces also offer the possibility of connecting to the CAN/CAN-FD network.

- Example: the two types of interfaces, connected respectively to USB 2.0 or 3.0: VN5610A and VN5640; the latter offers greater possibilities with 12 BroadR-Reach connections (see Figure 6.112).

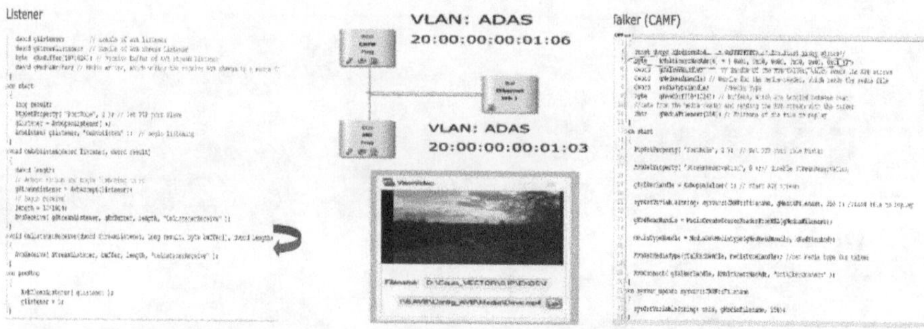

Figure 6.111 Examples of simulation of AVB/TSN Talkers and Listeners.

Figure 6.112 Network communication interfaces.

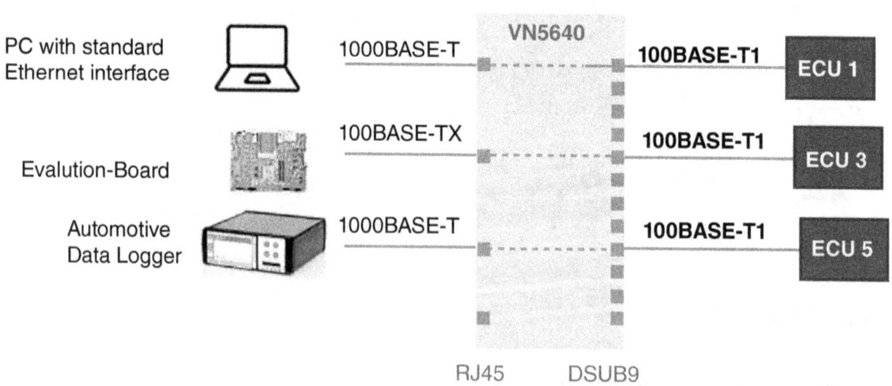

Figure 6.113 Configurable interfaces for simulation, measurement, eavesdropping, and recording.

These interfaces can be configured to provide multiple modes of operation: simulation, measurement, eavesdropping, or simultaneous recording of several networks using a data logger (see Figure 6.113).

To eavesdrop on an Ethernet link, we must have two physical connections. The latency introduced by the tool is very slight (Figure 6.114).

Interface for V2X radio communications: IEEE 802.11p

Today, a vehicle is an integral part of the ecosystem with which it is communicating (see Figure 6.115).

Remember that for V2X applications, there are currently two parallel options, which are not mutually compatible:

- DSRC (Dedicated Short-Range Communications), based on the IEEE 802.11p standard, rolled out in the United States, Europe, and Japan;
- C-V2X (Cellular-V2X), a technology built on the specifications of cellular radio networks (4G and 5G), defined by the 3GPP.

The network interface VN4610 is capable, firstly, of accessing DSRC, complying with IEEE 802.11p and, secondly, of accessing vehicle's onboard hardwired CAN and CAN-FD networks (Figure 6.116).

Coupled with the testing environment such as CANoe.Car2x from VECTOR, this solution allows developers working on computers that manage 802.11p radio

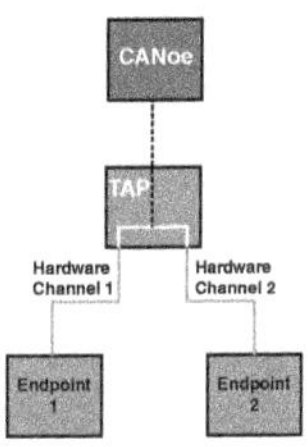

Physical Layer	Bypassing Latency Δt
BroadR-Reach <-> BroadR-Reach	1.9 µs
BroadR-Reach <-> IEEE 100 Mbit	1.5 µs
IEEE 100 Mbit <-> IEEE 100 Mbit	1.1 µs
IEEE 1000 Mbit <-> IEEE 1000 Mbit	1.5 µs

Figure 6.114 Bypassing latency due to the tool.

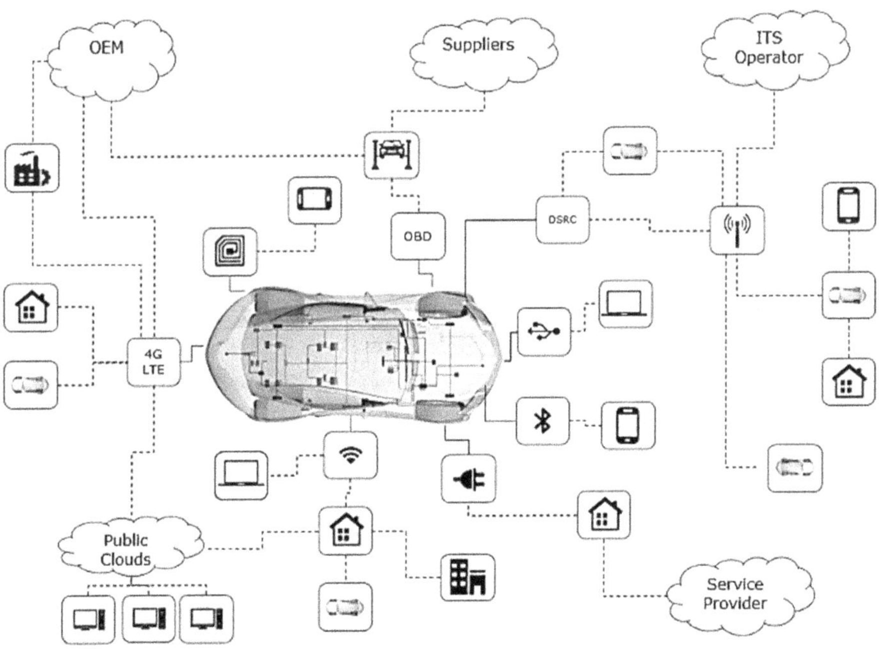

Figure 6.115 The ecosystem with which a vehicle communicates.

Figure 6.116 RF communication interface.

connections to move more swiftly on from prototyping to industrial production. The IEEE 802.11p acknowledgment and transmission messages pass from the USB interface to the development tools. The received messages are transferred to the application synchronously with the messages connected to the CAN bus. An integrated satellite receiver (GNSS) pinpoints the exact time and the position of the system being studied.

The testing toolkit CANoe.Car2x, in combination with the interface VN4610, offers an environment simulation solution capable of testing Car2x/V2X applications, using the VN4610 to send the transmitted messages and configure the communication parameters for different tests (see Figure 6.117).

The addressing of received messages over two unfiltered radio channels to a test toolkit offers the advantage, for developers, of being able to quickly analyze messages rejected by a computer due to a synchronization error, incorrect geographical information, or protocol errors. It also offers the possibility of measuring latency, because the time and date stamp of the messages is synchronized on the channels of the CAN bus (see Figure 6.118).

As with all protocols we have looked at previously, in an analytical tool such as CANoe, a database is used to decode the frames. The database format used is ASN.1 (Abstract Syntax Notation One). This is a standardized format and also includes notation that describes the rules and structures of representation, encoding, transmission, and decoding of data in telecommunications and computer networks.

Database for V2X

In telecommunications, an application that exchanges data with another needs to make itself understood by that other application. However, it is often the case that two applications represent the data they handle in different ways. For example, the way in which an integer is represented depends on the nature of the processor; the encoding of a file depends on the operating system; and the data structure depends on the applications themselves, or more specifically, on the programming language in which those applications are written. Each application on a given site has a particular form of

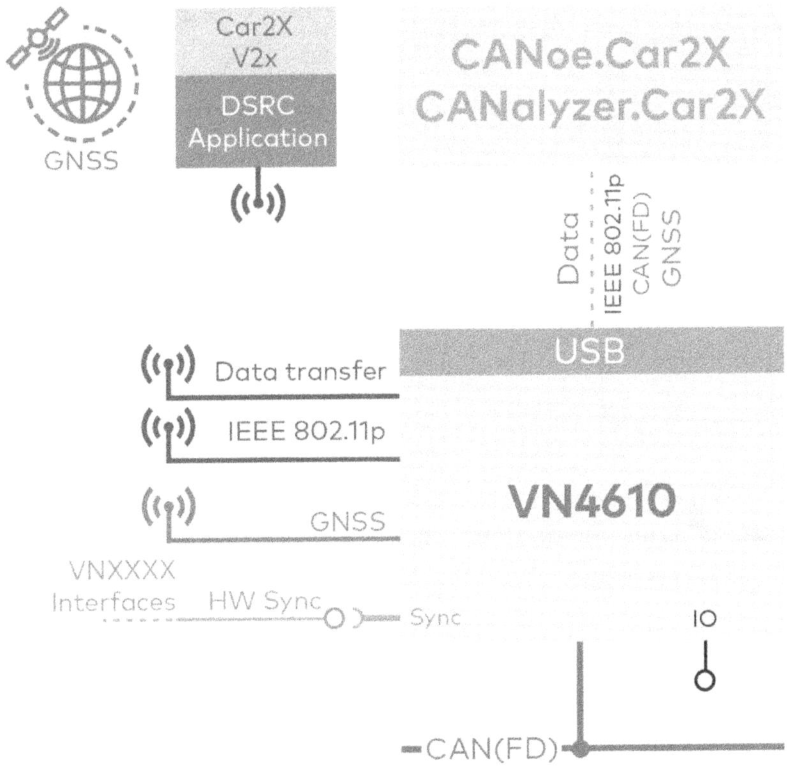

Figure 6.117 Environment simulation solution for Car2x/V2X applications.

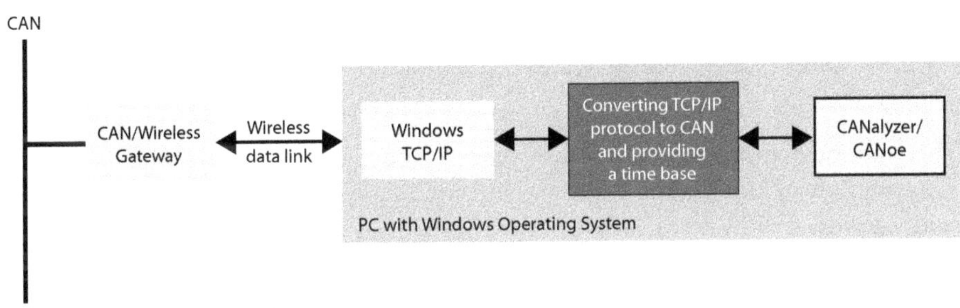

Figure 6.118 Example of a whole application.

syntax to represent information. One solution to this potential problem is to use a syntax that is shared by all applications. The transfer syntax defined by ASN.1 is shared and, therefore, facilitates communication between two applications (Figure 6.119).

Message type

In ASN.1 format, various types of messages, which are to be exchanged between vehicles, or between the vehicle and a piece of infrastructure, are described. These types are defined by the ETSI standard:

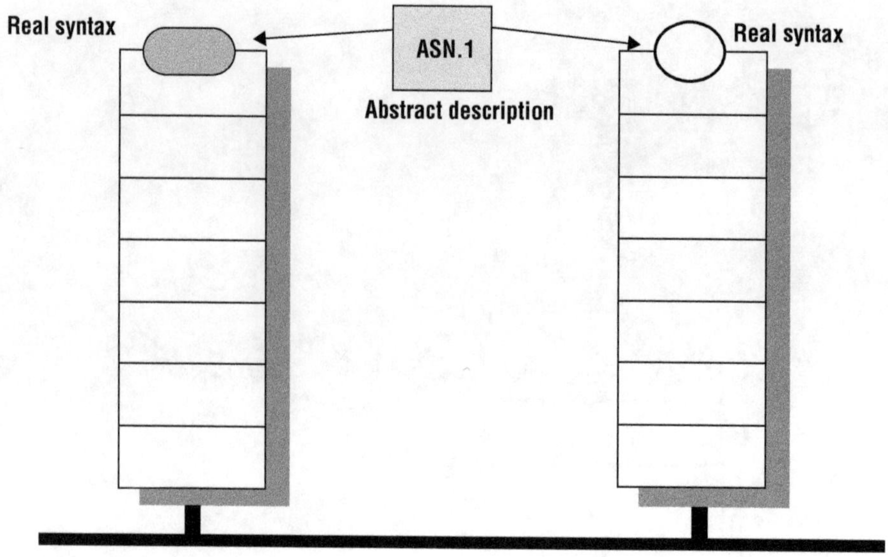

Figure 6.119 ASN.1 transfer syntax, facilitating communication between two applications.

- Firstly, the Cooperative Awareness Message (CAM) describes the basic features of the talker;
- Secondly, the Decentralized Environmental Notification Message (DENM) alerts the system to dangers on the road;
- Thirdly, a Signal Phase and Timing (SPAT) message is sent from a traffic light. This message contains various data, such as the current state of the traffic light (green, amber, or red), the time remaining before it changes, its position in relation to the vehicle and to an Intelligent Transport Station (ITS), in terms of its latitude and longitude.

This solution helps support security services and efficient traffic flow, which require a constant flow of information about the surrounding vehicles, or event-triggered alert notifications.

DENMs are generally transmitted cyclically. They contain, for example, information about position, dimensions, and road type, but also information to users on roadworks.

To transport these messages, a transport protocol is needed.

Transport protocol

The basic transport protocol (BTP) offers end-to-end transport without connection in the ITS network. The main purpose of the BTP is to manage the multiplexing and demultiplexing of messages such as CAMs or DENMs, for transmission of the packets via the GeoNetworking protocol (network layer). The BTP is used in the GeoNetwork protocol stack. A BTP packet is made up of protocol headers and the payload, as shown in Figure 6.120.

- The MAC header (Source Address);
- The GeoNetworking header;

MAC Header	GeoNetworking Header	GeoNetworking Security Header (optional)	BTP header	Payload (optional)

Figure 6.120 BTP packet header with protocol and payload.

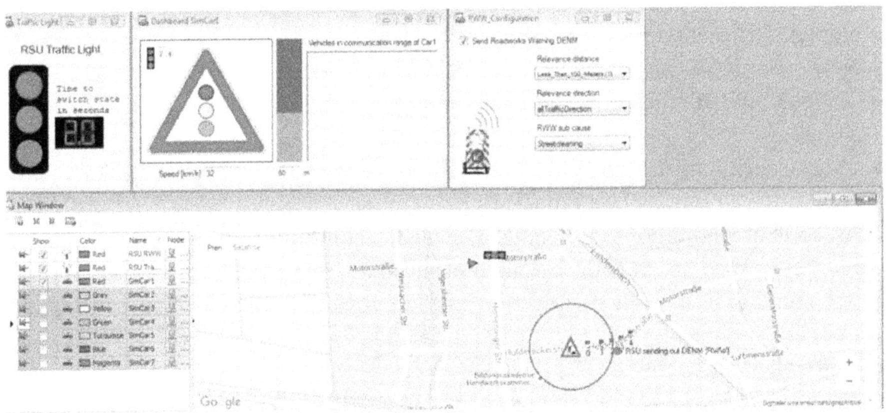

Figure 6.121 Example of simulation with CANoe.

- The GeoNetworking Security header;
- The BTP header;
- The payload, representing packets from the ITSs.

Validation of a V2X application

CANoe includes these various protocols to decode the different types of messages. This makes it an ideal environment for the analysis, testing, and validation of V2X communication, with full or partial simulation. Input/output HMIs can be used to interactively modify the road data in order to check the different functions – see Figure 6.121.

6.5 Conclusions

We come now to the end of this volume. We have attempted to give as full and as clear a picture as possible of the various (and lengthy) stages with which you need to be familiar, before venturing into the vast jungle of "autonomous and connected vehicles." It is true that certain proposed commercial solutions may make it appear that such-and-such an application is easy to implement. Be aware, though, of the complexity that often lies behind this façade.

In this book, we have set forth the regulations, standards, economic issues, etc., as well as applications, techniques, and technologies pertaining to autonomous vehicles, to provide an industry-level understanding of the architectures and networks involved. In actual fact, though, we are witnessing the very start of a revolution – a new age, in technical, technological, and industrial terms, which is likely to take place between 2022 and 2035.

Over that period of time, there is still an enormous amount of work to be done before we see "*level-5 vehicles driving on open roads, in all weathers, the world over.*"

- The Vienna Convention must be adapted as quickly as possible;
- Regulations must be put in place, both at national and international level;
- The texts of the highway codes need to be reviewed and applications built to enforce those codes;
- Signage and communication infrastructures must be rolled out on the ground;
- New types of insurance policies must be devised and fine-tuned;
- The operational safety of autonomous vehicles must be proven beyond a shadow of a doubt;
- Cybersecurity must be ensured, both within the vehicles and beyond them, guarding against all sorts of cyberattacks;
- All sorts of simulations (mechanical, acoustic, communications, networking, etc.) must be refined, to avoid the unthinkable need to drive millions of kilometers under test conditions to gain certification and approval for vehicles to be brought to market;
- And much, much more!

All these developments will take time and will generate information enough to fill a dozen further books just like this one!

Ultimately, our aim in writing this book was to equip readers to construct permanent gateways for autonomous, connected, and intelligent vehicles, connecting hardware and software, bridging the gap between electronics and software, reconciling economic and societal values, and so forth.

It is our sincere hope that we have fulfilled this aim. If readers have questions, or comments, the authors are very happy to receive correspondence via e-mail at the addresses below. Constructive exchanges can only enrich the knowledge of all involved.

Dominique Paret dp-consulting@orange.fr
Hassina Rebaine hassina.rebaine@vector.com

Index

Autonomous & Connected Vehicles: Network Architectures from Legacy Networks to Automotive Ethernet, First Edition. Dominique Paret and Hassina Rebaine.

www.ingramcontent.com/pod-product-compliance
Lightning Source LLC
LaVergne TN
LVHW062110200926
848484LV00007B/1630

* 9 7 8 1 1 1 9 8 1 6 1 2 6 *